AF746927

Biomaterials - Properties, Production and Devices

nova science publishers

www.novapublishers.com

Biomaterials - Properties, Production and Devices

Current Advances in Biopolymer Processing and Characterization
Martin Koller (Editor)
2017. ISBN: 978-1-53612-710-2 (Softcover)
2017. ISBN: 978-1-53612-711-9 (eBook)

Hydroxyapatite and Other Calcium Orthophosphates: Biocomposites, Self-Setting Formulations and Dissolution
Sergey V. Dorozhkin (Editor)
2017. ISBN: 978-1-53611-936-7 (Hardcover)
2017. ISBN: 978-1-53611-975-6 (eBook)

Hydroxyapatite and Other Calcium Orthophosphates: Nanodimensional, Multiphasic and Amorphous Formulations
Sergey V. Dorozhkin (Editor)
2017. ISBN: 978-1-53611-902-2 (Hardcover)
2017. ISBN: 978-1-53611-902-2 (eBook)

Hydroxyapatite and Other Calcium Orthophosphates: General Information and History
Sergey V. Dorozhkin (Editor)
2017. ISBN: 978-1-53611-901-5 (Hardcover)
2017. ISBN: 978-1-53611-921-3 (eBook)

Hydroxyapatite and Other Calcium Orthophosphates: Bioceramics, Coatings and Dental Applications
Sergey V. Dorozhkin (Editor)
2017. ISBN: 978-1-53611-897-1 (Hardcover)
2017. ISBN: 978-1-53611-916-9 (eBook)

More information about this series can be found at
https://novapublishers.com/product-category/series/biomaterials-properties-production-and-devices/

Isaac Lavoie
Editor

What to Know about Hydroxyapatite

nova
science publishers

www.novapublishers.com

Copyright © 2023 by Nova Science Publishers, Inc.

All rights reserved. No part of this book may be reproduced, stored in a retrieval system or transmitted in any form or by any means: electronic, electrostatic, magnetic, tape, mechanical photocopying, recording or otherwise without the written permission of the Publisher.

We have partnered with Copyright Clearance Center to make it easy for you to obtain permissions to reuse content from this publication. Simply navigate to this publication's page on Nova's website and locate the "Get Permission" button below the title description. This button is linked directly to the title's permission page on copyright.com. Alternatively, you can visit copyright.com and search by title, ISBN, or ISSN.

For further questions about using the service on copyright.com, please contact:
Copyright Clearance Center
Phone: +1-(978) 750-8400 Fax: +1-(978) 750-4470 E-mail: info@copyright.com.

NOTICE TO THE READER

The Publisher has taken reasonable care in the preparation of this book, but makes no expressed or implied warranty of any kind and assumes no responsibility for any errors or omissions. No liability is assumed for incidental or consequential damages in connection with or arising out of information contained in this book. The Publisher shall not be liable for any special, consequential, or exemplary damages resulting, in whole or in part, from the readers' use of, or reliance upon, this material. Any parts of this book based on government reports are so indicated and copyright is claimed for those parts to the extent applicable to compilations of such works.

Independent verification should be sought for any data, advice or recommendations contained in this book. In addition, no responsibility is assumed by the Publisher for any injury and/or damage to persons or property arising from any methods, products, instructions, ideas or otherwise contained in this publication.

This publication is designed to provide accurate and authoritative information with regard to the subject matter covered herein. It is sold with the clear understanding that the Publisher is not engaged in rendering legal or any other professional services. If legal or any other expert assistance is required, the services of a competent person should be sought. FROM A DECLARATION OF PARTICIPANTS JOINTLY ADOPTED BY A COMMITTEE OF THE AMERICAN BAR ASSOCIATION AND A COMMITTEE OF PUBLISHERS.

Additional color graphics may be available in the e-book version of this book.

Library of Congress Cataloging-in-Publication Data

ISBN: 979-8-88697-523-9

Published by Nova Science Publishers, Inc. † New York

Contents

Preface

This book is a collection of eight chapters which provide an overview of what one should know about hydroxyapatite (Ha). Chapter One reviews the modification of hydroxyapatite coatings for implant applications. Chapter Two discusses the removal of pollutants by means of hydroxyapatite composites. Chapter Three reviews the synthesis and characterization of hydroxyapatite. Chapter four explores the potential use of functionalized hydroxyapatite in bone regeneration. Chapter Five reviews the applications of modified hydroxyapatite in catalysis. Chapter Six examines hydroxyapatite in terms of sustainable product development and waste valorization. Chapter Seven reviews the deposition of Ha-based coatings on bio-inert substrates by various deposition methods for improved bio-mechanical properties. Chapter Eight explores 3D printed Ha ceramics and its composite scaffolds for tissue engineering.

Chapter 1 - Hydroxyapatites (HA) are ceramic coatings of the calcium phosphate group that researchers in dental and bone implants have highly regarded due to their high resemblance to bone and tooth constituents. Much research has been done on HA coatings applied to magnesium, stainless steel, NiTi alloy, and titanium implants to increase their compatibility with the body's physiological environment and accelerate bone tissue formation. Still, there are problems with the mechanical strength of this coating. And their adhesion has led to restrictions on their use. Moreover, the durability of these coatings in the physiological environment of the body is still discussed in many scientific articles. Therefore, many efforts have been made to increase corrosion resistance, adhesion, mechanical properties, and compatibility with the body environment. These efforts can be classified into three groups: the method of application of these coatings (electrodeposition, electrophoretic deposition. Sol-gel, dip coating, plasma spray, biomimetic coating, etc.), changes in the chemical structure of HA using the addition of different chemicals (cerium, magnesium, fluorine, carbon nanotubes, TiO_2, etc.), and post-treatment using various biocompatible polymers (chitosan, polyamides, polyvinyl alcohol, gelatin, collagen, etc.). In this chapter, the effects of the modifications on the corrosion resistance properties of HA coatings are collected to make an efficient comparison by examining the mechanism of action.

Chapter 2 - Water is an essential need for human life, which is polluted due to various reasons. The water pollution is a universal subject and the world community is facing problems caused by it. One of the main causes of environmental water pollution is anthropogenic actions and industrial activities in the world. Water pollutants include thousands of organic (pesticides, fertilizers, hydrocarbons, phenols, plasticizers, biphenyls, detergents, oils, greases, pharmaceuticals, etc.), inorganic (heavy metals, nitrate, sulfate, phosphate, fluoride, etc.), and biological pollutants (virus, bacteria, fungi, algae, amoebas, and planktons, etc.). Many studies were performed to remove emerging pollutants by the adsorption process. These processes do not add unfavorable by-products to the environment and have been proven to be superior to other wastewater treatment techniques in terms of the simplicity of design, operation, and insensitivity to toxic compounds. Due to the presence of both cationic and anionic groups in the structure of hydroxyapatite, this compound is a good candidate for absorbing cationic and anionic pollutants from water. Hydroxyapatite is widely used for adsorbing pollutants from water due to its properties such as low cost, ion exchange properties, high biocompatibility, and high adsorption capacity. Although, the use of pure HAP nanoparticles is very useful for removing toxic substances, but its preparation process is expensive. Also, the amount of HAP nanoparticles during filtration shows a significant drop due to their small size. A literature review shows how to overcome these problems as well as the applications and advantages of HAP nanoparticles. It is essential to prepare adsorbents based on HAP in a usable form by combining them with other inorganic materials such as synthetic and natural zeolite, Fe3O4, MnO2 and Al2O3.

Chapter 3 - Nanocrystalline hydroxyapatite (HAp) has been successfully used for biomedical applications during the last decades owing to its chemical similarity with natural apatite, its remarkable biocompatibility, bio-affinity, and bioactivity that can be related to its crystal structure. HAp is the most stable, least soluble of all calcium orthophosphates, and is one of the most promising calcium phosphate biomaterials which can be obtained from chemical reagent, natural sources, animal-like dental enamel, corals, etc. Thermally stable HAp powders have been prepared using wet chemical route and co-precipitation method from various precursors such as naturally abundant seashell waste, calcium hydroxide Ca(OH)6 and mono ammonium phosphate NH4H2PO4 as calcium and phosphate sources, respectively. The crystal structure, morphology, elemental analysis, molecular bonding, and electric properties of the synthesized HAp powders have been studied using X-ray diffraction (XRD), scanning electron microscopy (SEM), energy-dispersive X-ray spectroscopy (EDS), Raman spectroscopy, Fourier transform infrared spectroscopy (FT-IR), and complex impedance spectroscopy. The effect of simulated body fluids immersion on the molecular bending and electric properties is also reported.

Chapter 4 - From 1990 to 2019, there were globally 178 million new cases of fracture which accounts for a 33.4% increment. The epidemiological studies explored that in the last three decades the years lived with disabilities (YLD) have increased to 65.3%. The challenges of orthopedic surgery still deal with the dearth of satisfactory solutions for the healing of large osseous defects. Even though autografts are the gold standard, surgeons must rely on synthetic graft alternatives due to a lack of availability and donor site morbidity. Due to its close physio-biological and chemical similarity to natural bone, osteoconductive nature, and ease of adaptability with many other necessary elements for functionalization, calcium phosphate hydroxyapatite (HA, Ca10(PO4)6(OH)2) has gained considerable attention as a bone graft substitute since 1980. They consistently aid in bone growth on the implant surface. However, limited bioactivity is the main hindrance behind their wide use in orthopedic surgeries. In recent years, several strategies have been undertaken to improve the biological performances with the various ionic substitutions with the apatite. Functionalization is one of the key processes to alter the surface properties of materials aiming toward specific properties to be induced depending on the target tissue or to minimize any unwanted negative or hindrance factors. Functionalization of HA with specific biomolecules is the process by which the surface of HA is made more bioactive i.e., increment of bio integration and osteogenesis. The use of functionalized HA is not only restricted to bone tissue engineering, they are equally important for skin, cartilage, periodontal, vascular tissue, hepatic tissue, tendon, ligament as well as corneal tissue engineering. But, in this chapter, special emphasis has been given to the brief introduction of HA, basic properties, different synthesis techniques, summarized application in biomedical engineering in general, and especially the use and scope of clinical translational aspect of functionalized HA in bone tissue engineering.

Chapter 5 - Hydroxyapatite (HAP, Ca10(PO4)6(OH)2) and its applications in catalysis have attracted considerable attention in recent years.

Mild acid and basic sites uniformly distributed on the surface of HAP makes it an attractive solid catalyst for condensation reactions with high conversion, selectivity, and stability. Thus, HAP is the subject of increasing interest in producing chemicals from oxygenated molecules such as ethanol. This chapter addresses the various modified hydroxyapatites applications in catalysis. The first part of the review will include HAP as a catalyst totally or partially substituted with different ions. The unique structure of HAP allows it to accept many anionic and cationic substituents. The second part will present HAP as the active phase support. Different metals supported on HAP showed improvement in the many catalytic reactions. More recent findings and the current challenges owing to the development of modified hydroxyapatite are summarized.

Chapter 6 - Million tons of shell waste are dumped in oceans and landfills and considered nuisance waste. Rather than disposing of the shells, recycling and reusing are excellent alternatives. In general, shells are rich in calcium carbonate, so sustainable development of calcium-based products may help reduce waste shell accumulation. Hydroxyapatite is a calcium phosphate-based mineral having similar composition and structure to natural human bone and teeth. Excellent biocompatibility, bioactivity, biodegradability, and non-toxicity; are some properties that enable hydroxyapatite to use as a superior biomaterial. The demand to obtain hydroxyapatite in a simple, efficient, scalable, and environmentally friendly way is increasing daily. Not only from shells but hydroxyapatite can also be synthesized from animal bones, fish scales, plants, and algae. This review discussed the highlights of hydroxyapatite and its natural sources, synthesis, and biomedical applications. Moreover, it also explained the importance of hydroxyapatite synthesized from biogenic waste materials. From this review, researchers may get an idea of the synthesis of hydroxyapatite from various sources in the development of biomedical applications and the conservation of food and agricultural waste management in terms of waste valorization.

Chapter 7 - Hydroxyapatite (HA) is an attractive bioactive material in the biomedical field due to its chemical composition and structure which are similar to the mineral components of natural bone. It is a highly bio-compatible and osteo-conductive material. However, the low bending strength and fracture toughness of pure HA restrict it from implant applications. Although bio-inert materials such as titanium alloy and bio-grade stainless steel have excellent mechanical properties, they show poor osteo-conductivity. Coating of HA applied to these metallic materials can be a solution to the problem. The HA-coated metallic implant can have superior mechanical and biological characteristics. HA powder can be synthesized from chemical reagents or biogenic materials by using various routes. Chemical composition and other properties of biogenic material are preserved in the synthesized biogenic-derived hydroxyapatite. The HA powder is deposited as a coating on bio-inert metals for biomedical implant applications by using several deposition methods. In this chapter, several techniques such as plasma spray, electrophoretic deposition, sputtering deposition, pulsed laser deposition, spin coating, ion beam assisted deposition, sol-gel deposition, electrodeposition, cold spray, and suspension spray techniques which were used to deposit HA coatings are discussed thoroughly. The characteristics of the HA coatings in terms of their microstructure, phases, mechanical properties, and biological behavior are described in detail.

Chapter 8 - Hydroxyapatite (HA) ceramic has gained tremendous attention in tissue engineering owing to its excellent mechanical properties, biocompatibility, and osteoconductivity. In most tissue engineering cases, complex-shaped HA scaffolds are usually needed, which brings great challenges to their manufacturing. Recently, additive manufacturing, also known as 3D printing, has been widely studied for its fabrication ability of complex-shaped HA ceramic and its composite scaffolds. Herein, this chapter introduces the latest advances of 3D printed HA ceramic and its composite scaffolds, including HA, HA/TCP, and various HA/polymer composite scaffolds. The performances of these scaffolds, including their mechanical properties, and in vitro and in vivo biological behaviors, are further summarized. Finally, the challenges and opportunities involved in this field are discussed and forecasted. It is believed this chapter can give some guidance for researchers or graduate students in the field of 3D printing of bioceramic scaffolds for tissue engineering.

Chapter 1

Modification of Hydroxyapatite Coatings for Implant Applications

R. Ghamsarizade[1], H. Eivaz Mohammadloo[2,*] and Sh. Roshan[1]

[1]Department of Polymer Engineering and Color Technology,
Amirkabir University of Technology, Tehran, Iran
[2]Colour, Resin & Surface Coatings Department,
Iran Polymer and Petrochemical Institute, Tehran, Iran

Abstract

Hydroxyapatites (HA) are ceramic coatings of the calcium phosphate group that researchers in dental and bone implants have highly regarded due to their high resemblance to bone and tooth constituents. Much research has been done on HA coatings applied to magnesium, stainless steel, NiTi alloy, and titanium implants to increase their compatibility with the body's physiological environment and accelerate bone tissue formation. Still, there are problems with the mechanical strength of this coating. And their adhesion has led to restrictions on their use. Moreover, the durability of these coatings in the physiological environment of the body is still discussed in many scientific articles. Therefore, many efforts have been made to increase corrosion resistance, adhesion, mechanical properties, and compatibility with the body environment. These efforts can be classified into three groups: the method of application of these coatings (electrodeposition, electrophoretic deposition. Sol-gel, dip coating, plasma spray, biomimetic coating, etc.), changes in the chemical structure of HA using the addition of different chemicals (cerium, magnesium, fluorine, carbon nanotubes, TiO2, etc.), and post-treatment

*esponding Author's Email: h.mohammadloo@ippi.ac.ir.

ıat to Know about Hydroxyapatite
: Isaac Lavoie
979-8-88697-523-9
› Nova Science Publishers, Inc.

using various biocompatible polymers (chitosan, polyamides, polyvinyl alcohol, gelatin, collagen, etc.). In this chapter, the effects of the modifications on the corrosion resistance properties of HA coatings are collected to make an efficient comparison by examining the mechanism of action.

Keywords: hydroxyapatites, implants, corrosion

Introduction

Orthopedic implants are typically made of one of the following metals or metal alloys: titanium, stainless steel 316L, titanium, magnesium, or metal alloys centered on cobalt-chrome (Nasab, Hassan, and Sahari 2010). Due to their outstanding mechanical strength, longevity with wear, and corrosion resistance characteristics, they are in this spot. The mechanical features of metallic implants and the benefits and drawbacks linked with their use are outlined in the following. To overcome the corrosion, biocompatibility, and bone growth issues, various types of the coating were developed, and hydroxyapatite (HA) is one of the most promising coatings among them. The HA's shortcoming is the poor adhesion to the metallic surface of implants as well as the film uniformity, corrosion resistance, crystallinity, and thickness. Therefore, plenty of research was conducted to enhance all the mentioned factors to achieve implants with long-term durability. These attempts can be categorized into the method of HA deposition on the surface, pre/post treatments, and changing the chemical composition of HA by adding other minerals and biopolymers. In this chapter, firstly and foremost, the properties of the implant are presented and compared to the natural bone, then the standard coating for medical implants is stated. Next, the methods to apply HA are explained, and at the end, the attempts to chemically enhance the HA essential factors are provided. Some applications of HA are listed in Figure 1.

Implants Types and Properties

The natural bone factors are: density = 1.7 to 2 g/cm^3, compressive strength = 164 to 240 MPa[1], tensile strength = 80 to 150 MPa, Young's modulus = 3 to 30 GPa, and elongation at fracture = 3 to 4% (Córdoba 2016).

[1] Mega pascal.

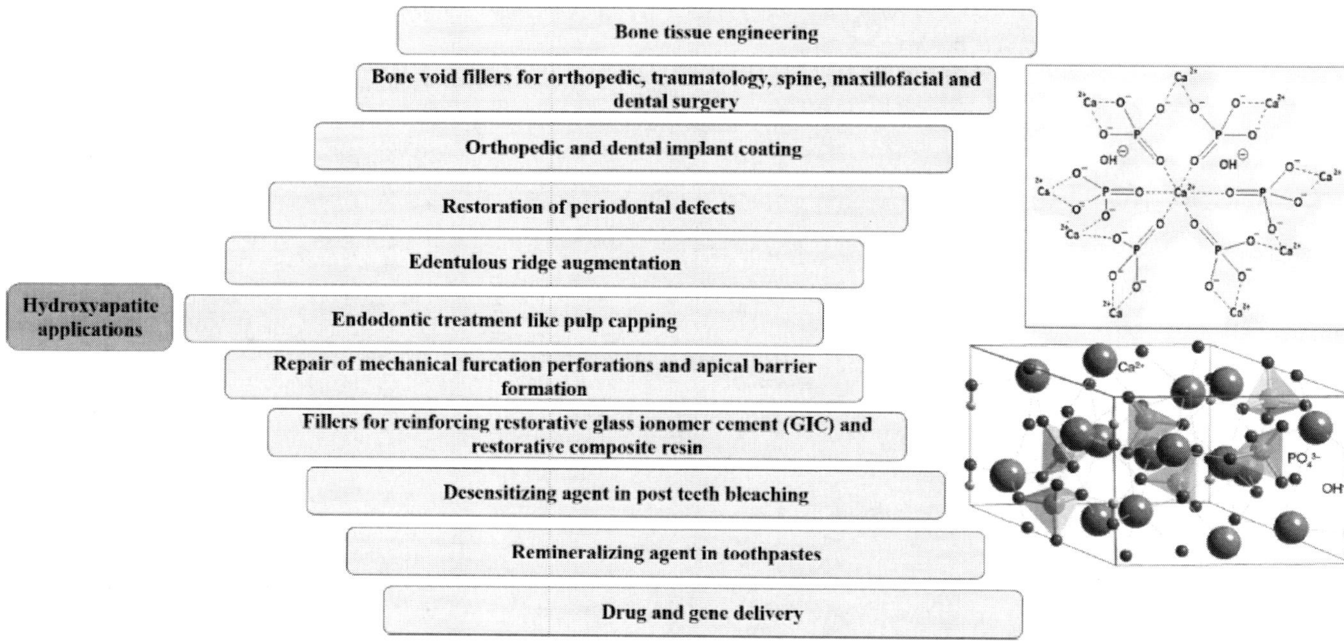

Figure 1. Applications of HA in the biomaterials.

Numerous metals and alloys have been introduced as the implants, including stainless steel, Co-based alloys, Ti-based alloys, Ta-based alloys, Mg-based alloys, and nitinol; some of them are detailed in the following.

Stainless steel 316L has a density of 7.9 to 8.1 g/cm^3 and an elastic modulus of 165 to 200 GPa[1]. The compressive strength is in the range of 170 to 310 MPa, the tensile strength is in the range of 480 to 620 MPa, and its elongation at break is 30 to 40%. The advantages of these implants are suitable corrosion resistance and wear resistance, but high elastic modulus is its disadvantage. Moreover, the high elastic modulus hinders bone regeneration. The main use of this type of implant is in temporary devices such as hip nails, screws, plates, and total hip replacement (Manivasagam et al. 2009). The corrosion reactions taking place for this implant is given in the Equations (1-4).

$$Fe \rightarrow Fe^{2+} + 2e^- \quad (1)$$

$$Fe^{2+} \rightarrow Fe^{3+} + e^- \quad (2)$$

$$Fe + nH_2O + Cl^- \rightarrow FeClOH^- + (n-1)H_2O + H^+ + e^- \quad (3)$$

$$2Fe^{2+} + O_2 + 2H_2O + 2e^- \rightarrow 2FeOOH + 2H^+ \quad (4)$$

Titanium alloy (Ti-6Al-4V) with a density of 4.43 g/cm^3 has an elastic modulus of 100 to 110 GPa. The compressive strength of this alloy varies from 758 to 1117 MPa, the tensile strength of it ranges from 930 to 1140 MPa, and its elongation at break is recorded between 8 to 15%. Compared to other implants, it has higher biocompatibility and osseointegration (Manivasagam et al. 2009). Remarkable tensile strength, fatigue stress, and fracture toughness are the advantages of these implants. In contrast, it may be toxic for the body because of the V and Al elements. It is costly compared to the other implants.

Moreover, the corrosion resistance and wear characterizations are lower than stainless steel and Co-Cr alloys but higher than magnesium implants. These implants are usually used in permanent devices such as nails and pacemakers. The corrosion reactions taking place for this implant are given in Equations (5-10).

$$Ti \rightarrow Ti^{4+} + 4e^- \quad (5)$$

[1] Giga pascal.

$$Ti^{4+} + H_2O \rightarrow Ti(OH)^{3+} + \mathrm{H}^+ \quad (6)$$

$$Ti(OH)^{3+} + H_2O \rightarrow Ti(OH)_2^{2+} + \mathrm{H}^+ \quad (7)$$

$$Ti(OH)_2^{2+} + H_2O \rightarrow Ti(OH)_3^{+} + \mathrm{H}^+ \quad (8)$$

$$Ti(OH)_3^{+} + H_2O \rightarrow Ti(OH)_4 + \mathrm{H}^+ \quad (9)$$

$$4H^+ + O_2 + 4e^- \rightarrow 2H_2O \quad (10)$$

Co-Cr alloys, as another sort of implant, are heavier and stronger than the other implants since they got a density of 9.2 g/cm^3 and an elastic modulus of 210 GPa. The tensile strength of this alloy is ranged from 908 to 1282 MPa. The elongation at break of this material is recorded between 8 to 41%. Regrettably, due to the existence of Co, Cr, and Ni elements, these implants might be dangerous for in vivo employment. In addition, these implants suffer from low corrosion, wear, and friction resistance. The main use of this type of implant is in total hip replacement and dentistry casting (Manivasagam et al. 2009). The galvanic cell between the Co and Cr phases causes the corrosion of this implant (Jansen van Vuuren, Odendaal, and Pistorius 2008).

Finally, magnesium alloys are introduced as the lightest implants, with a density of 1.74 g/cm^3. The elastic modulus of these implants varies from 41 to 45 on average. The tensile strength of this material is 240 MPa, and the elongation at the break of it is recorded at 3%. The advantages of these types of implants are: that the characteristics of these implants are incredibly close to the natural bone. These implants decrease the stress shielding effects.

Moreover, they are easy to produce in complex forms of geometry. They support load-bearing applications due to their lightweight, and Mg^{2+} is necessary for the human body. It is worth mentioning that these implants provide stimulatory impacts for bone regeneration. Conversely, they suffer from low corrosion resistance, hydrogen gas release, and early mechanical strength loss. These implants are now being used in biodegradable orthopedic implants (Sillekens and Bormann 2012). The corrosion reactions taking place for this implant are given in Equations (11-14).

$$Mg \rightarrow Mg^{2+} + 2e^- \quad (11)$$

$$2H_2O + 2e^- \rightarrow H_2 + 2\mathrm{OH}^- \quad (12)$$

$$Mg^{2+} + 2\text{OH}^- \rightarrow Mg(OH)_2 \tag{13}$$

$$Mg(OH)_2 + 2\text{Cl}^- \rightarrow MgCl_2 + 2\text{OH}^- \tag{14}$$

Implants Coatings

For tissue regeneration, magnesium-based materials are thought to be as biologically active, biodegradable, and bio-tolerant as possible (Maier et al. 2020). These metals and alloys are typically shaped as fracture plates, bones, hip nails, pins, wires, joint caps, and screws (Hanawa 1999). Such implants are inserted to keep the bones in their proper position and aid in osteogenesis with the bone cells. When implants are used to support tissue, they need to be biocompatible and provide cues to the host cells to initiate the process of natural cell recovery as quickly as possible. Coatings applied to biomedical implants are a potentially useful method for enhancing the implant-tissue interactions, promoting their biocompatibility and bifunctionality, and doing so without affecting the properties of the underlying substance (Priyadarshini et al. 2019; Jiang et al. 2019). Bioactive coating materials include HA[2], calcium phosphate salts, and bio-active glasses. Inactive coatings like zirconia are also used as a covering for the implants. Interestingly, HA (density = 3.05 to 3.15 g/cm^3, compressive strength = 100 to 900 MPa, tensile strength = 40 to 200 MPa, and young's modulus = 70 to 120 GPa) is particularly popular among the mentioned coatings. Ceramics are the material category that encompasses these substances, which have excellent chemical interoperability with physiological medium and rigid tissues, such as bones and teeth (Arcos and Vallet-Regí 2020). Moreover, excluding bioactive glass, all these materials show chemical and structural similarities with natural apatite (Niinomi 2002; Sridhar 2010). Using a precursor that is a hydrated calcium phosphate compound is by far the most widespread technique for the preparation of HA (Francis and Webb 1970), although there are other ways to create HA.

In 1786, Werner was the first person to refer to HA as a mineral. As a phosphatic substance on the earth, HA can be found in very high concentrations. HA is very similar to natural bone in terms of its microstructure and its chemical composition. Crystalline in appearance, and it has a density of 3.22 g/cm^3. Most importantly, though, Ca^{2+}, PO_4^{3-}, and OH^-

[2] Hydroxyapatite.

ions are tightly packed together to form the apatite structure in each HA unit cell (GRAFTS and SUBSTITUTES 2004). This results in the formation of apatite. Dental enamel and dentin both contain a significant amount of carbonated HA lacking in calcium (Martin and Brown 1995). Calcium and phosphorus are combined in the HA in a molecular ratio of 1.67. This solid apatite is primarily steady to the calcium phosphate salt under environmental conditions and an extensive array of pH ranging from 4 to 12. Changes in HA crystal's chemical composition, crystallinity, size, and shape can significantly shift the material's attributes (Surmenev, Surmeneva, and Ivanova 2014; Prakasam et al. 2015). Compared to HA in its bulk form, HA nanostructures have superior mechanical properties alongside increased interaction and osteoclast capacity (Lu et al. 2020). This is due to the fact that HA nanostructures provide greater surface energy than the other nanoparticle (Sadat-Shojai et al. 2013). Because of these and other significant characteristics, HA is currently the most favorable choice for the coating of orthopaedical implants. The requirements from a HA coating to possess the mechanical, biological, and anti-corrosion factors are Ca/P ratio=1.67 to 1.76, phase purity= 95% in minimum, crystallinity= 62% in minimum, tensile strength more than 50.8 MPa, shear strength more than 22 MPa, and presence of the heavy metals lower than 50 ppm. There are various calcium phosphate structures which are all represented in Table 1 (Berndt et al. 2014). The reactions of HA formation are outlined below (Equations (15-18)).

$$3Ca^{2+} + 2PO_4^{3-} + 4H_2O \rightarrow Ca_3(PO_4)_2.4H_2O \downarrow \quad (15)$$

$$5Ca^{2+} + 3PO_4^{3-} + OH^- \rightarrow 0.5Ca_{10}(PO_4)_6(OH)_2 \downarrow \quad (16)$$

$$2H_2O + 10Ca^{2+} + 6HPO_4^{2-} \rightarrow Ca_{10}(PO_4)_6(OH)_2 \downarrow +8H^+ \quad (17)$$

$$2H_2O + Ca^{2+} + HPO_4^{2-} \rightarrow CaHPO_4.2H_2O \downarrow \quad (18)$$

In addition, HA decomposition reactions are provided in Equations (19-22).

$$Ca_{10}(PO_4)_6(OH)_2 + 2H^+ \rightarrow Ca_{10}(PO_4)_6(H_2O)_2^+ \quad (19)$$

$$Ca_{10}(PO_4)_6(H_2O)_2^+ \rightarrow 3Ca_3(PO_4)_2 + Ca^{2+} + 2H_2O \quad (20)$$

$$Ca_3(PO_4)_2 + 2H^+ \rightarrow Ca^{2+} + 2CaHPO_4 \tag{21}$$

$$CaHPO_4 + H^+ \rightarrow Ca^{2+} + H_2PO_4^- \tag{22}$$

As a result of HA's less-than-perfect crystalline structure, the HA film connection on the metallic surface begins to weaken and finally breaks down (Lacefield 1988). As a consequence of this breakdown, metallic ions are released into the body (Mahapatro 2015) as the metal surface becomes exposed to the surrounding environment. To improve the adhesion of HA films, surface adjusting agents are necessary. These agents contribute to creating a long-lasting film atop the metallic surface, which helps to achieve the desired goal. Because of its biocompatibility, HA, the primary inorganic component of hard tissues (bones), has been utilized in various biomedical applications over the past fifty years.

Table 1. Types of calcium phosphate crystals, their solubility in water, and pH range of their stability

Chemical formula	Chemical definition	Symbol	Ca/P ratio	Solubility in water at 25°C (g/L)	Stability range pH
$Ca(H_2PO_4)_2$	Monocalcium phosphate hydrate	MCP	0.50	18	0.0-2.0
$CaHPO_4.2H_2O$	Dicalcium phosphate dihydrate	DCPD	1.00	0.088	2.0-6.0
$Ca_8H_2(PO_4)_6.5H_2O$	Octa-calcium phosphate	OCP	1.33	0.0081	5.5-7.0
α-$Ca_3(PO_4)_2$	α-tricalcium phosphate	α-TCP	1.50	0.0025	-
β-$Ca_3(PO_4)_2$	β-tricalcium phosphate	β-TCP	1.50	0.0005	-
$Ca_4(PO_4)_2O$	Tetra-calcium phosphate	TTCP	2.00	0.0007	-
$Ca_{10}(PO_4)_6F_2$	Fluorapatite	FA	1.67	0.0002	7.0-12.0
$Ca_{10}(PO_4)_6O$	Oxy-apatite	OA	1.67	0.087	-
$Ca_{10}(PO_4)_6(OH)_2$	hydroxyapatite	HA	1.67	0.0003	9.5-12.0

On the other hand, research conducted in the past found that HA possesses the characteristics of brittle ceramics that cannot handle the load of pressure (Abu Bakar et al. 2003). Because of earlier research findings, HA was chosen to serve as the bioactive coating material to improve mechanical qualities. Due to the bioactivity of this layer, the wound is given the assistance it needs to regenerate more effectively (Roy et al. 2008). Only a handful of research projects have been conducted employing a mixture of HA and other mineral substances to produce composite products with improved mechanical performance (Kannan 2015). In their work (Witte et al. 2007), Witte and colleagues developed a metallic composite by combining HA particles with AZ91 magnesium alloy as a reinforced filler in a metallic matrix. The particle size distribution of HA significantly impacts the mechanical characteristics of composite materials.

Moreover, biological uses of the combination of chitosan and HA are being investigated. The preliminary study has examined the adhesive bond between HA/chitosan film and metallic surfaces. Compared to a standard HA film, the HA/chitosan film possesses superior tensile strength. Including the active ingredient calcium silicate, which forms pores with HA on the film surface, contributes to the increased bioactivity and biocompatibility offered by the composite coating compared to that of the HA film (Kezhi et al. 2017).

The adhesion of adhesive on the substrate will decide the surface and mechanical properties of metals used for biomedical applications. The improper coating on the substrate due to variation in process parameters increases the chances of sudden collapse when used as the bone supporting implants (Bauer et al. 2013; Tao et al. 2015). The poor coating layers will exfoliate and create serious effects during tissue healing. The coating pieces will leave the surface due to poor strength and affect the surrounding body parts (Huang et al. 2016; Zhao et al. 2016). The coating on metallic substrates is widely used to make it bio-compatible and bio-active. The crystalline nature of HA offers better mechanical behavior against load and stress. The stability of HA film is the biggest challenge in these coating implants. The complete degradation of the HA occurred within 4–5 years by natural cell mechanism with the assistance of body environments like pH, water content, and osteoclast cells. In vivo degradation of HA can be started by dissolution or cell mediation. The speed of degradation also depends upon the chemical properties of HA, Ca/P ratio, crystal size, crystallinity, and porosity (Hannink and Arts 2011).

After this time, the malfunction (dissolution) can occur starting from the surface and penetrating to the metallic substrate. Thus, stable HA film will provide a reliable metallic implant for load-bearing applications. Operational difficulties during the coating process also create problems for successful HA coating with uniformity on metallic implants. The porosity and development of cracks on the coating surface are also issues for the biomedical implant. Porosity is an essential parameter for judging the quality of ceramic coatings on biomedical implants. The porosity in coating layers dictates its bioactivity for biomedical implants. HA offers more bioactivity and biocompatibility due to active agents such as calcium and phosphate, similar to human bone minerals (Mediaswanti et al. 2013; Berndt et al. 2014). Further enhancement of adhesive strength can be achieved by increasing the surface roughness and pores. The porous implants coated with HA provide dual benefits to enhance bioactivity further. These implants have Young's Modulus closer to that of bone, addressing the stress shielding effect and the increased surface area coated with HA for good osseointegration (Ma et al. 2016; Almeida, Gama, and Vueva 2011). A stress shielding effect arises when the modulus of the implant is greater than the young's modulus of bone. The support of the implant to the damaged bone is reduced due to this phenomenon. Due to this, the implant takes the loads and is shielded from going to the damaged bones (Mi et al. 2007). According to Wolff's law, a bone develops its structure depending on its force. The area of bone that experiences a higher load will increase bone density. The bone mass decrease results in the implant loosening (Bugbee, Sychterz, and Engh 1996). Porous coating for bone regeneration allows the migration and proliferation of osteoblast cells and vascularization.

Additionally, the pores on the surface facilitate better mechanical interlocking between implant and bone. This interlocking provides higher mechanical stability and reduces stress shielding (Hannink and Arts 2011). Most pores are open and interconnected with each other. In vivo tests have shown that the pores facilitate cell migration, tissue growth, and transport of waste products (Bellucci et al. 2019).

The history behind HA uses in bioimplants goes back to the 1950s, when the HA was firstly synthesized for biomedical implants. In 1980, the HA coating was commercialized in the US, Europe, and Japan. The first hip prosthesis implant was coated with 200 microns of HA and used in-vivo in 1985. Utilizing the nano HA was done for the first time in 1994. Since today, numerous developments have been done in HA coatings for biomedical implants, as stated in the following sections.

Methods of HA Coatings on the Implants

There are a variety of coating methodologies that can be used to apply HA to metallic implants (Awasthi et al. 2021). Utilizing HA is a straightforward method for compensating for the non-bioactive nature of metals. Because of the substantial interaction between the HA layer and the host tissue, the metallic implant with the HA coating motivates the growth of new bone (Song,

Shan, and Han 2008). Furthermore, HA coating even serves the function of a corrosion control film against risky biological fluids for these functions. This HA film further decreases the rate at which metallic ions dissolve, reducing the likelihood of any leaching occurring (Zhong, Qin, and Ma 2015). Several different coating techniques can be used to prepay HA on metallic implants. Sol-gel, dip coating, electrochemical deposition, chemical vapor deposition, thermal spraying, radio frequency magnetron sputtering, micro-arc oxidation, high-velocity suspension flame spraying, plasma spraying, and pulsed laser deposition are some cases of the methodologies that fall into this category (Surmenev and Surmeneva 2019). Because it forms a coating layer that is homogeneous all over the surface of the metal, thermal spray coating is the method that is currently considered to be the most effective for use on metallic implants (Cizek and Matejicek 2018). The ineffective adhesion of HA to the metallic surface is the primary source of worry during the application process of HA coatings. This is because the adhesive bond between the metallic load-bearing locations and the HA layer is not particularly strong (Mohseni, Zalnezhad, and Bushroa 2014).

In the following section, the process of applying HA on implants by each of the methods mentioned above is discussed. Also, the benefits and difficulties of each are stated.

Sol-Gel and Dip Coating

A substantial amount of literature utilizing sol-gels and the utilization in metallic implants through the use of the dip-coating process has been conducted until this point. The coatings that can be formed using this method have the greatest bonding strength of any other techniques now known (Zhang, Guan, and Zhang 2011; Yuan and Golden 2009; Suwanprateeb et al. 2018). It is possible to fabricate a wide range of products using the sol-gel process (Almeida, Gama, and Vueva 2011), contributing to the method's achievements. This allows the materials to have a governed porous morphology. Most calcium and phosphate-based compounds are included in the precursors employed to create sol-gel. For sol-gel fabrication, ethanol and water are frequently used (Zhang et al. 2006; Combes and Rey 2010; Costa, Dixon, and Rizkalla 2012). Water is considered the widespread solvent. Ethanol is used to dissolve organic phosphorus substances, while water is used for most sol synthesis processes (Liu, Troczynski, and Tseng 2001; Alves Cardoso, Jansen, and G. Leeuwenburgh 2012). The phosphorus solutions are gradually and carefully added to the calcium salt solutions (Ben-Nissan and Choi 2006; Choi and Ben-Nissan 2007; Davar and Shayan 2017). Calcium is typically obtained from calcium nitrate salt when used in HA here as a source of calcium. After being generated individually, all these solutions are combined and then warmed at a range of temperatures corresponding to different reactions. Evaporating the extra solvents that exist in the solution mixture is one of the steps taken in preparing sol-gel suspension. The viscosity of the combination increases as a result of the evaporation of the solvent, which ends in a quite viscous mixture (Shadanbaz and Dias 2012). Both the creation of the apatite phase and the characteristics of the sol-gel material are greatly sensitive to the type of salts containing calcium and phosphorus utilized and the temperature at which the material is operated. The formation of sol-gel occurs after removing water, aging, drying, and calcination at extreme temperatures have been carried out.

Because of how straightforward and affordable the sol-gel method is, it has garnered a lot of support and has found widespread application; this approach is beneficial to others because it can cover objects with various geometric patterns, which is a common challenge in the surface coating. Furthermore, it has the potential to deliver increased coating strength combined with advanced adhesion (Yuan and Golden 2009; Qu and Wei 2008; Bakan, Laçin, and Sarac 2013). The dip-coating technique is combined with the sol-gel method to produce an even coating layer on metallic bio-implants. The dip-coating way has three distinct stages: the first is the dipping process, followed by the removing stage, and finally, the drying phase. This method has many affecting factors, including the pH adjustment, the temperature of the solution, time of immersion, the concentration of the solution, and pre-treatments; therefore, this method is complicated, although many researchers have used it. By this method, even complex shapes can be coated.

Biomimetic Method

The biomimetic method involves producing results similar to those produced by the body's natural bone-building mechanisms. In this manner, HA can be utilized to improve the osseointegration of natural bone as well as a manufactured implant that has been coated (Kokubo et al. 1990; Mali, Nune, and Misra 2016). Since it imitates the characteristics of natural bone tissues, the technique of biomimetic coating encourages the binding and multiplication of osteoblast cells. For a biomimetic strategy to work, the implant's surface must include hydroxyl groups, as it does for other chemical coating procedures. During the pre-treatment process, either an acid or an alkali can be used to facilitate a simple attachment of these substituents. These hydroxyl groups on the surface made CaP[3] nucleation easier, which then accelerated the crystal growth that led to the creation of apatite. The methods of the biomimetic coating were carried out under typical settings in terms of pH and temperature (Escada et al. 2011; Stigter, de Groot, and Layrolle 2002; Bharati, Sinha, and Basu 2005; Arrés et al. 2020).

Chemical Vapor Deposition

Chemical vapor deposition, also known as CVD, is a method of coating that employs volatile precursors to cover pre-heated surfaces through reaction or decomposing on the surface (Gunputh and Le 2020; Goto and Katsui 2015). Coating metal implants with HA and calcium phosphate-based coating is often accomplished via CVD (Cabañas and Vallet-Regí 2003). The ability of CVD to control the creation of crystalline phases and microstructures during coating is one of its most exciting and potentially valuable properties. Furthermore, it can homogeneously cover complicated metallic forms (Goto and Katsui 2015).

Spraying with Thermal Energy

The approach of thermal spray is attracting considerable interest and is being accepted as the current approach to coating since it gives bio-medical metallic implants admirable features. During the thermal spray process, precursors are applied to the appropriate sample in the form of a solution or a suspension to cover it and give the necessary qualifications. Flame spraying, plasma arc spraying, and electrical arc spraying are the three sub-groups that fall under the category of thermal spraying. When doing thermal spraying on surfaces, a few distinct operating strategies can be applied. These techniques include

[3] Calcium phosphate.

spraying with APS[4], VPS[5], LPS[6], SPS[7], HVOF[8], HVSFS[9], DGS[10], and GTPS[11]. These approaches were utilized to cover the metallic implants with HA (Rath et al. 2012; Albayrak, El-Atwani, and Altintas 2008; Gadow, Killinger, and Stiegler 2010; Heimann 2013; Saadati et al. 2020; Santos, Santos, and Carmezim 2019).

4.4.1. High-Velocity Suspension Flame Spray Coating

The high-velocity suspension flame spraying (HVSFS) approach was the first thermal spray process ever produced, and it was created in 1910 (Liu et al. 2018). Out of all the ways to apply a thermal spray coating, FS is the one that is the easiest and most cost-effective. Flame spray coating is initiated by igniting an oxygen torch to melt HA particles, which results in a porous and composite layer on metal substrates. In contrast to the various other thermal spray coating methods, FS has many drawbacks. Some of these downsides are a larger size of the microstructure and larger pores and fractures on the coating layer. As was discussed previously, FS is a cost-effective solution that is also simple to operate in terms of commercial procedures. The upgraded flames allow for a particle velocity ranging from 200 to 300 meters per second (m/s). To reach a more significant flame temperature of approximately 2600°C (Guo et al. 1995; Yang et al. 2017; Sun, Berndt, and Gross 2002; Antala and Rathod 2017), the fuel utilized to run the torch is a mixture of oxygen and acetylene. A zinc-doped HA was deposited using the flame spray technique on Ti-6Al-4V surfaces to improve its biocompatibility and antibacterial activity. This was done to increase its surface area (Yang et al. 2017). By employing flame spraying, Liu and his colleagues placed a porous HA coating on the titanium implant. The generation of porous coating requires wetting the metallic substrate before coating it. The wetting process promotes the proliferation and differentiation of pre-osteoblast cells to a greater extent. The coatings formed in moist conditions had several fractures ranging in size from nanometers to one hundred nanometers (Liu et al. 2018). By using the flame spray method, Monsalve et al. (Monsalve et al. 2015) were able to spray 316L steel and titanium alloys with a bioactive glass layer. The crystallinity of the coating

---

[4] Atmospheric plasma.
[5] Vacuum plasma.
[6] Liquid plasma.
[7] Suspension plasma.
[8] High-velocity oxy-fuel.
[9] High-velocity suspension flame spraying.
[10] Detonation gun spraying.
[11] Gas tunnel type plasma.

layer is impacted by the magnesium concentration present in the bioactive glass. Because of the increased concentration, the coating layer tends to be more crystallized, which encourages the creation of holes.

Moreover, titanium alloys have a lower thermal conductivity, which facilitates increased crystallinity in conjunction with a slower cooling rate. This contributes to the formation of various crystal structures. The construction of a layer of hydroxy carbonate apatite, which proved the bioactivity of the surface, takes place upon immersion of the coated substrate in the SBF solution.

4.4.2. Plasma Spraying

Plasma spray is a method that is frequently utilized in today's society for coating bio-active HA on bio-medical implants (Morks 2008; Fauchais and Vardelle 2011; Ročňáková et al. 2018). In order to melt and spray HA onto the metallic implants, this method utilized an electric arc that operated at an elevated temperature and pressure. Dry HA is used as the raw material for the coating process, and with the assistance of a heated plasma jet, it is transformed into plasma. Afterward, the high-temperature plasma that was formed interacts with the surface, and it then binds to it. Plasma depositing can also be done by a founded method known as air or vacuum spray, which is yet another sort of plasma deposit. When a surface is coated with HA using this method, the resulting coating is more durable and possesses unique advantages. The temperature of jets can range anywhere from 10,000 K to 30,000 K, and it will get cooler as the distance from the jet nozzle gets ever further away (Bakshi et al. 2009; Levingstone et al. 2015). This specific method is utilized to apply bioactive coatings on various biomedical implant materials. The plasma spray procedure has only one shortcoming: the HA structure deformation that occurs due to the high-temperature coating operation. Hot temperatures can occasionally reduce the interfacial adhesion between the HA coating and the metal surface (Fielding et al. 2012). Using secondary heat treatment, coated HA's structural drawbacks can be overcome in various ways. The coated HA was heated for 90 hours at 400°C. Consequently, the structure of HA is altered, and the crystalline phase of the molecule is elevated (Vahabzadeh et al. 2015). A greater crystalline structure favors strong adhesion strength between the injured tissues and the implant (Singh, Singh, and Prakash 2010).

On the other hand, heat treatment at a high temperature can help reduce the fatigue stresses present on the surface. This happens because the film thickness gets thinner as the color of the coating gets lighter. This post-coating heat treatment is carried out at a temperature of 700°C for one hour to remove any excess water and impurities (Popa et al. 2011; Chen, Lu, and Li 2006). This helps to make the coating purer. Due to the increased adhesion, the osteoconductivity of the metallic implants that have been coated with HA via plasma spray has been improved.

Pulsed Laser Deposition

Pulsed laser deposition (PLD) is a coating method that is used to coat various substrates with the assistance of a highly accelerated laser beam in the existence of a vacuum. This method is used to coat a wide range of materials. This laser beam makes contact with the surface we want to coat with the compound we have specified. The coating material evaporates, resulting in plasma formation at a high temperature. A very thin film is produced when this plasma comes into contact with the target material. A high vacuum atmosphere is preferred because it allows for superior coating and flawless film deposition of thin layers. The surface is also recommended to be heated to achieve homogeneity in the coating (Cotell, Chrisey, and Grabowski 1991). The PLD technique has garnered much interest thanks to its adaptable coating properties and its wide range of processing conditions. These factors can be categorized as homogeneity, thickness, crystallinity, and strength (Suchanek and Yoshimura 1998). The coating breakdown is connected with the residual tensions left over from this method, which is one of the technique's drawbacks. Because of the high temperature operating condition, these stresses are created. (Yang and Chang 2001) The variation in crystal structure between the HA coating and the metallic surface is another factor contributing to residual stresses. Because of these residual tensions, cracks can appear on coating films, although they are impossible to spot on very thick layers (1 um thickness). During the process of bone regeneration, surface roughness is indeed a significant factor. PLD results in a more homogenous covering that influences the responses of surrounding bone tissues for tissue regeneration. PLD results in broadly consistent coatings with a microscopically minor porous structure and a surface roughness that varies from a few to a few hundred nanometers (Zhao et al. 2008). Post-heat treatment operations are typically conducted to reduce the stresses as much as possible. Because of the greater surface area, the presence of surface imperfections and porosity on the bio-implant stimulates cell attachment and growth (Xiong et al. 2010). The particle size, texture, morphologies, and proportion of surface area that is supplied all have an impact on how the body reacts to HA coating. Rough spots on the coating make it more wettable, which is necessary for growth and interaction with body fluids (Ninomiya et al. 2001). The chemistry and morphology of the coating surface are also essential for successful bone regeneration (Dohan Ehrenfest et al. 2010). The process by which broken bone heals in the presence of a biomedical implant is comparable to the process by which primary bone heals. In the beginning, blood flowed, and over time, it changed into a clot in the space between the implant and the bone. The presence of phagocytic cells is necessary for the completion of this process. The speed of the response caused by the bioactive coatings is determined by whether or not the implant has already been covered with them. Compared to metallic implants without coatings, this reactivity allowed for more rapid bone repair. The bioactive coatings stimulate various biological answers from the body, including adhesion to tissues and the formation of tissues. There are two distinct kinds of substances to choose from in most cases. The first one has the quality of osteo-conductivity, and the second one has the quality of osteo-productivity. As soon as these coating layers came into contact with the bodily fluid, they began to provide stimulation, focusing on supporting the bone formation process along the surface or beyond the biomedical implant (Hench and Jones 2005).

Hydrothermal

The hydrothermal approach involves heating and pressurizing an aqueous phase as a reactive system in a particular sealed reaction chamber to generate a high heating, high-pressure reaction state (Morey and Niggli 1913). Hydrothermal crystallization is attributed to the following stages: Hydrothermal environment is used to dissolve the substances entering the mixture as ions or molecular groupings. Next, the temperature gradient among the top and the bottom parts of the pot separates the ions or molecules. In order to generate a supersaturation, the ions or molecules are moved to a lower temperature. Then at the growing contact, ions or molecules are adsorbed, degraded, and desorbed. Just at the surface, the adsorbed material flows. Eventually, the dissolved substance precipitates out of the water. In hydrothermal circumstances, the crystalline shape of hydrothermal crystals is directly linked to the environmental condition. Under various hydrothermal circumstances, the same crystals can take on a variety of microstructures (Penn and Banfield 1999; Polsongkram et al. 2008; Sekiguchi et al. 2000).

Electrochemical Deposition

The surface modification approach electrodeposition involves the application of an external circuit to the anode material, coupled with motion. Anodes can be either sacrificing or inactive. Ions can be reduced at the cathode/electrolyte contact, and a desirable coating is generated on the cathode's region. Solution oxidizing is sometimes used to carry out electrodeposition. The cell's electrodes are connected by an electrolyte circuit (Rasooli, Safavi, and Kasbkar Hokmabad 2018; Safavi and Rasooli 2019b, a; Safavi et al. 2019; Rasooli et al. 2020; Fathi et al. 2020; Safavi, Rasooli, and Sorkhabi 2020). Electrodeposition is a potential option for dry procedures, such as thermal spraying, to create HA films. The electrolyte is prepared by dissolving calcium and phosphorus salts in water before the electrodeposition of HA. The pH-dependent solubility of calcium phosphate salts is used in this approach. Many modern initiatives have been aimed at improving operational metrics (Ducheyne 2015). First, the following processes might happen at the cathode surface immersed in the salt-containing aqueous electrolyte throughout electrodeposition (Chakraborty et al. 2016; Wang et al. 2010; Drevet et al. 2011). The production of hydrogen bubbles at the working electrode results from the reduction of H_2O. Due to the hydroxide produced in Equation (23), the electrolyte solution's pH rises near the cathode. Acid-base reactions might occur because of the higher pH level, as indicated in Equation (24-27). HPO_4^{2-} and PO_4^{3-} reduction processes may produce hydrogen phosphate and phosphate ions. When the pH suddenly rises, CaP crystals form and expand. The movement of electrons between electrodes and electrolytes occurs during HPO_4^{2-} ion dissociation. Other reactions may be triggered as a result of the raised pH (Lobo et al. 2011; Manso et al. 2000; Pezzatini et al. 2006).

$$2H_2O + 2e^- \rightarrow H_2 + 2OH^- \tag{23}$$

$$H_2PO_4^- + OH^- \rightarrow HPO_4^{2-} + H_2 \tag{24}$$

$$HPO_4^{2-} + OH^- \rightarrow PO_4^{3-} + H_2 \tag{25}$$

$$H_2PO_4^- + OH^- \rightarrow HPO_4^{2-} + H_2O \tag{26}$$

$$HPO_4^{2-} + OH^- \rightarrow PO_4^{3-} + H_2O \tag{27}$$

Ions in the solution, including calcium and hydrogen ions, migrate towards the cathode owing to the electric field gradients. Existing HPO_4^{2-} ions stay in the diffusion layers because of a concentration gradient among them and their surroundings. The concentration of these ions may grow as the distances from the electrode surface increase. HPO_4^{2-} ions may be propelled by a concentration difference to migrate through the electrolyte bulk and reach the diffusion layer on the cathode's surface (Equations (28-31)) (Seyedraoufi and Mirdamadi 2015; Chen et al. 2007; Fornell et al. 2017).

$$2H_2O + Ca^{2+} + HPO_4^{2-} \rightarrow CaHPO_4.2H_2O \downarrow \quad (28)$$

$$4Ca^{2+} + 2PO_4^{3-} + HPO_4^{2-} + 2.5H_2O \rightarrow Ca_4H(PO_4)_3.2.5H_2O \downarrow \quad (29)$$

$$4Ca^{2+} + 2PO_4^{3-} \rightarrow Ca_3(PO_4)_2 \downarrow \quad (30)$$

$$5Ca^{2+} + 3PO_4^{3-} + OH^- \rightarrow 0.5Ca_{10}(PO_4)_6(OH)_2 \downarrow \quad (31)$$

The primary calcium phosphates in as-deposited coatings are DCPD, OCP, and HA, with TCP as an inclusion. DCPD creation leads to HA formation early in the crystallization process for two main reasons: 1) There are insufficient OH^- ions to produce HA. Only in the existence of a low OH^- concentration in the solution could the conversion of $H_2PO_4^-$ to HPO_4^{2-} be performed. PO_4^{3-} and OH^- ions are required to produce HA. 2) When the pH is below 5, DCPD is steady, and its precipitate process gets promoted in the existence of slight concentrations of OH^- ions (Equation (32)).

$$9.5Ca^{2+} + 5.5PO_4^{3-} + 0.5OH^- + 0.5HPO_4^{2-} \rightarrow Ca_{9.5}(HPO_4)_{0.5}(PO_4)_{5.5}(OH)_{0.5} \downarrow \quad (32)$$

Electrodeposited CaP films under acidic conditions are mainly composed of DCPD, as seen by available research. The most stable CaP phase in essential circumstances is HA, and DCPD is also converted to HA at high pH values. Research (Yuan and Golden 2009) has proven that DCPD and OCP are intermediates to produce HA at high pH values.

These crystals depend on alkaline conditions for their formation and development, and hydroxyl ions are essential for this mechanism. Post-alkaline treatment is included after acid treatment to meet the thermodynamic and Hydroxyl ions saturation needs. Some calcium and phosphate ions may also be produced whenever a DCPD-containing film is submerged in an alkaline medium. Calcium and phosphate do not require to be added to the solution. Ca^{2+} and PO_4^{3-}ions in the alkaline environment, as well as a decent amount of Hydroxyl, work to support the synthesis of HA from DCPD (Equations (33-34)) (Seyedraoufi and Mirdamadi 2015; Seyedraoufi and Mirdamadi 2014; Zhang et al. 2011; Chen, Birbilis, and Abbott 2011).

$$10CaHPO_4.2H_2O + 12OH^- \rightarrow Ca_{10}(PO_4)_6(OH)_2 \downarrow +3H_2O + 4PO_4^{3-} \quad (33)$$

$$20Ca_4H(PO_4)_3.2.5H_2O \rightarrow 8Ca_{10}(PO_4)_6(OH)_2 + 34H_2O + 12H_3PO_4 \quad (34)$$

A pulsed reversing current may be used throughout electrodeposition to create a HA precipitate. Adsorption of OH^- ions occurs due to double-layer charges due to proton desorption during the reverse current stage. As an outcome, there may be an increase in the amount of the HA phase. Notably, in several experiments, the dissolving of ambient CO_2 in the aqueous electrolyte has produced carbonated HA in the FTIR spectra of coatings. Carbonate HA is formed from dissolved CO_2 in water (Li et al. 2019).

Electrophoretic Deposition

Electrophoretic deposition is typically carried out in a 2-electrode cell, with the voltages maintained at a fixed value. It is a famous colloid-based approach for producing ceramic coatings, wherein electrically suspended nanoparticles in a liquid media settle upon a metal substrate. This method is used for the manufacturing of ceramic coatings. Generally, the scale of the colloidal particles must not be more than thirty micrometers (Ma et al. 2018; Besra and Liu 2007). Through the use of Electrophoretic deposition, it is now available to apply HA as well as composite coatings that are centered in HA on biomedical implants. HA particles, a solution, and a surfactant are the three types of things that are most often used in the Electrophoretic deposition process to produce HA solution (Horandghadim, Khalil-Allafi, and Urgen 2020; Horandghadim and Khalil-Allafi 2019).

All in all, each HA coating method's advantages and disadvantages are summarized as follows.

The thickness of the HA coating produced in the Sol-Gel method is lower than 1 micrometer. Easy reaction circumstances, inexpensive, smooth layers of HA, high purity of the HA film, and lax process for producing complicated structures are the pros of this method. On the other hand, some reactions require a close system which might be considered a con for this method. In addition, this method requires expensive chemicals sometimes (Ballarre et al. 2007; Ballarre et al. 2010; Ballarre et al. 2011; Hsieh, Perng, and Chin 2002; Liu, Yang, and Troczynski 2002; Stoch et al. 2005; Wang et al. 2014).

In the Dip-coating method, the HA layer thickness varies from 0.05 to 15 millimeters. This method's pros are inexpensiveness, and an uncomplicated HA apply process, and uniformity of the layer. The difficulty of this method is the high finishing temperature, which might damage the HA film (Metikoš-Huković et al. 2003; Li et al. 2010; Zhang, Guan, and Zhang 2011; Bryington et al. 2013; Rojaee, Fathi, and Raeissi 2013; Catauro et al. 2015; Fu et al. 2015; Zhao et al. 2015).

The electrochemical deposition method provides an HA layer with a thickness of 0.05 to 0.5 millimeters. This method is doing well in terms of economy and ease of the process. Like the above techniques, electrodeposition also creates a uniform layer of HA. The weak bonding between the surface and the HA layer is the drawback of this method (Ma, Ye, and Wang 2008; Wang et al. 2006; Qiu, Wang, and Yin 2010; Wang and Luo 2011).

Electrophoretic deposition can provide 0.1 to 2 millimeters thickness of HA film. The uniformity of the layer, easy process, and high coating rate are the benefits of this method, but it needs post-treatments at high temperatures. Moreover, some cracks might develop at low temperatures (Qiu et al. 2011; Tahmasbi Rad et al. 2014; Rojaee et al. 2014; Sun, Ma, and Zhang 2014).

The thickness of HA produced by the bio-mimetic technique is lower than 30 microns. This method needs lower temperatures, and it can easily coat complex shapes. The issues with this method are its time-taking nature, the requirement for unchanging pH values, and solution makeup (Bigi et al. 2005).

In the plasma spraying method, the thickness of the HA coating is limited to 20 microns. This method is inexpensive in process and fast. The resulted coating is smooth and provides interconnected pores for multilayer coatings. The difficulty of this method is the fluctuations of HA coating density which might result in unevenness. If the temperature rises during the process, some grains may start to form. In addition, the adhesion strength of HA coating to the surface is poor (Schrooten and Helsen 2000; Yang, Ong, and Tian 2003; Yang and Chou 2007; Grandfield et al. 2011; Vencl et al. 2011; Hung et al. 2013; Latifi et al. 2014; Hameed et al. 2019).

In the HVSFS technique, the thickness of the coating is 50 microns at maximum. The final coating is uniform with nanometric pores, the process is economical, and there is no need for post-treatment. The con of this method is the requirement of high-temperature levels (Rauch et al. 2009; Bolelli et al. 2014; Bolelli et al. 2015).

Pulse laser deposition provides 0.05 to 5 microns of HA coating, which is adjustable, meaning that porous, dense, crystalline, or amorphous coating can be achieved. Moreover, the final coating is uniform. But the main shortcoming of this method is the high-cost process, and the pre-treatment of the substrate is needed (Clèries et al. 2000; Fernández-Pradas et al. 2001; Khandelwal et al. 2013).

In the flame spraying technique, the resulting coating has a range of thicknesses between 100 and 250 microns. This method is better than other thermal spraying methods in terms of economy. The process is adaptable, and a porous coating can be achieved. The drawbacks of this technique are the requirement of post-treatment, crack developments at low temperatures, and the presence of the melted particles in the microstructure (Das, Bandyopadhyay, and Nath 2018; Monsalve et al. 2015).

It is possible to construct crystal structures that are unstable at the melting temperature using the hydrothermal approach. The hydrothermal process can even be used to create materials with high vapor pressures around the melting temperatures. With this technology, high-quality crystals may be grown at the same time as they can be controlled. Because of the use of a metal cylinder, it is impossible to see the crystals develop, which is a drawback of thc approach.

Table 2 summarizes the categorized research of enhancements of HA by different methods and for different metallic implants.

Inventive Approaches to the Coating Process

Yuan et al. utilized electro-deposition to apply a coating of HA on stainless steel 316L (Yuan and Golden 2009). In order to reduce the amount of time that the implant is in touch with bodily fluid, the surfaces were given a double coating. Following the coating application, the samples were subjected to a heat treatment in either a vacuum at 800°C or in the presence of air at 600°C. The advantages of using a bi-layer coating were increased bioactivity and homogeneity thanks to the strong bonds that were formed between the surface and the HA. A further cutting-edge approach is placing an oxide layer between the metallic implant and the HA. This is an example of an innovative approach.

Table 2. List of the research focused on the various metallic implants and the methods to coat HA on them

Materials used	Coating method	Operation conditions	Results	Year	Ref
Stainless steel implants					
Triethylephosphite and calcium nitrate	Sol-Gel	Dried for 15 minutes at 80°C- annealed at 375°C, 400°C, and 500°C in air- 5 cm/min dipping speed	Annealing at 500°C caused nano crystals and micro cracks.	2002	(Liu, Yang, and Troczynski 2002)
Bovine serum albumin and calcium nitrate	biomimetic	Dipping in diammonium hydrogen phosphate for 24 hours and dried at 50°C	Optimum crystallinity akin to bone	2005	(Nayar et al. 2006)
HA	Pulse laser deposition	Substrate temperature: 570- gas medium: oxygen- laser frequency: 50 Hz- standoff distance: 35 mm- voltage: 19.7 kV- number of shots: 40000- laser energy: 300 and 500 kJ	500 kJ laser energy provided more crystalline and uniform coating	2013	(Khandelwal et al. 2013)
HA, ZrO_2, and TiO_2	Electrochemical deposition	pH: 4.2- temperature : 85°C - time: 2 hours	Good corrosion resistance and bioactivity	2018	(Poorraeisi and Afshar 2018)
HA, Al_2O_3, and tricalcium phosphate	Plasma spraying	Torch speed: 50, 200, and 500 mm/s- arc current 530 A- standoff distance: 85 mm- suspension pressure: 1 bar- suspension rate of feeding: 25 g/min	Porous coating in the range of 0.2 to 6 microns	2018	(Ročňáková et al. 2018)
HA, TiO_2	High-velocity suspension flame spray	Pre-heating of substrate: 150°C to 200°C - spray distance 100 mm- compressed air for cooling- isopropanol as solvent	Enhanced tensile strength and war resistance compared to HA	2018	(Yao, Wang, et al. 2018)

Table 2. (Continued)

Materials used	Coating method	Operation conditions	Results	Year	Ref
Magnesium alloys implants					
Calcium nitrate Tetra hydrate, phosphorous penta oxide	Sol-Gel	Reaction at 26°C for 5 hours- withdrawn speed: 0.1 mm/s- dried at 60°C for 24 hours- calcination and sintering at 400°C for 6 hours	Improved corrosion resistance	2013	(Rojaee, Fathi, and Raeissi 2013)
Calcium nitrate, Ammonium di hydrogen phosphate	Electrochemical deposition	Volts applied: 90 V, 100 V, 110 V, and 120 V- time intervals: 20 minutes, 30 minutes, 40 minutes, and 60 minutes- temperature: 80°C - drying at 140°C for 4 hours in an autoclave	Bioactivity in SBF and corrosion resistance improved	2015	(Furko et al. 2016)
Calcium nitrate, sodium dihydrogen phosphate, and sodium hydrogen carbonate	biomimetic	Kept at 37°C for 2 days- dried in air at room temperature	Uniform HA coating- promoted cell growth- improved corrosion performance	2015	(Bao et al. 2015)
HA	Plasma spraying	Voltage: 500 V- current: 70 A- rotation speed: 35 rpm- thickness: 0.1 mm	Bioactivity in SBF and corrosion resistance improved. Contact angle decreased, and blood compatibility enhanced	2018	(Gao, Liu, and Song 2018)
HA	Pulse laser deposition	Temperature: 100°C, 200°C, 300°C, and 400°C	The anti-corrosion performance in SBF increased for samples produced at 200 and 300	2018	(Rau et al. 2018)
HA/TiO_2 and HA/ZrO_2	High-velocity suspension flame spray	Substrate pre-heating: 200°C - spray distance: 100 mm- suspension flow rate: 40 ml/min- torch traverse speed: 200 mm/s	Rough, porous surface with nanoscale agglomerates- well-bonded coating/surface interfaces- improved corrosion resistance	2018	(Yao, Hu, et al. 2018)

Materials used	Coating method	Operation conditions	Results	Year	Ref
Co-Cr alloys implants					
Calcium nitrate, sodium hydrogen phosphate, tris-hydroxy-methyl-amino-methane	Electrochemical deposition	CoCrMo as cathode and platinum as an anode- pH:7- stirring rate: 200 rpm- dried at room temperature	Excellent bonding to the substrate	2011	(Wang and Luo 2011)
HA	biomimetics	Soaking time: 15 minutes- NaOH 5 molar- soaking in SBF: 18 hours, 24 hours, and 36 hours- temperature: 37°C	Refined grains of pure HA crystals	2016	(Charlena, Sukaryo, and Fajar 2016)
Calcium chloride, sodium phosphate, titania nanotube	Dip coating	Dip coater rate: 200 mm/min- oven drying temperature: 70°C – kept for 24 hours at room temperature	Low porosity- enhanced adhesion strength and corrosion resistance at sintering temperature: 669.11°C	2016	(Shirdar and Taheri 2017)
Calcium nitrate, ammonium dihydrogen phosphate	Electrophoretic deposition	Current: 12 mA/cm^2- time: 20 minutes- temperature: 25°C - anode: graphite- cathode: substrate	Fairy uniform microstructure with low porosity- low corrosion rate	2017	(Rezazadeh Shirdar et al. 2016)
HA, niobium	Plasma spray coating	Arc current: 600 A- arc voltage: 63 V- primary gas: Argon- Secondary gas: H_2- standoff distance: 75 mm	Crystallinity and surface hardness increased by the addition of Nb- no microcracks- improved corrosion resistance in the absence of Nb	2019	(Singh, Singh, and Sidhu 2019)
Titanium alloys implants					
Calcium nitrate tetra hydrate, tri-ammonium phosphate tri hydrate, gelatin	Sol-Gel	Reaction at ambient temperature- pH: 6 to 7.8- under argon gas annealed at 460°C to 750°C	Bioactivity increased	2005	(Stoch et al. 2005)
Sodium hydroxide, calcium phosphate	biomimetic	Surface activation: 10 M NaOH- room temperature, voltage: 10 V- Time: 30 minutes- pH values just above 7	Coating promotes bone growth	2008	(Xie and Luan 2008)

Table 2. (Continued)

Materials used	Coating method	Operation conditions	Results	Year	Ref
HA	Pulse laser deposition	Temperature: ambient- annealed in air at 290°C to 310°C	Increased purity after post-treatment- crystalline HA showed no dissolution in SBF	2009	(Dinda, Shin, and Mazumder 2009)
Bis-dipivaloy imethanocalcium triphenyl phosphate	Chemical vapor deposition	Pressure: 800 Pa- temperature: 973 K	HA showed excellent mechanical and bioactivity	2010	(Tsutsumi et al. 2010)
HA, Al_2O_3	Plasma spraying	Traverse speed: 38 mm/s- spray time: 35 seconds- spray gun passes 15	Osteogenic response rose	2017	(Fielding et al. 2012)
HA	High-velocity suspension flame spray	Temperature: 357°C to 616°C - torch cycle: 2 to 4	More dense and reactive coating due to calcium phosphate	2015	(Bolelli et al. 2015)
Calcium nitrate, ammonium dihydrogen phosphate	Electrochemical deposition	Anode: platinum plate- cathode: substrate- temperature: 90°C - deposition time: 60 minutes- current density: 1.25 mA/cm^2, 1.87 mA/cm^2, 2.50 mA/cm^2, 3.12 mA/cm^2, and 3.61 mA/cm^2	Increased bioactivity	2016	(He et al. 2016)

This oxide film shields the metallic surface and prevents harmful ions from the top metallic layer from being released (Rojaee et al. 2014; Mumith et al. 2020; Ullah et al. 2020). The oxide layers that are coated in HA coatings make the adhesion between the oxide and the HA even stronger (Zieliński, Antoniuk, and Krzysztofowicz 2014). Additionally, the metallic covering lessens the impacts of cytotoxicity while simultaneously improving the implant's ability to function biologically (Vu et al. 2019; Furko et al. 2018). In addition to this, a ceramic material such as Zr was put in the space between the HA and the metallic surface. This ceramic layer performed the functions of a strong adhesive and a composite material. This particular layer proved to be beneficial for implants that were exposed to repeated loading (Kurzweg et al. 1998). Improving the binding strength between the metallic implant and the coating is one of the most critical factors contributing to the implant's longevity. The super-high-speed blasting process is an innovative and fresh approach that improves the bond strength and does away with the exfoliating of HA layers. The HA film that is produced using the super-high-speed blasting process has superior wettability capabilities and better adhesion strength than its counterpart (Jung et al. 2012).

Chemical Treatments of HA to Enhance Corrosion Resistance, Adhesion to the Substrate, and Biocompatibility

The porous structure of the HA coating and their low adhesion to the metal substrate in an aqueous solution may decrease their lifetime and corrosion resistance during prolonged exposure to the corrosive environment. So, finding different ways for chemical treatment of this type of coating is necessary. Up to now, many efforts on chemical treatment of HA coatings by pre-treatment, post-treatment, and direct addition of chemical compounds were performed to achieve a HA coating with high corrosion resistance and lifetime in an aqueous solution.

Pre-Treatments

Surface Activation

Rahman et al. used sodium hydroxide to anodize WE43 alloy to investigate surface pretreatment on the surface characteristics of further layers. Then they applied dicalcium phosphate dihydrate ($CaHPO_4.2H_2O$, DCPD) and HA on the pretreated substrates. They reported that by surface pretreatment, DCPD coatings changed to HA with dense and plate-like structures. Besides, the adhesion strength and microhardness of prepared HA coating were significantly improved (Viswanath and Ravishankar 2008).

Inorganic Layer

Applying an inorganic layer as a pre-treatment before applying HA coating could help limit the access of electrolyte coming to the substrate as a result of the porous structure of HA coatings. The microstructure and anti-corrosion performance of HA coatings was studied by applying them to fluoride-coated Mg-2.1Zn-0.22Ca alloy and the non-coated one. The obtained coating on the pretreated substrate showed a uniform, nano-rod structure. Also, the measured corrosion current density was decreased from 5.72×10^{-5} A/cm^2 for the non-coated substrate to 4.32×10^{-7} A/cm^2, showing visual enhancement in the anti-corrosion performance of the HA coating (Feng, Zhu, et al. 2017).

Micro-arc oxidation (MAO) is usually used to prepare a metal oxide film to enhance the anti-corrosion performance of the metal substrates (Simchen et al. 2020). Chaharmahali et al. applied HA coating with different concentrations (5, 10, 15 g/L) on the MAO Mg alloy substrate. Results showed that the HA coating with a concentration of 15 g/L had the lowest corrosion current density (1.99×10^{-6} A/cm^2), and its polarization resistance was 62% higher than the uncoated substrate (Chaharmahali, Fattah-Alhosseini, and Esfahani 2020). TiO_2 compounds are another inorganic layer applied before HA coating by Guo et al. As expected, the presence of the TiO_2 layer improved the corrosion resistance. They also used the extraction method to evaluate the cell viability of the HA/TiO_2 coated substrate. The cell viability value of the HA/TiO_2 coated substrate was up to 86.99%. The cell adherence to each other caused conductivity and cell diffusion on the surface of the coating.

Moreover, in vitro antibacterial test was done on the prepared coatings. Because accumulation and adhesion of bacteria on the implant's surface, a biofilm and implantation failure formed. Results showed that the antibacterial rate of HA/TiO_2 coating on Staphylococcus aureus was much higher than on uncoated substrate (Guo et al. 2018).

Organic Layer

The lack of adsorption sites on the surface of Mg alloys limits the deposition of HA coating in some approaches like hydrothermal methods (Hiromoto and Tomozawa 2010). To overcome this shortcoming, organic compounds, especially the ones with film formation ability, were utilized to improve HA deposition on the Mg alloys. Zhou et al. used PDA as an interlayer of AZ31 Mg alloy and HA coating. So, the presence of PDA could significantly decrease the corrosion rate and increase the cells' survival rate to 120% after 5 days (Zhou et al. 2020).

Ahadi Parsa et al. treated Mg AZ31 surface with vinyl tri-ethoxy silane via a sol-gel route via two hydrolyzation times (1, 24 h) and three levels of immersion time (30, 60, 180 s) (AhadiParsa et al. 2022). In the second step, HA coating is applied on the silane-treated surface. Results showed that both hydrolyzation and immersion time has an observable effect on the morphology and electrochemical behavior of the prepared coating. The Si/HA coating with one-hour hydrolyzation and 60 s immersion time had a homogenous and smooth surface. The results of EIS measurements during different immersion times in SBF solution confirmed the regeneration of coating for a long exposure time.

Direct Addition Treatments

6.2.1. Utilizing Inorganic Compounds

By changing the number of nucleation sites, rate of deposition, and blocking the active sites, inorganic compounds can affect the thermal stability, microhardness, electrochemical properties, and so on (Singh, Singh, and Sidhu 2018; Singh, Singh, and Prakash 2011). Dehghani et al. prepared HA-Si coatings varying different concentrations of SiO_4^{4-} on an Mg-5Zn-0.3 Ca alloy. According to their report, SiO_4^{4-} could be substituted with PO_4^{3-} and increase the local pH, which provides a proper condition for high-density coating, resulting in higher nucleation sites. The obtained coating with higher density and uniformity could effectively prevent SBF penetration into the coating and improve the barrier characteristic of the HA coating (Dehghanian, Aboudzadeh, and Shokrgozar 2018).

HA-bioglass coating application in orthopedics and tissue engineering has received significant attention from many researchers. Singh et al. prepared HA-bioglass composite with different Fe_2O_3 content (1, 3, 5 wt.%) on the AZ91 Mg alloy substrate. The prepared crack-free coating had good adhesion to the substrate with a porous structure. Also, the homolysis rate of the optimized coating was less than 5%, overcoming the implant field requirements. The red blood cell morphology remains round in the presence of Fe_2O_3, (Singh, Singh, and Bala 2019). Strontium was introduced as an influential element in promoting bone regeneration and preventing osteoporosis (Hiromoto and Tomozawa 2010; Marie, Felsenberg, and Brandi 2011; Ni et al. 2006). Yang et al. induced Sr in HA coating and applied it on the AZ91D alloy. The effect of Sr presence is observed by reducing the Mg^{2+} release. It also enhanced the biocompatibility of the substrate (Reginster et al. 2009). ZnO is an inorganic compound that has attracted researchers' attention to be used to reinforce HA coating (Singh, Singh, and Sidhu 2018; Singh, Singh, and Prakash 2011). In HA coating preparation, ZnO was used with different concentrations (4, 8, 12 wt.%). The microhardness of the HA/ZnO coating reached 280 HV at a concentration of 12 wt.%. The HA/ZnO coatings gathered proper hydrophilicity and good corrosion resistance (Bansal, Singh, and Sidhu 2020). In a further study, HA was reinforced by Nb at different levels (4, 12, 18 wt.%). With an increase in Nb content, the surface protection ability was increased. However, it has no positive effect on the hemolysis rate and erythrocytes (Singh et al. 2019).

6.2.2. Utilizing Organic Compounds

Organic compounds could improve the coatings' adhesion, biological activity, and mechanical properties. The single-step method was used to precipitate HA/graphene oxide (GO) coating on the surface of the Mg alloy substrate by Wen et al. The adhesion strength of the prepared coating to the substrate was 40.7 MPa, above the one measured for natural cortical bone (35 MPa). Based on the results of SEM observation, HA/GO composite could successfully seal the pores and reduce their size (Wen et al. 2017). Additionally, the measured corrosion current density on HA/GO composite was two orders of magnitude lower than the one measured on Mg alloy substrate, indicating an enhancement in corrosion resistance of the HA/GO composite.

Afshari et al. used poly vinyl alcohol (PVA) at three levels (0.01, 0.05, 0.1 wt.%) in HA coating to enhance corrosion and bio-corrosion resistance of HA coating (Afshari et al. 2021). They reported that the presence of PVA significantly affects the morphology and crystallinity of the HA coating. The best anti-corrosion performance was obtained for the HA/PVA coating with a concentration of 0.05 wt.% PVA. Compared with the Mg alloy substrate, the corrosion current density decreased from 36 to 0.46 $\mu A/cm^2$.

Post-Treatments

Post-treatment of the HA coating usually is done to enhance the surface protection against corrosive species. Feng et al. used polydopamine (PDA) in the post-treatment step on the HA coating. Since it could uniformly cover the surface of the HA coating, the corrosion potential was shifted to 150 mV positively. They also reported that by overlaying PDA on the HA coating, the hydrogen evolution rate decreased to 4.4 mm/year, resulting in higher corrosion resistance (Feng, Ma, et al. 2017). Chitosan is an organic compound with film formation ability that can reduce the porosity of HA coating. Roshan et al. used different concentrations of chitosan to enhance the barrier ability of HA coating against the corrosive species (Roshan et al. 2022). Based on their research, different concentration of chitosan (0.05, 0.1, 0.5 wt.%) was used for post-treating of HA coating. Chitosan with 0.5 wt.% concentration is uniformly overlayed on HA coating. This thin and sticky layer significantly improved the surface protection even after 7 days of exposure to the salt spray test.

Other Attempts to Reinforce the HA Coatings

Here is the list of researches done to increase the durability and biocompatibility of the HA coatings on the metallic implants either by incorporation of ions in the chemical formula of HA, or by biocompatible polymers as a primary layer that HA can be coated on (Table 3).

Table 3. Attempts to increase the stability and compatibility of HA coatings

	Material	Note	Ref
	F^-	Lower dissociation than HA	(Okazaki 1995; Huang et al. 2013)
	Cu^{2+}	Increased corrosion resistance, antibacterial, and cell biocompatibility	(Zhang et al. 2021)
Polymers	Collagen	Better cell proliferation behaviors	(Teng et al. 2008)
	Albumin	Rougher than HA coating which is good for cell adhesion	(Vasilescu et al. 2011)
Ions	Ag^+	Strong corrosion and wear resistance, antibacterial and biocompatible	(Zhou et al. 2014)
	Zr^{4+}	Higher bioactivity and corrosion resistance	(Indira, KamachiMudali, and Rajendran 2014)

Table 3. (Continued)

	Material	Note	Ref
	Sr^{2+}	Lower dissociation than HA	(LeGeros et al. 1988; Huang et al. 2013)
	Zn^{2+}	Superior corrosion resistance compared to HA	(LeGeros et al. 1988; Ding et al. 2015)
	Mg^{2+}	Excellent corrosion resistance in Ringer's solution	(B et al. 2019)
	Ce^{3+}	Increased inhibition against pathogens, better corrosion resistance than HA	(Priyadarshini et al. 2021)
	Gelatin	Possess biocompatibility, excellent bioactivity, and proper mechanical strength compared to non-coated HA	(Zandi et al. 2010)
	Alginate	Promising biomaterial for orthopedics	(Yin et al. 2020)
	Hyaluronic acid	Bone-implant contact and interface shear strength of BMP-2 implants were lower than HA implants	(Aebli et al. 2005)
	Cyclodextrins	Flexibility to choose any drug after implantation	(Taha et al. 2013)
	Poly lactic acid	coatings could be efficiently applied to titanium dental implants	(Karacan et al. 2017)
	Poly phosphate	Increased corrosion resistance	(Jiang et al. 2022)
	Poly dopamine	Increased cell spreading, proliferation and adhesion	(Xu et al. 2018)

Publications from the Last Year

In this section, a brief review of the most recent studies is listed.

- *HA/MgO/Chitosan:* Excellent apatite forming on Ti implants, Released Ca^{2+} and Mg^{2+}, Anti-bacterial rate of 96% (Zhang et al. 2022).
- *HA/Metformin/Chitosan:* Excellent corrosion resistance for Mg implants, High osteogenic activity, and biocompatibility (Li et al. 2022).
- *HA/Mg(OH)$_2$/Graphene oxide:* Promotes the regeneration of bones for Mg implants (Yuan et al. 2022).
- *HA/Zn^{2+}/F^-:* 70% better anti-bacterial performance in Ti implants compared to HA, 1.5 times higher osteogenic activity (Bhattacharjee, Bandyopadhyay, and Bose 2022).

- *HA/Graphene oxide:* Excellent adhesion to stainless steel implant, Improved anti-bacterial and anti-corrosion resistance (Nizami, Campéon, and Nishina 2022).
- *HA/Ag*$^{+}$*:* Improved biocompatibility of HA (Iconaru et al. 2022).
- *HA/Mg*$^{2+}$*/Sr*$^{2+}$*:* Increased the protection efficiency against the corrosive saliva medium (Kadhim et al. 2022).
- *HA/Mg*$^{2+}$*/Sr*$^{2+}$*/*TiO_2*:* Greater hydrophilicity and roughness compared to HA, Great adhesion to Ti implants and increased bioactivity (Praharaj et al. 2022).
- *HA/ZnO:* Anti-bacterial, Increased viability of the cells in NiTi implants (Mehrvarz, Khalil-Allafi, and Khosrowshahi 2022).
- *HA/Honeycomb:* Increased cell viability in Ti implants (Sari et al. 2022).
- *HA/Zn*$^{2+}$*/Sr*$^{2+}$*/F*$^{-}$ *and HA/Zn*$^{2+}$*/Sr*$^{2+}$*/Ag*$^{+}$*:* Enhanced osseointegration and corrosion resistance in SBF of Ti implants (Ratha et al. 2022).
- *HA/hydroxypropyltrimethyl ammonium chloride chitosan:* Remarkable hydrophilicity which decreased the adhesion of bacteria in Mg implants, Biocompatible and anti-bacterial due to the presence of chloride (Tian et al. 2022).
- *HA/Polypyrrole:* Improved adhesion strength to stainless steel implants, Reduced the corrosion current density (Mohandesnezhad et al. 2022).
- *HA/Sr*$^{2+}$*/Nb*$^{5+}$*:* Increased the roughness of the surface, better wettability and protein adsorption, Positive-shift in corrosion potential and negative shift in corrosion current density in Ti implants (Panda et al. 2022).
- *HA/SiC/Ag:* Increased corrosion resistance and hydrophilicity in Ti implants (Pana et al. 2022).
- *HA/Si/Sr:* Polarization resistance increased and corrosion potential shifted to the more positive values in Ti implants (Yu, Cho, and Choe 2022).

Conclusion

There does seem to be evidence that HA coatings are one of the main biocompatible coatings that are common in metallic implants categories. There are various ways of applying HA coatings on the implants, but, with regards to the future purpose of the coating and application, proper method must be chosen. Pre-treatments, post-treatments, and direct addition of minerals in the HA coatings have been studied in detailed by countless researches, however, regardless of how sufficient they are in enhancing the anti-corrosion and adhesion to the substrate, still lots of steps must be taken to achieve the HA coatings with greater anti-corrosion performance as well as the well-adhesion properties. The combination of applying methods might be a good idea in the future to reach the desired HA coatings. Furthermore, introduction of new biocompatible compounds to the HA formulation can be a way to increase the performance of this coating.

References

Abu Bakar, M. S., M. H. W. Cheng, S. M. Tang, S. C. Yu, K. Liao, C. T. Tan, K. A. Khor, and P. Cheang. 2003. "Tensile properties, tension–tension fatigue and biological response of polyetheretherketone–hydroxyapatite composites for load-bearing orthopedic implants." *Biomaterials* 24 (13):2245-2250. doi: https://doi.org/10.1016/S0142-9612(03)00028-0.

Aebli, Nikolaus, Hermann Stich, Peter Schawalder, Jean-Claude Theis, and Jörg Krebs. 2005. "Effects of bone morphogenetic protein-2 and hyaluronic acid on the osseointegration of hydroxyapatite-coated implants: An experimental study in sheep." *Journal of Biomedical Materials Research Part A* 73A (3):295-302. doi: https://doi.org/10.1002/jbm.a.30299.

Afshari, M., H. Eivaz Mohammadloo, A. A. Sarabi, and Sh Roshan. 2021. "Modification of hydroxyapatite-based coating in the presence of polyvinylalcohol (PVA) for implant application: Corrosion, structure and surface study." *Corrosion Science* 192:109859. doi: https://doi.org/10.1016/j.corsci.2021.109859.

Ahadi Parsa, M., H. Eivaz Mohammadloo, S. M. Mirabedini, and Sh Roshan. 2022. "Bio-corrosion assessment and surface study of hydroxyapatite-coated AZ31 Mg alloy pre-treated with vinyl tri-ethoxy silane." *Materials Chemistry and Physics* 287:126147. doi: https://doi.org/10.1016/j.matchemphys.2022.126147.

Albayrak, Onder, Osman El-Atwani, and Sabri Altintas. 2008. "Hydroxyapatite coating on titanium substrate by electrophoretic deposition method: Effects of titanium dioxide inner layer on adhesion strength and hydroxyapatite decomposition." *Surface and Coatings Technology* 202 (11):2482-2487. doi: https://doi.org/10.1016/j.surfcoat.2007.09.031.

Almeida, R. M., A. Gama, and Y. Vueva. 2011. "Bioactive sol–gel scaffolds with dual porosity for tissue engineering." *Journal of Sol-Gel Science and Technology* 57 (3):336-342. doi: 10.1007/s10971-009-2134-8.

Alves Cardoso, D., J. A. Jansen, and S. C. G. Leeuwenburgh. 2012. "Synthesis and application of nanostructured calcium phosphate ceramics for bone regeneration."

Journal of Biomedical Materials Research Part B: Applied Biomaterials 100B (8):2316-2326. doi: https://doi.org/10.1002/jbm.b.32794.

Antala, Nikunj, and PA Rathod. 2017. "Review of the Coating Characteristics Achieved Employing Thermal Flame-spray Coating Method." *Int. J. Eng. Technol. Manag. Appl. Sci* 5.

Arcos, Daniel, and María Vallet-Regí. 2020. "Substituted hydroxyapatite coatings of bone implants." *Journal of Materials Chemistry B* 8 (9):1781-1800. doi: https://doi.org/10.1039/C9TB02710F.

Arrés, Mar, Mariana Salama, Diogo Rechena, Patrizia Paradiso, Luis Reis, Marta M. Alves, Ana M. Botelho do Rego, Maria J. Carmezim, Maria Fátima Vaz, Augusto M. Deus, and Catarina Santos. 2020. "Surface and mechanical properties of a nanostructured citrate hydroxyapatite coating on pure titanium." *Journal of the Mechanical Behavior of Biomedical Materials* 108:103794. doi: https://doi.org/10.1016/j.jmbbm.2020.103794.

Awasthi, Shikha, Sarvesh Kumar Pandey, E. Arunan, and Chandan Srivastava. 2021. "A review on hydroxyapatite coatings for the biomedical applications: experimental and theoretical perspectives." *Journal of Materials Chemistry B* 9 (2):228-249. doi: 10.1039/D0TB02407D.

Bakan, Feray, Oral Laçin, and Hanifi Sarac. 2013. "A novel low temperature sol–gel synthesis process for thermally stable nano crystalline hydroxyapatite." *Powder Technology* 233:295-302. doi: https://doi.org/10.1016/j.powtec.2012.08.030.

Bakshi, Srinivasa R., Virendra Singh, Sudipta Seal, and Arvind Agarwal. 2009. "Aluminum composite reinforced with multiwalled carbon nanotubes from plasma spraying of spray dried powders." *Surface and Coatings Technology* 203 (10):1544-1554. doi: https://doi.org/10.1016/j.surfcoat.2008.12.004.

Ballarre, Josefina, Damián A. López, Wido H. Schreiner, Alicia Durán, and Silvia M. Ceré. 2007. "Protective hybrid sol–gel coatings containing bioactive particles on surgical grade stainless steel: Surface characterization." *Applied Surface Science* 253 (17):7260-7264. doi: https://doi.org/10.1016/j.apsusc.2007.03.007.

Ballarre, Josefina, Inderchand Manjubala, Wido H Schreiner, Juan Carlos Orellano, Peter Fratzl, and Silvia Ceré. 2010. "Improving the osteointegration and bone–implant interface by incorporation of bioactive particles in sol–gel coatings of stainless steel implants." *Acta biomaterialia* 6 (4):1601-1609.

Ballarre, Josefina, Rocío Seltzer, Emigdio Mendoza, Juan Carlos Orellano, Yiu-Wing Mai, Claudia García, and Silvia M. Ceré. 2011. "Morphologic and nanomechanical characterization of bone tissue growth around bioactive sol–gel coatings containing wollastonite particles applied on stainless steel implants." *Materials Science and Engineering: C* 31 (3):545-552. doi: https://doi.org/10.1016/j.msec.2010.11.030.

Bansal, Puneet, Gurpreet Singh, and Hazoor Singh Sidhu. 2020. "Investigation of surface properties and corrosion behavior of plasma sprayed HA/ZnO coatings prepared on AZ31 Mg alloy." *Surface and Coatings Technology* 401:126241. doi: https://doi.org/10.1016/j.surfcoat.2020.126241.

Bao, Quan He, Chao Sun, Chen Zhang, and Jie Qing Zhang. 2015. "Dip Coated Magnesium-Substituted Hydroxyapatite Coatings on Magnesium Alloy for

Biomedical Applications." *Journal of Biomimetics, Biomaterials and Biomedical Engineering* 25:83-89. doi: https://10.4028/www.scientific.net/JBBBE.25.83.

Bauer, Sebastian, Patrik Schmuki, Klaus von der Mark, and Jung Park. 2013. "Engineering biocompatible implant surfaces: Part I: Materials and surfaces." *Progress in Materials Science* 58 (3):261-326. doi: https://doi.org/10.1016/j.pmatsci.2012.09.001.

Bellucci, Devis, Elena Veronesi, Valentina Strusi, Tiziana Petrachi, Alba Murgia, Ilenia Mastrolia, Massimo Dominici, and Valeria Cannillo. 2019. "Human Mesenchymal Stem Cell Combined with a New Strontium-Enriched Bioactive Glass: An ex-vivo Model for Bone Regeneration." *Materials* 12 (21). doi: https://10.3390/ma12213633.

Ben-Nissan, Besim, and Andy H. Choi. 2006. "Sol-gel production of bioactive nanocoatings for medical applications. Part 1: an introduction." *Nanomedicine* 1 (3):311-319. doi: https://10.2217/17435889.1.3.311.

Berndt, C. C., Fahad Hasan, U. Tietz, and K. P. Schmitz. 2014. "A Review of Hydroxyapatite Coatings Manufactured by Thermal Spray." In *Advances in Calcium Phosphate Biomaterials*, edited by Besim Ben-Nissan, 267-329. Berlin, Heidelberg: Springer Berlin Heidelberg.

Besra, Laxmidhar, and Meilin Liu. 2007. "A review on fundamentals and applications of electrophoretic deposition (EPD)." *Progress in Materials Science* 52 (1):1-61. doi: https://doi.org/10.1016/j.pmatsci.2006.07.001.

Bharati, S., M. K. Sinha, and D. Basu. 2005. "Hydroxyapatite coating by biomimetic method on titanium alloy using concentrated SBF." *Bulletin of Materials Science* 28 (6):617-621. doi: https://10.1007/BF02706352.

Bhattacharjee, Arjak, Amit Bandyopadhyay, and Susmita Bose. 2022. "Plasma sprayed fluoride and zinc doped hydroxyapatite coated titanium for load-bearing implants." *Surface and Coatings Technology* 440:128464. doi: https://doi.org/10.1016/j.surfcoat.2022.128464.

Bigi, Adriana, Elisa Boanini, Barbara Bracci, Alessandro Facchini, Silvia Panzavolta, Francesco Segatti, and Luigina Sturba. 2005. "Nanocrystalline hydroxyapatite coatings on titanium: a new fast biomimetic method." *Biomaterials* 26 (19):4085-4089. doi: https://doi.org/10.1016/j.biomaterials.2004.10.034.

Bolelli, Giovanni, Devis Bellucci, Valeria Cannillo, Luca Lusvarghi, Antonella Sola, Nico Stiegler, Philipp Müller, Andreas Killinger, Rainer Gadow, Lina Altomare, and Luigi De Nardo. 2014. "Suspension thermal spraying of hydroxyapatite: Microstructure and in vitro behaviour." *Materials Science and Engineering: C* 34:287-303. doi: https://doi.org/10.1016/j.msec.2013.09.017.

Bolelli, Giovanni, Devis Bellucci, Valeria Cannillo, Rainer Gadow, Andreas Killinger, Luca Lusvarghi, Philipp Müller, and Antonella Sola. 2015. "Comparison between Suspension Plasma Sprayed and High Velocity Suspension Flame Sprayed bioactive coatings." *Surface and Coatings Technology* 280:232-249. doi: https://doi.org/10.1016/j.surfcoat.2015.08.039.

Bryington, Matthew S., Mariko Hayashi, Yusuke Kozai, Stefan Vandeweghe, Martin Andersson, Ann Wennerberg, and Ryo Jimbo. 2013. "The influence of nano hydroxyapatite coating on osseointegration after extended healing periods." *Dental Materials* 29 (5):514-520. doi: https://doi.org/10.1016/j.dental.2013.02.004.

Bugbee, W. D., C. J. Sychterz, and C. A. Engh. 1996. "Bone remodeling around cementless hip implants." *Southern medical journal* 89 (11):1036-1040. doi: https://10.1097/00007611-199611000-00002.

Cabañas, M. V., and M. Vallet-Regí. 2003. "Calcium phosphate coatings deposited by aerosol chemical vapour deposition." *Journal of Materials Chemistry* 13 (5):1104-1107. doi: https://10.1039/B301435E.

Catauro, Michelina, Flavia Bollino, Ferdinando Papale, and Giuseppe Lamanna. 2015. "TiO2/PCL Hybrid Layers Prepared via Sol-Gel Dip Coating for the Surface Modification of Titanium Implants: Characterization and Bioactivity Evaluation." *Applied Mechanics and Materials* 760:353-358. doi: https://10.4028/www.scientific.net/AMM.760.353.

Chaharmahali, Razieh, Arash Fattah-Alhosseini, and Hamid Esfahani. 2020. "Increasing the in-vitro corrosion resistance of AZ31B-Mg alloy via coating with hydroxyapatite using plasma electrolytic oxidation." *Journal of Asian Ceramic Societies* 8 (1):39-49. doi: 10.1080/21870764.2019.1698143.

Chakraborty, Rajib, Srijan Sengupta, Partha Saha, Karabi Das, and Siddhartha Das. 2016. "Synthesis of calcium hydrogen phosphate and hydroxyapatite coating on SS316 substrate through pulsed electrodeposition." *Materials Science and Engineering: C* 69:875-883. doi: https://doi.org/10.1016/j.msec.2016.07.044.

Charlena, S. G. Sukaryo, and M. Fajar. 2016. "Hidroxyapatite Coating on CoCrMo Alloy Titanium Nitride Coated Using Biomimetic Method." *Journal of Physics: Conference Series* 776:012056. doi: https://10.1088/1742-6596/776/1/012056.

Chen, X. B., N. Birbilis, and T. B. Abbott. 2011. "A simple route towards a hydroxyapatite–Mg(OH)2 conversion coating for magnesium." *Corrosion Science* 53 (6):2263-2268. doi: https://doi.org/10.1016/j.corsci.2011.03.008.

Chen, Xing-yu, Zhong-wei Zhao, Ai-liang Chen, and Hong-gui Li. 2007. "Pulsed electrodeposition of hydroxyapatite on titanium substrate in solution containing hydrogen peroxide." *Transactions of Nonferrous Metals Society of China* 17 (3):617-621. doi: https://doi.org/10.1016/S1003-6326(07)60144-7.

Chen, Y. M., Y. P. Lu, and M. S. Li. 2006. "Surface changes of plasma sprayed hydroxyapatite coatings before and after heat treatment." *Surface Engineering* 22 (6):462-467. doi: https://10.1179/174329406X150413.

Choi, Andy H., and Besim Ben-Nissan. 2007. "Sol-gel production of bioactive nanocoatings for medical applications. Part II: current research and development." *Nanomedicine* 2 (1):51-61. doi: https://10.2217/17435889.2.1.51.

Cizek, J., and J. Matejicek. 2018. "Medicine Meets Thermal Spray Technology: A Review of Patents." *Journal of Thermal Spray Technology* 27 (8):1251-1279. doi: https://10.1007/s11666-018-0798-8.

Clèries, L., E. Martínez, J. M. Fernández-Pradas, G. Sardin, J. Esteve, and J. L. Morenza. 2000. "Mechanical properties of calcium phosphate coatings deposited by laser ablation." *Biomaterials* 21 (9):967-971. doi: https://doi.org/10.1016/S0142-9612(99)00240-9.

Combes, C., and C. Rey. 2010. "Amorphous calcium phosphates: Synthesis, properties and uses in biomaterials." *Acta Biomaterialia* 6 (9):3362-3378. doi: https://doi.org/10.1016/j.actbio.2010.02.017.

Córdoba, Laura. 2016. "Magnesium-based biodegradable materials: from surface functionalization to cellular evaluation."

Costa, Daniel O., S. Jeffrey Dixon, and Amin S. Rizkalla. 2012. "One- and Three-Dimensional Growth of Hydroxyapatite Nanowires during Sol–Gel–Hydrothermal Synthesis." *ACS Applied Materials & Interfaces* 4 (3):1490-1499. doi: https://10.1021/am201735k.

Cotell, C. M., D. B. Chrisey, and K. S. Grabowski. 1991. "Pulsed Laser Deposition of Biocompatible Thin Films: Calcium Hydroxylapatrte and Other Calcium Phosphates." *MRS Proceedings* 252:3. doi: https://10.1557/PROC-252-3.

Das, Biswajit, P. P. Bandyopadhyay, and A. K. Nath. 2018. "An investigation on corrosion resistance and mechanical properties of laser remelted flame sprayed coating." *Advances in Materials and Processing Technologies* 4 (4):660-668. doi: https://10.1080/2374068X.2018.1489583.

Davar, Fatemeh, and Nasrin Shayan. 2017. "Preparation of zirconia-magnesia nanocomposite powders and coating by a sucrose mediated sol-gel method and investigation of its corrosion behavior." *Ceramics International* 43 (3):3384-3392. doi: https://doi.org/10.1016/j.ceramint.2016.11.184.

Dehghanian, Changiz, Neda Aboudzadeh, and Mohammad Ali Shokrgozar. 2018. "Characterization of silicon- substituted nano hydroxyapatite coating on magnesium alloy for biomaterial application." *Materials Chemistry and Physics* 203:27-33. doi: https://doi.org/10.1016/j.matchemphys.2017.08.020.

Dinda, G. P., J. Shin, and J. Mazumder. 2009. "Pulsed laser deposition of hydroxyapatite thin films on Ti–6Al–4V: Effect of heat treatment on structure and properties." *Acta Biomaterialia* 5 (5):1821-1830. doi: https://doi.org/10.1016/j.actbio.2009.01.027.

Ding, Qiongqiong, Xuejiao Zhang, Yong Huang, Yajing Yan, and Xiaofeng Pang. 2015. "In vitro cytocompatibility and corrosion resistance of zinc-doped hydroxyapatite coatings on a titanium substrate." *Journal of Materials Science* 50 (1):189-202. doi: 10.1007/s10853-014-8578-4.

Dohan Ehrenfest, David M., Paulo G. Coelho, Byung-Soo Kang, Young-Taeg Sul, and Tomas Albrektsson. 2010. "Classification of osseointegrated implant surfaces: materials, chemistry and topography." *Trends in Biotechnology* 28 (4):198-206. doi: https://doi.org/10.1016/j.tibtech.2009.12.003.

Drevet, R., F. Velard, S. Potiron, D. Laurent-Maquin, and H. Benhayoune. 2011. "In vitro dissolution and corrosion study of calcium phosphate coatings elaborated by pulsed electrodeposition current on Ti6Al4V substrate." *Journal of Materials Science: Materials in Medicine* 22 (4):753-761. doi: 10.1007/s10856-011-4251-5.

Ducheyne, Paul. 2015. *Comprehensive biomaterials*. Vol. 1: Elsevier.

Escada, A. L. A., J. P. B. Machado, S. G. Schneider, M. C. R. Alves Rezende, and A. P. R. Alves Claro. 2011. "Biomimetic calcium phosphate coating on Ti-7.5Mo alloy for dental application." *Journal of Materials Science: Materials in Medicine* 22 (11):2457-2465. doi: https://10.1007/s10856-011-4434-0.

Fathi, Mehdad, Mir Saman Safavi, Sahand Mirzazadeh, Alireza Ansariyan, and Iraj Ahadzadeh. 2020. "A Promising Horizon in Mechanical and Corrosion Properties Improvement of Ni-Mo Coatings Through Incorporation of Y2O3 Nanoparticles."

Metallurgical and Materials Transactions A 51 (2):897-908. doi: 10.1007/s11661-019-05559-5.

Fauchais, P., and A. Vardelle. 2011. "Innovative and emerging processes in plasma spraying: from micro- to nano-structured coatings." *Journal of Physics D: Applied Physics* 44 (19):194011. doi: https://10.1088/0022-3727/44/19/194011.

Feng, Yashan, Shijie Zhu, Liguo Wang, Lei Chang, Bingbing Yan, Xiaozhe Song, and Shaokang Guan. 2017. "Characterization and corrosion property of nano-rod-like HA on fluoride coating supported on Mg-Zn-Ca alloy." *Bioactive Materials* 2 (2):63-70. doi: https://doi.org/10.1016/j.bioactmat.2017.05.001.

Feng, Yashan, Xun Ma, Lei Chang, Shijie Zhu, and Shaokang Guan. 2017. "Characterization and cytocompatibility of polydopamine on MAO-HA coating supported on Mg-Zn-Ca alloy." *Surface and Interface Analysis* 49 (11):1115-1123. doi: https://doi.org/10.1002/sia.6286.

Fernández-Pradas, J. M., L. Clèries, E. Martínez, G. Sardin, J. Esteve, and J. L. Morenza. 2001. "Influence of thickness on the properties of hydroxyapatite coatings deposited by KrF laser ablation." *Biomaterials* 22 (15):2171-2175. doi: https://doi.org/10.1016/S0142-9612(00)00408-7.

Fielding, Gary A., Mangal Roy, Amit Bandyopadhyay, and Susmita Bose. 2012. "Antibacterial and biological characteristics of silver containing and strontium doped plasma sprayed hydroxyapatite coatings." *Acta Biomaterialia* 8 (8):3144-3152. doi: https://doi.org/10.1016/j.actbio.2012.04.004.

Fornell, J., Y. P. Feng, E. Pellicer, S. Suriñach, M. D. Baró, and J. Sort. 2017. "Mechanical behaviour of brushite and hydroxyapatite coatings electrodeposited on newly developed FeMnSiPd alloys." *Journal of Alloys and Compounds* 729:231-239. doi: https://doi.org/10.1016/j.jallcom.2017.09.187.

Francis, Marion D., and Ned C. Webb. 1970. "Hydroxyapatite formation from a hydrated calcium monohydrogen phosphate precursor." *Calcified Tissue Research* 6 (1):335-342. doi: https://10.1007/BF02196214.

Fu, Tao, Jian-min Sun, Zafer Alajmi, and Feng Wu. 2015. "Sol-gel preparation, corrosion resistance and hydrophilicity of Ta-containing TiO2 films on Ti6Al4V alloy." *Transactions of Nonferrous Metals Society of China* 25 (2):471-476. doi: https://doi.org/10.1016/S1003-6326(15)63626-3.

Furko, M., Y. Jiang, T. A. Wilkins, and C. Balázsi. 2016. "Electrochemical and morphological investigation of silver and zinc modified calcium phosphate bioceramic coatings on metallic implant materials." *Materials Science and Engineering: C* 62:249-259. doi: https://doi.org/10.1016/j.msec.2016.01.060.

Furko, Monika, Viktor Havasi, Zoltán Kónya, Alina Grünewald, Rainer Detsch, Aldo R. Boccaccini, and Csaba Balázsi. 2018. "Development and characterization of multi-element doped hydroxyapatite bioceramic coatings on metallic implants for orthopedic applications." *Boletín de la Sociedad Española de Cerámica y Vidrio* 57 (2):55-65. doi: https://doi.org/10.1016/j.bsecv.2017.09.003.

Gadow, R., A. Killinger, and N. Stiegler. 2010. "Hydroxyapatite coatings for biomedical applications deposited by different thermal spray techniques." *Surface and Coatings Technology* 205 (4):1157-1164. doi: https://doi.org/10.1016/j.surfcoat.2010.03.059.

Gao, Ya Li, Yu Liu, and Xue Ying Song. 2018. "Plasma-Sprayed Hydroxyapatite Coating for Improved Corrosion Resistance and Bioactivity of Magnesium Alloy." *Journal of Thermal Spray Technology* 27 (8):1381-1387. doi: https://10.1007/s11666-018-0760-9.

Gayathri, B., Muthukumarasamy N., Shanthi Bhupathi Santhosh, and Dhayalan Velauthapillai. 2019. "Investigation of Corrosion protection performance of Magnesium incorporated Hydroxyapatite coating on surgical grade Titanium." *Materials Today: Proceedings* 18:1678-1685. doi: https://doi.org/10.1016/j.matpr.2019.05.264.

Goto, Takashi, and Hirokazu Katsui. 2015. "Chemical vapor deposition of Ca–P–O film coating." In *Interface oral health science 2014*, 103-115. Springer, Tokyo.

GRAFTS, I Bone, and Bone Substitutes. 2004. "Three-dimensionally engineered hydroxyapatite ceramics with interconnected pores as a bone substitute and tissue engineering scaffold." *Biomaterials in orthopedics* 287.

Grandfield, Kathryn, Anders Palmquist, Stéphane Gonçalves, Andy Taylor, Mark Taylor, Lena Emanuelsson, Peter Thomsen, and Håkan Engqvist. 2011. "Free form fabricated features on CoCr implants with and without hydroxyapatite coating in vivo: a comparative study of bone contact and bone growth induction." *Journal of Materials Science: Materials in Medicine* 22 (4):899-906. doi: https://10.1007/s10856-011-4253-3.

Gunputh, Urvashi F., and Huirong Le. 2020. "A Review of In-Situ Grown Nanocomposite Coatings for Titanium Alloy Implants." *Journal of Composites Science* 4 (2). doi: https://10.3390/jcs4020041.

Guo, D. Z., F. L. Li, J. Y. Wang, and J. S. Sun. 1995. "Effects of post-coating processing on structure and erosive wear characteristics of flame and plasma spray coatings." *Surface and Coatings Technology* 73 (1):73-78. doi: https://doi.org/10.1016/0257-8972(94)02364-6.

Guo, Yunting, Yingchao Su, Siqi Jia, Guixun Sun, Rui Gu, Donghui Zhu, Guangyu Li, and Jianshe Lian. 2018. "Hydroxyapatite/Titania Composite Coatings on Biodegradable Magnesium Alloy for Enhanced Corrosion Resistance, Cytocompatibility and Antibacterial Properties." *Journal of The Electrochemical Society* 165 (14):C962-C972. doi: 10.1149/2.1171814jes.

Hameed, Pearlin, Vasanth Gopal, Stefan Bjorklund, Ashish Ganvir, Dwaipayan Sen, Nicolaie Markocsan, and Geetha Manivasagam. 2019. "Axial Suspension Plasma Spraying: An ultimate technique to tailor Ti6Al4V surface with HAp for orthopaedic applications." *Colloids and Surfaces B: Biointerfaces* 173:806-815. doi: https://doi.org/10.1016/j.colsurfb.2018.10.071.

Hanawa, Takao. 1999. "In vivo metallic biomaterials and surface modification." *Materials Science and Engineering: A* 267 (2):260-266. doi: https://doi.org/10.1016/S0921-5093(99)00101-X.

Hannink, Gerjon, and J. J. Chris Arts. 2011. "Bioresorbability, porosity and mechanical strength of bone substitutes: What is optimal for bone regeneration?" *Injury* 42:S22-S25. doi: https://doi.org/10.1016/j.injury.2011.06.008.

He, Dai-Hua, Pu Wang, Ping Liu, Xin-Kuan Liu, Feng-Cang Ma, and Jun Zhao. 2016. "HA coating fabricated by electrochemical deposition on modified Ti6Al4V alloy." *Surface and Coatings Technology* 301:6-12. doi: https://doi.org/10.1016/j.surfcoat.2016.07.005.

Heimann, Robert B. 2013. "Structure, properties, and biomedical performance of osteoconductive bioceramic coatings." *Surface and Coatings Technology* 233:27-38. doi: https://doi.org/10.1016/j.surfcoat.2012.11.013.

Hench, L, and J Jones. 2005. *Biomaterials, artificial organs and tissue engineering*: Elsevier.

Hiromoto, Sachiko, and Masanari Tomozawa. 2010. "Corrosion behavior of magnesium with hydroxyapatite coatings formed by hydrothermal treatment." *Materials transactions* 51 (11):2080-2087. doi: https://doi.org/10.2320/matertrans.M2010192.

Horandghadim, Nazila, and Jafar Khalil-Allafi. 2019. "Characterization of hydroxyapatite-tantalum pentoxide nanocomposite coating applied by electrophoretic deposition on Nitinol superelastic alloy." *Ceramics International* 45 (8):10448-10460. doi: https://doi.org/10.1016/j.ceramint.2019.02.105.

Horandghadim, Nazila, Jafar Khalil-Allafi, and Mustafa Urgen. 2020. "Influence of tantalum pentoxide secondary phase on surface features and mechanical properties of hydroxyapatite coating on NiTi alloy produced by electrophoretic deposition." *Surface and Coatings Technology* 386:125458. doi: https://doi.org/10.1016/j.surfcoat.2020.125458.

Hsieh, Ming-Fa, Li-Hsiang Perng, and Tsung-Shune Chin. 2002. "Hydroxyapatite coating on Ti6Al4V alloy using a sol–gel derived precursor." *Materials Chemistry and Physics* 74 (3):245-250. doi: https://doi.org/10.1016/S0254-0584(01)00474-6.

Huang, Yong, Min Hao, Xiaofeng Nian, Haixia Qiao, Xuejiao Zhang, Xiaoyun Zhang, Guiqin Song, Jiachi Guo, Xiaofeng Pang, and Honglei Zhang. 2016. "Strontium and copper co-substituted hydroxyapatite-based coatings with improved antibacterial activity and cytocompatibility fabricated by electrodeposition." *Ceramics International* 42 (10):11876-11888. doi: https://doi.org/10.1016/j.ceramint.2016.04.110.

Huang, Yong, Qiongqiong Ding, Xiaofeng Pang, Shuguang Han, and Yajing Yan. 2013. "Corrosion behavior and biocompatibility of strontium and fluorine co-doped electrodeposited hydroxyapatite coatings." *Applied Surface Science* 282:456-462. doi: https://doi.org/10.1016/j.apsusc.2013.05.152.

Hung, Kuo-Yung, Sung-Cheng Lo, Chung-Sheng Shih, Yung-Chin Yang, Hui-Ping Feng, and Yi-Chih Lin. 2013. "Titanium surface modified by hydroxyapatite coating for dental implants." *Surface and Coatings Technology* 231:337-345. doi: https://doi.org/10.1016/j.surfcoat.2012.03.037.

Iconaru, Simona L., Daniela Predoi, Carmen S. Ciobanu, Mikael Motelica-Heino, Régis Guegan, and Coralia Bleotu. 2022. "Development of Silver Doped Hydroxyapatite Thin Films for Biomedical Applications." *Coatings* 12 (3). doi: 10.3390/coatings12030341.

Indira, K., U. KamachiMudali, and N. Rajendran. 2014. "In vitro bioactivity and corrosion resistance of Zr incorporated TiO2 nanotube arrays for orthopaedic applications." *Applied Surface Science* 316:264-275. doi: https://doi.org/10.1016/j.apsusc.2014.08.001.

Jansen van Vuuren, Ludwig, J. Odendaal, and P. Pistorius. 2008. "Galvanic corrosion of dental cobalt-chromium alloys and dental amalgam in artificial saliva." *SADJ: journal of the South African Dental Association = tydskrif van die Suid-Afrikaanse Tandheelkundige Vereniging* 63:034-8.

Jiang, Dan, Qi-Ke Li, Yun-Zhi Liu, Zhe Xiao, Bao-Hong Xia, Shuo-Qi Li, Fen Zhang, Lan-Yue Cui, and Rong-Chang Zeng. 2022. "Polyphosphate assisted hydrothermal synthesis of hydroxyapatite coating on Mg alloys: Enhanced mechanical properties and corrosion resistance." *Surface and Coatings Technology* 432:128033. doi: https://doi.org/10.1016/j.surfcoat.2021.128033.

Jiang, Jiawei, Gui Han, Xuesong Zheng, Gang Chen, and Peizhi Zhu. 2019. "Characterization and biocompatibility study of hydroxyapatite coating on the surface of titanium alloy." *Surface and Coatings Technology* 375:645-651. doi: https://doi.org/10.1016/j.surfcoat.2019.07.067.

Jung, Ui-Won, Ji-Wan Hwang, Da-Yae Choi, Kyung-Seok Hu, Mi-Kyung Kwon, Seong-Ho Choi, and Hee-Jin Kim. 2012. "Surface characteristics of a novel hydroxyapatite-coated dental implant." *JPIS* 42 (2):59-63. doi: https://10.5051/jpis.2012.42.2.59.

Kadhim, Mustafa M., Haider Abdulkareem AlMashhadani, Raid D. Hashim, Anees A. Khadom, Khulood Abid Salih, and Abbas Washeel Salman. 2022. "Effect of Sr/Mg co-substitution on corrosion resistance properties of hydroxyapatite coated on Ti–6Al–4V dental alloys." *Journal of Physics and Chemistry of Solids* 161:110450. doi: https://doi.org/10.1016/j.jpcs.2021.110450.

Kannan, M. B. 2015. "13 - Hydroxyapatite coating on biodegradable magnesium and magnesium-based alloys." In *Hydroxyapatite (Hap) for Biomedical Applications*, edited by Michael Mucalo, 289-306. Woodhead Publishing.

Karacan, Ipek, Innocent Jacob Macha, Gina Choi, Sophie Cazalbou, and Besim Ben-Nissan. 2017. "Antibiotic Containing Poly Lactic Acid/Hydroxyapatite Biocomposite Coatings for Dental Implant Applications." *Key Engineering Materials* 758:120-125. doi: 10.4028/www.scientific.net/KEM.758.120.

Kezhi, Li, Guo Qian, Zhang Leilei, Zhang Yulei, Liu Shoujie, Guo Kebing, and Li Shaoxian. 2017. "Synthesis and characterization of Si-substituted hydroxyapatite bioactive coating for SiC-coated carbon/carbon composites." *Ceramics International* 43 (1, Part B):1410-1414. doi: https://doi.org/10.1016/j.ceramint.2016.10.102.

Khandelwal, Himanshu, Gurbhinder Singh, Khelendra Agrawal, Satya Prakash, and R. D. Agarwal. 2013. "Characterization of hydroxyapatite coating by pulse laser deposition technique on stainless steel 316 L by varying laser energy." *Applied Surface Science* 265:30-35. doi: https://doi.org/10.1016/j.apsusc.2012.10.072.

Kokubo, T., S. Ito, Z. T. Huang, T. Hayashi, S. Sakka, T. Kitsugi, and T. Yamamuro. 1990. "Ca, P-rich layer formed on high-strength bioactive glass-ceramic A-W." *Journal of Biomedical Materials Research* 24 (3):331-343. doi: https://doi.org/10.1002/jbm.820240306.

Kurzweg, H., R. B. Heimann, T. Troczynski, and M. L. Wayman. 1998. "Development of plasma-sprayed bioceramic coatings with bond coats based on titania and zirconia." *Biomaterials* 19 (16):1507-1511. doi: https://doi.org/10.1016/S0142-9612(98)00067-2.

Lacefield, W. R. 1988. "Hydroxyapatite coatings." *Annals of the New York Academy of Sciences* 523:72-80. doi: https://10.1111/j.1749-6632.1988.tb38501.x.

Latifi, Afrooz, Mohammad Imani, Mohammad Taghi Khorasani, and Morteza Daliri Joupari. 2014. "Plasma surface oxidation of 316L stainless steel for improving adhesion strength of silicone rubber coating to metal substrate." *Applied Surface Science* 320:471-481. doi: https://doi.org/10.1016/j.apsusc.2014.09.084.

LeGeros, R. Z., R. Kijowska, W. Jia, and J. P. LeGeros. 1988. "Fluoride-cation interactions in the formation and stability of apatites." *Journal of Fluorine Chemistry* 41 (1):53-64. doi: https://doi.org/10.1016/S0022-1139(00)83016-X.

Levingstone, Tanya J., Malika Ardhaoui, Khaled Benyounis, Lisa Looney, and Joseph T. Stokes. 2015. "Plasma sprayed hydroxyapatite coatings: Understanding process relationships using design of experiment analysis." *Surface and Coatings Technology* 283:29-36. doi: https://doi.org/10.1016/j.surfcoat.2015.10.044.

Li, Hanyang, Zainen Qin, Yiqiang Ouyang, Bo Zheng, Hong Wei, Jun Ou, and Chong Shen. 2022. "Hydroxyapatite/chitosan-metformin composite coating enhances the biocompatibility and osteogenic activity of AZ31 magnesium alloy." *Journal of Alloys and Compounds* 909:164694. doi: https://doi.org/10.1016/j.jallcom.2022.164694.

Li, Ting-Ting, Lei Ling, Mei-Chen Lin, Qian Jiang, Qi Lin, Jia-Horng Lin, and Ching-Wen Lou. 2019. "Properties and Mechanism of Hydroxyapatite Coating Prepared by Electrodeposition on a Braid for Biodegradable Bone Scaffolds." *Nanomaterials* 9 (5). doi: 10.3390/nano9050679.

Li, Yunfeng, Qing Li, Songsong Zhu, En Luo, Jihua Li, Ge Feng, Yunmao Liao, and Jing Hu. 2010. "The effect of strontium-substituted hydroxyapatite coating on implant fixation in ovariectomized rats." *Biomaterials* 31 (34):9006-9014. doi: https://doi.org/10.1016/j.biomaterials.2010.07.112.

Liu, Dean-Mo, T. Troczynski, and Wenjea J. Tseng. 2001. "Water-based sol–gel synthesis of hydroxyapatite: process development." *Biomaterials* 22 (13):1721-1730. doi: https://doi.org/10.1016/S0142-9612(00)00332-X.

Liu, Dean-Mo, Quanzu Yang, and Tom Troczynski. 2002. "Sol–gel hydroxyapatite coatings on stainless steel substrates." *Biomaterials* 23 (3):691-698. doi: https://doi.org/10.1016/S0142-9612(01)00157-0.

Liu, Yu-Cheng, Geng Sheng Lin, Jheng-Yang Wang, Che-Shun Cheng, Yung-Chin Yang, Bor-Shiunn Lee, and Kuo-Lun Tung. 2018. "Synthesis and characterization of porous hydroxyapatite coatings deposited on titanium by flame spraying." *Surface and Coatings Technology* 349:357-363. doi: https://doi.org/10.1016/j.surfcoat.2018.06.010.

Lobo, AO, FR Marciano, I Regiani, JT Matsushima, SC Ramos, and EJ Corat. 2011. "Influence of temperature and time for direct hydroxyapatite electrodeposition on superhydrophilic vertically aligned carbon nanotube films." *J. Nanomedic. Nanotechnol. S* 8 (2).

Lu, Minxun, Hongjie Chen, Bo Yuan, Yong Zhou, Li Min, Zhanwen Xiao, Xiangdong Zhu, Chongqi Tu, and Xingdong Zhang. 2020. "Electrochemical Deposition of Nanostructured Hydroxyapatite Coating on Titanium with Enhanced Early Stage Osteogenic Activity and Osseointegration." *International journal of nanomedicine* 15:6605-6618. doi: https://10.2147/IJN.S268372.

Ma, J., C. Z. Wang, C. L. Ban, C. Z. Chen, and H. M. Zhang. 2016. "Pulsed laser deposition of magnesium-containing bioactive glass film on porous Ti–6Al–4V substrate pretreated by micro-arc oxidation." *Vacuum* 125:48-55. doi: https://doi.org/10.1016/j.vacuum.2015.12.005.

Ma, Menghan, Wei Ye, and Xiao-Xiang Wang. 2008. "Effect of supersaturation on the morphology of hydroxyapatite crystals deposited by electrochemical deposition on titanium." *Materials Letters* 62 (23):3875-3877. doi: https://doi.org/10.1016/j.matlet.2008.05.009.

Ma, Yifei, Jiemin Han, Mei Wang, Xuyuan Chen, and Suotang Jia. 2018. "Electrophoretic deposition of graphene-based materials: A review of materials and their applications." *Journal of Materiomics* 4 (2):108-120. doi: https://doi.org/10.1016/j.jmat.2018.02.004.

Mahapatro, Anil. 2015. "Bio-functional nano-coatings on metallic biomaterials." *Materials Science and Engineering: C* 55:227-251. doi: https://doi.org/10.1016/j.msec.2015.05.018.

Maier, Petra, Benjamin Clausius, Jens Wicke, and Norbert Hort. 2020. "Characterization of an Extruded Mg-Dy-Nd Alloy during Stress Corrosion with C-Ring Tests." *Metals* 10 (5):584.

Mali, S. A., K. C. Nune, and R. D. K. Misra. 2016. "Biomimetic nanostructured hydroxyapatite coatings on metallic implant materials." *Materials Technology* 31 (13):782-790. doi: https://10.1080/10667857.2016.1224609.

Manivasagam, Geetha, A. K. Singh, Asokamani Rajamanickam, and Ashok Gogia. 2009. "Ti based biomaterials, the ultimate choice for orthopaedic implants–A review." *Progress in Materials Science* 54:397-425. doi: 10.1016/j.pmatsci.2008.06.004.

Manso, Miguel, Carmen Jiménez, Carmen Morant, Pilar Herrero, and J. M. Martínez-Duart. 2000. "Electrodeposition of hydroxyapatite coatings in basic conditions." *Biomaterials* 21 (17):1755-1761. doi: https://doi.org/10.1016/S0142-9612(00)00061-2.

Marie, P. J., D. Felsenberg, and M. L. Brandi. 2011. "How strontium ranelate, via opposite effects on bone resorption and formation, prevents osteoporosis." *Osteoporosis International* 22 (6):1659-1667. doi: 10.1007/s00198-010-1369-0.

Martin, R. I., and P. W. Brown. 1995. "Mechanical properties of hydroxyapatite formed at physiological temperature." *Journal of Materials Science: Materials in Medicine* 6 (3):138-143. doi: https://10.1007/BF00120289.

Mediaswanti, Kun, Cuie Wen, Elena Ivanova, Christopher Berndt, Francois Malherbe, Vy Pham, and James Wang. 2013. "A review on bioactive porous metallic biomaterials." *Journal of Biomimetics Biomaterials and Tissue Engineering* 18 (1):1-8. doi: https://10.4172/1662-100X.1000104.

Mehrvarz, Alireza, Jafar Khalil-Allafi, and Amir Kahaie Khosrowshahi. 2022. "Biocompatibility and antibacterial behavior of electrochemically deposited Hydroxyapatite/ZnO porous nanocomposite on NiTi biomedical alloy." *Ceramics International* 48 (11):16326-16336. doi: https://doi.org/10.1016/j.ceramint.2022.02.183.

Metikoš-Huković, M., E. Tkalčec, A. Kwokal, and J. Piljac. 2003. "An in vitro study of Ti and Ti-alloys coated with sol–gel derived hydroxyapatite coatings." *Surface and*

Coatings Technology 165 (1):40-50. doi: https://doi.org/10.1016/S0257-8972(02)00732-6.

Mi, Z Ridzwan, Solehuddin Shuib, AY Hassan, AA Shorki, and MM Ibrahim. 2007. "Problem of stress shielding and improvement to the hip Implat designs: a review." *J. Med. Sci* 7 (3):460-467.

Mohandesnezhad, Sadaf, Mohamadreza Etminanfar, Soheil Mahdavi, and Mir Saman Safavi. 2022. "Enhanced bioactivity of 316L stainless steel with deposition of polypyrrole/hydroxyapatite layered hybrid coating: Orthopedic applications." *Surfaces and Interfaces* 28:101604. doi: https://doi.org/10.1016/j.surfin.2021.101604.

Mohseni, E., E. Zalnezhad, and A. R. Bushroa. 2014. "Comparative investigation on the adhesion of hydroxyapatite coating on Ti–6Al–4V implant: A review paper." *International Journal of Adhesion and Adhesives* 48:238-257. doi: https://doi.org/10.1016/j.ijadhadh.2013.09.030.

Monsalve, M., E. Lopez, H. Ageorges, and F. Vargas. 2015. "Bioactivity and mechanical properties of bioactive glass coatings fabricated by flame spraying." *Surface and Coatings Technology* 268:142-146. doi: https://doi.org/10.1016/j.surfcoat.2014.08.041.

Morey, George W., and Paul Niggli. 1913. "The Hydrothermal Formation of Silicates, A Review." *Journal of the American Chemical Society* 35 (9):1086-1130. doi: 10.1021/ja02198a600.

Morks, M. F. 2008. "Fabrication and characterization of plasma-sprayed HA / SiO2 coatings for biomedical application." *Journal of the Mechanical Behavior of Biomedical Materials* 1 (1):105-111. doi: https://doi.org/10.1016/j.jmbbm.2007.04.003.

Mumith, Aadil, Vee San Cheong, Paul Fromme, Melanie J Coathup, and Gordon W Blunn. 2020. "The effect of strontium and silicon substituted hydroxyapatite electrochemical coatings on bone ingrowth and osseointegration of selective laser sintered porous metal implants." *PloS one* 15 (1):e0227232. doi: https://doi.org/10.1371/journal.pone.0227232.

Nasab, Marjan Bahrami, Mohd Roshdi Hassan, and B Bin Sahari. 2010. "Metallic biomaterials of knee and hip-a review." *Trends Biomater. Artif. Organs* 24 (1):69-82.

Nayar, Suprabha, Ashit Kumar Pramanick, Binay Kumar Sharma, Rahul Kumar Mishra, Satyendra Kumar Bansal, Amardeep Prajapati, Kaushal Kishor Sahu, Swapan Kr Das, Lokesh Pathak, and Arvind Sinha. 2006. "Hydroxyapatite coating on stainless steel pre-coated with bovine serum albumin at ambient conditions." *Colloids and Surfaces B: Biointerfaces* 48 (2):183-187. doi: https://doi.org/10.1016/j.colsurfb.2005.09.006.

Ni, G. X., W. W. Lu, B. Xu, K. Y. Chiu, C. Yang, Z. Y. Li, W. M. Lam, and K. D. K. Luk. 2006. "Interfacial behaviour of strontium-containing hydroxyapatite cement with cancellous and cortical bone." *Biomaterials* 27 (29):5127-5133. doi: https://doi.org/10.1016/j.biomaterials.2006.05.030.

Niinomi, Mitsuo. 2002. "Recent metallic materials for biomedical applications." *Metallurgical and Materials Transactions A* 33 (3):477. doi: https://10.1007/s11661-002-0109-2.

Ninomiya, James T., Janine A. Struve, Cary T. Stelloh, Jeffrey M. Toth, and Kevin E. Crosby. 2001. "Effects of hydroxyapatite participate debris on the production of cytokines and proteases in human fibroblasts." *Journal of Orthopaedic Research* 19 (4):621-628. doi: https://doi.org/10.1016/S0736-0266(00)00061-9.

Nizami, M. Z. I., B. D. L. Campéon, and Y. Nishina. 2022. "Electrodeposition of hydroxyapatite and graphene oxide improves the bioactivity of medical grade stainless steel." *Materials Today Sustainability* 19:100193. doi: https://doi.org/10.1016/j.mtsust.2022.100193.

Okazaki, M. 1995. "Crystallographic properties of heterogeneous Mg-containing fluoridated apatites synthesized with a two-step supply system." *Biomaterials* 16 (9):703-707. doi: https://doi.org/10.1016/0142-9612(95)99698-L.

Pana, Iulian, Viorel Braic, Alina Vladescu, Raluca Ion, Anca C. Parau, Nicolae C. Zoita, Mihaela Dinu, Adrian E. Kiss, Anisoara Cimpean, and Mariana Braic. 2022. "SiC- and Ag-SiC-Doped Hydroxyapatite Coatings Grown Using Magnetron Sputtering on Ti Alloy for Biomedical Application." *Coatings* 12 (2). doi: 10.3390/coatings12020195.

Panda, Sikta, Tanmay Bharadwaj, Devendra Verma, Chandan Kumar Biswas, and Subhankar Paul. 2022. "Influence of strontium and niobium on the physical and biological performance of hydroxyapatite as a bioactive coating on implant materials." *Ceramics International*. doi: https://doi.org/10.1016/j.ceramint.2022.07.268.

Penn, R. Lee, and Jillian F. Banfield. 1999. "Morphology development and crystal growth in nanocrystalline aggregates under hydrothermal conditions: insights from titania." *Geochimica et Cosmochimica Acta* 63 (10):1549-1557. doi: https://doi.org/10.1016/S0016-7037(99)00037-X.

Pezzatini, Silvia, Raffaella Solito, Lucia Morbidelli, Stefania Lamponi, Elisa Boanini, Adriana Bigi, and Marina Ziche. 2006. "The effect of hydroxyapatite nanocrystals on microvascular endothelial cell viability and functions." *Journal of Biomedical Materials Research Part A* 76A (3):656-663. doi: https://doi.org/10.1002/jbm.a.30524.

Polsongkram, D., P. Chamninok, S. Pukird, L. Chow, O. Lupan, G. Chai, H. Khallaf, S. Park, and A. Schulte. 2008. "Effect of synthesis conditions on the growth of ZnO nanorods via hydrothermal method." *Physica B: Condensed Matter* 403 (19):3713-3717. doi: https://doi.org/10.1016/j.physb.2008.06.020.

Poorraeisi, Morteza, and Abdollah Afshar. 2018. "The study of electrodeposition of hydroxyapatite-ZrO2-TiO2 nanocomposite coatings on 316 stainless steel." *Surface and Coatings Technology* 339:199-207. doi: https://doi.org/10.1016/j.surfcoat.2018.02.030.

Popa, Mihai V., Jose Maria Calderon Moreno, Monica Popa, Ecaterina Vasilescu, Paula Drob, Cora Vasilescu, and Silviu I. Drob. 2011. "Electrochemical deposition of bioactive coatings on Ti and Ti–6Al–4V surfaces." *Surface and Coatings Technology* 205 (20):4776-4783. doi: https://doi.org/10.1016/j.surfcoat.2011.04.040.

Praharaj, Rinmayee, Snigdha Mishra, R. D. K. Misra, and Tapash R. Rautray. 2022. "Biocompatibility and adhesion response of magnesium-hydroxyapatite/strontium-titania (Mg-HAp/ Sr-TiO2) bilayer coating on titanium." *Materials Technology* 37 (4):230-239. doi: 10.1080/10667857.2020.1825898.

Prakasam, Mythili, Janis Locs, Kristine Salma-Ancane, Dagnija Loca, Alain Largeteau, and Liga Berzina-Cimdina. 2015. "Fabrication, Properties and Applications of Dense Hydroxyapatite: A Review." *Journal of Functional Biomaterials* 6 (4). doi: https://10.3390/jfb6041099.

Priyadarshini, B., M. Rama, Chetan, and U. Vijayalakshmi. 2019. "Bioactive coating as a surface modification technique for biocompatible metallic implants: a review." *Journal of Asian Ceramic Societies* 7 (4):397-406. doi: https://10.1080/21870764.2019.1669861.

Priyadarshini, B., S. Ramya, E. Shinyjoy, L. Kavitha, D. Gopi, and U. Vijayalakshmi. 2021. "Structural, morphological and biological evaluations of cerium incorporated hydroxyapatite sol–gel coatings on Ti–6Al–4V for orthopaedic applications." *Journal of Materials Research and Technology* 12:1319-1338. doi: https://doi.org/10.1016/j.jmrt.2021.03.009.

Qiu, Deliang, Aiping Wang, and Yansheng Yin. 2010. "Characterization and corrosion behavior of hydroxyapatite/zirconia composite coating on NiTi fabricated by electrochemical deposition." *Applied Surface Science* 257 (5):1774-1778. doi: https://doi.org/10.1016/j.apsusc.2010.09.014.

Qiu, Deliang, Lejiao Yang, Yansheng Yin, and Aiping Wang. 2011. "Preparation and characterization of hydroxyapatite/titania composite coating on NiTi alloy by electrochemical deposition." *Surface and Coatings Technology* 205 (10):3280-3284. doi: https://doi.org/10.1016/j.surfcoat.2010.11.049.

Qu, Haibo, and Mei Wei. 2008. "Improvement of bonding strength between biomimetic apatite coating and substrate." *Journal of Biomedical Materials Research Part B: Applied Biomaterials* 84B (2):436-443. doi: https://doi.org/10.1002/jbm.b.30889.

Rasooli, Ali, Mir Saman Safavi, Farid Babaei, and Alireza Ansarian. 2020. "Electrodeposited Ni–Fe–Cr2O3 nanocomposite coatings: A survey of influences of Cr2O3 nanoparticles loadings in the electrolyte." *Journal of Alloys and Compounds* 822:153725. doi: https://doi.org/10.1016/j.jallcom.2020.153725.

Rasooli, Ali, Mir Saman Safavi, and Mehdi Kasbkar Hokmabad. 2018. "Cr2O3 nanoparticles: A promising candidate to improve the mechanical properties and corrosion resistance of Ni-Co alloy coatings." *Ceramics International* 44 (6):6466-6473. doi: https://doi.org/10.1016/j.ceramint.2018.01.044.

Rath, Purna C., Laxmidhar Besra, Bimal P. Singh, and Sarama Bhattacharjee. 2012. "Titania/hydroxyapatite bi-layer coating on Ti metal by electrophoretic deposition: Characterization and corrosion studies." *Ceramics International* 38 (4):3209-3216. doi: https://doi.org/10.1016/j.ceramint.2011.12.026.

Ratha, Itishree, Pradyot Datta, Nimu Chand Reger, Himanka Das, Vamsi Krishna Balla, K. Bavya Devi, Mangal Roy, Samit Kumar Nandi, and Biswanath Kundu. 2022. "In vivo osteogenesis of plasma sprayed ternary-ion doped hydroxyapatite coatings on Ti6Al4V for orthopaedic applications." *Ceramics International* 48 (8):11475-11488. doi: https://doi.org/10.1016/j.ceramint.2022.01.004.

Rau, J. V., I. Antoniac, M. Filipescu, C. Cotrut, M. Fosca, L. C. Nistor, R. Birjega, and M. Dinescu. 2018. "Hydroxyapatite coatings on Mg-Ca alloy prepared by Pulsed Laser Deposition: Properties and corrosion resistance in Simulated Body Fluid." *Ceramics*

International 44 (14):16678-16687. doi: https://doi.org/10.1016/j.ceramint.2018.06.095.

Rauch, J., G. Bolelli, A. Killinger, R. Gadow, V. Cannillo, and L. Lusvarghi. 2009. "Advances in High Velocity Suspension Flame Spraying (HVSFS)." *Surface and Coatings Technology* 203 (15):2131-2138. doi: https://doi.org/10.1016/j.surfcoat.2008.12.002.

Reginster, J. Y., O. Bruyère, A. Sawicki, A. Roces-Varela, P. Fardellone, A. Roberts, and J. P. Devogelaer. 2009. "Long-term treatment of postmenopausal osteoporosis with strontium ranelate: Results at 8 years." *Bone* 45 (6):1059-1064. doi: https://doi.org/10.1016/j.bone.2009.08.004.

Rezazadeh Shirdar, Mostafa, Mohammad Mahdi Taheri, Hossein Moradifard, Ali Keyvanfar, Arezou Shafaghat, Tolou Shokuhfar, and Sudin Izman. 2016. "Hydroxyapatite–Titania nanotube composite as a coating layer on Co–Cr-based implants: Mechanical and electrochemical optimization." *Ceramics International* 42 (6):6942-6954. doi: https://doi.org/10.1016/j.ceramint.2016.01.080.

Ročňáková, I., K. Slámečka, E. B. Montufar, M. Remešová, L. Dyčková, A. Břínek, D. Jech, K. Dvořák, L. Čelko, and J. Kaiser. 2018. "Deposition of hydroxyapatite and tricalcium phosphate coatings by suspension plasma spraying: Effects of torch speed." *Journal of the European Ceramic Society* 38 (16):5489-5496. doi: https://doi.org/10.1016/j.jeurceramsoc.2018.08.007.

Rojaee, Ramin, Mohammadhossein Fathi, and Keyvan Raeissi. 2013. "Controlling the degradation rate of AZ91 magnesium alloy via sol–gel derived nanostructured hydroxyapatite coating." *Materials Science and Engineering: C* 33 (7):3817-3825. doi: https://doi.org/10.1016/j.msec.2013.05.014.

Rojaee, Ramin, Mohammadhossein Fathi, Keyvan Raeissi, and Ali Sharifnabi. 2014. "Biodegradation assessment of nanostructured fluoridated hydroxyapatite coatings on biomedical grade magnesium alloy." *Ceramics International* 40 (9, Part B):15149-15158. doi: https://doi.org/10.1016/j.ceramint.2014.06.129.

Roshan, Sh, H. Eivaz Mohammadloo, A. A. Sarabi, and M. Afshari. 2022. "Biocompatible hybrid chitosan/hydroxyapatite coating applied on the AZ31 Mg alloy substrate: In-vitro corrosion, surface and structure studies." *Materials Today Communications* 30:103153. doi: https://doi.org/10.1016/j.mtcomm.2022.103153.

Roy, Mangal, B. Vamsi Krishna, Amit Bandyopadhyay, and Susmita Bose. 2008. "Laser processing of bioactive tricalcium phosphate coating on titanium for load-bearing implants." *Acta Biomaterialia* 4 (2):324-333. doi: https://doi.org/10.1016/j.actbio.2007.09.008.

Saadati, Ahmad, Hamid Hesarikia, Mohammad Reza Nourani, and Ramezan Ali Taheri. 2020. "Electrophoretic deposition of hydroxyapatite coating on biodegradable Mg–4Zn–4Sn–0.6Ca–0.5Mn alloy." *Surface Engineering* 36 (9):908-918. doi: https://10.1080/02670844.2019.1661145.

Sadat-Shojai, Mehdi, Mohammad-Taghi Khorasani, Ehsan Dinpanah-Khoshdargi, and Ahmad Jamshidi. 2013. "Synthesis methods for nanosized hydroxyapatite with diverse structures." *Acta Biomaterialia* 9 (8):7591-7621. doi: https://doi.org/10.1016/j.actbio.2013.04.012.

Safavi, M. S., A. Rasooli, and F. A. Sorkhabi. 2020. "Electrodeposition of Ni-P/Ni-Co-Al2O3 duplex nanocomposite coatings: towards improved mechanical and corrosion properties." *Transactions of the IMF* 98 (6):320-327. doi: 10.1080/00202967.2020.1802106.

Safavi, Mir Saman, Farid Babaei, Alireza Ansarian, and Iraj Ahadzadeh. 2019. "Incorporation of Y2O3 nanoparticles and glycerol as an appropriate approach for corrosion resistance improvement of Ni-Fe alloy coatings." *Ceramics International* 45 (8):10951-10960. doi: https://doi.org/10.1016/j.ceramint.2019.02.177.

Safavi, Mir Saman, and Ali Rasooli. 2019a. "Ni-P-TiO2 nanocomposite coatings with uniformly dispersed Ni3Ti intermetallics: Effects of current density and post heat treatment." *Surface and Coatings Technology* 372:252-259. doi: https://doi.org/10.1016/j.surfcoat.2019.05.058.

Safavi, Mir Saman, and Ali Rasooli. 2019b. "Ni-P-TiO2 nanocomposite coatings with uniformly dispersed Ni3Ti intermetallics: effects of TiO2 nanoparticles concentration." *Surface Engineering* 35 (12):1070-1080. doi: 10.1080/02670844.2018.1564475.

Santos, M., C. Santos, and M. J. Carmezim. 2019. "Production of bioactive hydroxyapatite coating by coblast process for orthopedic implants." *2019 IEEE 6th Portuguese Meeting on Bioengineering (ENBENG),* 22-23 Feb. 2019.

Sari, Mona, Chotimah, Ika D. Ana, and Yusril Yusuf. 2022. "Cell Viability Assay and Surface Morphology Analysis of Carbonated Hydroxyapatite/Honeycomb/Titanium Alloy Coatings for Bone Implant Applications." *Bioengineering* 9 (7). doi: 10.3390/bioengineering9070325.

Schrooten, J., and J. A. Helsen. 2000. "Adhesion of bioactive glass coating to Ti6Al4V oral implant." *Biomaterials* 21 (14):1461-1469. doi: https://doi.org/10.1016/S0142-9612(00)00027-2.

Sekiguchi, T., S. Miyashita, K. Obara, T. Shishido, and N. Sakagami. 2000. "Hydrothermal growth of ZnO single crystals and their optical characterization." *Journal of Crystal Growth* 214-215:72-76. doi: https://doi.org/10.1016/S0022-0248(00)00065-8.

Seyedraoufi, Z. S., and S. Mirdamadi. 2014. "Effects of pulse electrodeposition parameters and alkali treatment on the properties of nano hydroxyapatite coating on porous Mg–Zn scaffold for bone tissue engineering application." *Materials Chemistry and Physics* 148 (3):519-527. doi: https://doi.org/10.1016/j.matchemphys.2014.06.067.

Seyedraoufi, Z. S., and Sh Mirdamadi. 2015. "In vitro biodegradability and biocompatibility of porous Mg-Zn scaffolds coated with nano hydroxyapatite via pulse electrodeposition." *Transactions of Nonferrous Metals Society of China* 25 (12):4018-4027. doi: https://doi.org/10.1016/S1003-6326(15)64051-1.

Shadanbaz, Shaylin, and George J. Dias. 2012. "Calcium phosphate coatings on magnesium alloys for biomedical applications: A review." *Acta Biomaterialia* 8 (1):20-30. doi: https://doi.org/10.1016/j.actbio.2011.10.016.

Shirdar, Mostafa Rezazadeh, and Mohammad Mahdi Taheri. 2017. "Surface Morphology and Corrosion Behavior of Hydroxyapatite-Coated Co-Cr Implant: Effect of Sintering Conditions." *JOM* 69 (12):2831-2837. doi: https://10.1007/s11837-017-2598-x.

Sillekens, W. H., and D. Bormann. 2012. "13 - Biomedical applications of magnesium alloys." In *Advances in Wrought Magnesium Alloys*, edited by Colleen Bettles and Matthew Barnett, 427-454. Woodhead Publishing.

Simchen, Frank, Maximilian Sieber, Alexander Kopp, and Thomas Lampke. 2020. "Introduction to Plasma Electrolytic Oxidation—An Overview of the Process and Applications." *Coatings* 10 (7). doi: 10.3390/coatings10070628.

Singh, Balraj, Gurpreet Singh, and Buta Singh Sidhu. 2018. "Analysis of Corrosion Behavior and Surface Properties of Plasma-Sprayed HA/Ta Coating on CoCr Alloy." *Journal of Thermal Spray Technology* 27 (8):1401-1413. doi: 10.1007/s11666-018-0786-z.

Singh, Balraj, Gurpreet Singh, and Buta Singh Sidhu. 2019. "Investigation of the in vitro corrosion behavior and biocompatibility of niobium (Nb)-reinforced hydroxyapatite (HA) coating on CoCr alloy for medical implants." *Journal of Materials Research* 34 (10):1678-1691. doi: https://10.1557/jmr.2019.94.

Singh, Balraj, Gurpreet Singh, Buta Singh Sidhu, and Nitish Bhatia. 2019. "In-vitro assessment of HA-Nb coating on Mg alloy ZK60 for biomedical applications." *Materials Chemistry and Physics* 231:138-149. doi: https://doi.org/10.1016/j.matchemphys.2019.04.037.

Singh, Gurbhinder, Surendra Singh, and Satya Prakash. 2010. "Role of post heat treatment of plasma sprayed pure and Al2O3-TiO2 reinforced hydroxyapatite coating on the microstructure and mechanical properties." *Journal of Minerals & Materials Characterization & Engineering* 9 (12):1059-1069.

Singh, Gurbhinder, Surendra Singh, and Satya Prakash. 2011. "Surface characterization of plasma sprayed pure and reinforced hydroxyapatite coating on Ti6Al4V alloy." *Surface and Coatings Technology* 205 (20):4814-4820. doi: https://doi.org/10.1016/j.surfcoat.2011.04.064.

Singh, Sandeep, Gurpreet Singh, and Niraj Bala. 2019. "Corrosion behavior and characterization of HA/Fe3O4/CS composite coatings on AZ91 Mg alloy by electrophoretic deposition." *Materials Chemistry and Physics* 237:121884. doi: https://doi.org/10.1016/j.matchemphys.2019.121884.

Song, Y. W., D. Y. Shan, and E. H. Han. 2008. "Electrodeposition of hydroxyapatite coating on AZ91D magnesium alloy for biomaterial application." *Materials Letters* 62 (17):3276-3279. doi: https://doi.org/10.1016/j.matlet.2008.02.048.

Sridhar, T. M. 2010. "Nanobioceramic coatings for biomedical applications." *Materials Technology* 25 (3-4):184-195. doi: https://10.1179/175355510X12723642365449.

Stigter, M., K. de Groot, and P. Layrolle. 2002. "Incorporation of tobramycin into biomimetic hydroxyapatite coating on titanium." *Biomaterials* 23 (20):4143-4153. doi: https://doi.org/10.1016/S0142-9612(02)00157-6.

Stoch, A., W. Jastrze̜bski, E. Długoń, W. Lejda, B. Trybalska, G. J. Stoch, and A. Adamczyk. 2005. "Sol–gel derived hydroxyapatite coatings on titanium and its alloy Ti6Al4V." *Journal of Molecular Structure* 744-747:633-640. doi: https://doi.org/10.1016/j.molstruc.2004.10.080.

Suchanek, Wojciech, and Masahiro Yoshimura. 1998. "Processing and properties of hydroxyapatite-based biomaterials for use as hard tissue replacement implants." *Journal of Materials Research* 13 (1):94-117. doi: https://10.1557/JMR.1998.0015.

Sun, Guangfei, Jun Ma, and Shengmin Zhang. 2014. "Electrophoretic deposition of zinc-substituted hydroxyapatite coatings." *Materials Science and Engineering: C* 39:67-72. doi: https://doi.org/10.1016/j.msec.2014.02.023.

Sun, Limin, Christopher C. Berndt, and Karlis A. Gross. 2002. "Hydroxyapatite/polymer composite flame-sprayed coatings for orthopedic applications." *Journal of Biomaterials Science, Polymer Edition* 13 (9):977-990. doi: https://10.1163/156856202760319135.

Surmenev, Roman A., and Maria A. Surmeneva. 2019. "A critical review of decades of research on calcium phosphate–based coatings: How far are we from their widespread clinical application?" *Current Opinion in Biomedical Engineering* 10:35-44. doi: https://doi.org/10.1016/j.cobme.2019.02.003.

Surmenev, Roman A., Maria A. Surmeneva, and Anna A. Ivanova. 2014. "Significance of calcium phosphate coatings for the enhancement of new bone osteogenesis – A review." *Acta Biomaterialia* 10 (2):557-579. doi: https://doi.org/10.1016/j.actbio.2013.10.036.

Suwanprateeb, Jintamai, Waraporn Suvannapruk, Watchara Chokevivat, Siripong Kiertkrittikhoon, Nara Jaruwangsanti, and Prakit Tienboon. 2018. "Bioactivity of a sol–gel-derived hydroxyapatite coating on titanium implants in vitro and in vivo." *Asian Biomedicine* 12 (1):35-44. doi: https://10.1515/abm-2018-0029.

Taha, M., F. Chai, N. Blanchemain, M. Goube, B. Martel, and H. F. Hildebrand. 2013. "Validating the poly-cyclodextrins based local drug delivery system on plasma-sprayed hydroxyapatite coated orthopedic implant with toluidine blue O." *Materials Science and Engineering: C* 33 (5):2639-2647. doi: https://doi.org/10.1016/j.msec.2013.02.022.

Tahmasbi Rad, Armin, Mehran Solati-Hashjin, Noor Azuan Abu Osman, and Shahab Faghihi. 2014. "Improved bio-physical performance of hydroxyapatite coatings obtained by electrophoretic deposition at dynamic voltage." *Ceramics International* 40 (8, Part B):12681-12691. doi: https://doi.org/10.1016/j.ceramint.2014.04.116.

Tao, Yuan, Genshui Ke, Yan Xie, Yigang Chen, Siqi Shi, and Haibo Guo. 2015. "Adhesion strength and nucleation thermodynamics of four metals (Al, Cu, Ti, Zr) on AlN substrates." *Applied Surface Science* 357:8-13. doi: https://doi.org/10.1016/j.apsusc.2015.08.243.

Teng, Shu-Hua, Eun-Jung Lee, Chee-Sung Park, Won-Young Choi, Du-Sik Shin, and Hyoun-Ee Kim. 2008. "Bioactive nanocomposite coatings of collagen/hydroxyapatite on titanium substrates." *Journal of Materials Science: Materials in Medicine* 19 (6):2453-2461. doi: 10.1007/s10856-008-3370-0.

Tian, Meng, Shu Cai, Lei Ling, You Zuo, Zhongyan Wang, Pengbo Liu, Xiaogang Bao, and Guohua Xu. 2022. "Superhydrophilic hydroxyapatite/hydroxypropyltrimethyl ammonium chloride chitosan composite coating for enhancing the antibacterial and corrosion resistance of magnesium alloy." *Progress in Organic Coatings* 165:106745. doi: https://doi.org/10.1016/j.porgcoat.2022.106745.

Tsutsumi, Harumi, Mitsuo Niinomi, Masaaki Nakai, Tatsuya Gozawa, Toshikazu Akahori, Kazumi Saito, Rong Tu, and Takashi Goto. 2010. "Fabrication of hydroxyapatite film on Ti-29Nb-13Ta-4.6 Zr using a MOCVD technique." *Materials transactions* 51 (12):2277-2283.

Ullah, Ihsan, Muhammad Ali Siddiqui, Hui Liu, Sharafadeen Kunle Kolawole, Ji Zhang, Shuyuan Zhang, Ling Ren, and Ke Yang. 2020. "Mechanical, Biological, and Antibacterial Characteristics of Plasma-Sprayed (Sr,Zn) Substituted Hydroxyapatite Coating." *ACS Biomaterials Science & Engineering* 6 (3):1355-1366. doi: https://10.1021/acsbiomaterials.9b01396.

Vahabzadeh, Sahar, Mangal Roy, Amit Bandyopadhyay, and Susmita Bose. 2015. "Phase stability and biological property evaluation of plasma sprayed hydroxyapatite coatings for orthopedic and dental applications." *Acta Biomaterialia* 17:47-55. doi: https://doi.org/10.1016/j.actbio.2015.01.022.

Vasilescu, C., P. Drob, E. Vasilescu, I. Demetrescu, D. Ionita, M. Prodana, and S. I. Drob. 2011. "Characterisation and corrosion resistance of the electrodeposited hydroxyapatite and bovine serum albumin/hydroxyapatite films on Ti–6Al–4V–1Zr alloy surface." *Corrosion Science* 53 (3):992-999. doi: https://doi.org/10.1016/j.corsci.2010.11.033.

Vencl, Aleksandar, Saioa Arostegui, Gregory Favaro, Fatima Zivic, Mihailo Mrdak, Slobodan Mitrović, and Vladimir Popovic. 2011. "Evaluation of adhesion/cohesion bond strength of the thick plasma spray coatings by scratch testing on coatings cross-sections." *Tribology International* 44 (11):1281-1288. doi: https://doi.org/10.1016/j.triboint.2011.04.002.

Viswanath, B., and N. Ravishankar. 2008. "Controlled synthesis of plate-shaped hydroxyapatite and implications for the morphology of the apatite phase in bone." *Biomaterials* 29 (36):4855-4863. doi: https://doi.org/10.1016/j.biomaterials.2008.09.001.

Vu, Ashley A., Samuel Ford Robertson, Dongxu Ke, Amit Bandyopadhyay, and Susmita Bose. 2019. "Mechanical and biological properties of ZnO, SiO2, and Ag2O doped plasma sprayed hydroxyapatite coating for orthopaedic and dental applications." *Acta Biomaterialia* 92:325-335. doi: https://doi.org/10.1016/j.actbio.2019.05.020.

Wang, H. X., S. K. Guan, X. Wang, C. X. Ren, and L. G. Wang. 2010. "In vitro degradation and mechanical integrity of Mg–Zn–Ca alloy coated with Ca-deficient hydroxyapatite by the pulse electrodeposition process." *Acta Biomaterialia* 6 (5):1743-1748. doi: https://doi.org/10.1016/j.actbio.2009.12.009.

Wang, Hao, Noam Eliaz, Zhou Xiang, Hu-Ping Hsu, Myron Spector, and Linn W. Hobbs. 2006. "Early bone apposition in vivo on plasma-sprayed and electrochemically deposited hydroxyapatite coatings on titanium alloy." *Biomaterials* 27 (23):4192-4203. doi: https://doi.org/10.1016/j.biomaterials.2006.03.034.

Wang, Lu-Ning, and Jing-Li Luo. 2011. "Preparation of hydroxyapatite coating on CoCrMo implant using an effective electrochemically-assisted deposition pretreatment." *Materials Characterization* 62 (11):1076-1086. doi: https://doi.org/10.1016/j.matchar.2011.08.002.

Wang, Xuexin, Shu Cai, Tielong Liu, Mengguo Ren, Kai Huang, Ruiyue Zhang, and Huan Zhao. 2014. "Fabrication and corrosion resistance of calcium phosphate glass-ceramic coated Mg alloy via a PEG assisted sol–gel method." *Ceramics International* 40 (2):3389-3398. doi: https://doi.org/10.1016/j.ceramint.2013.09.093.

Wen, Cuilian, Xiaozhang Zhan, Xiaogui Huang, Feng Xu, Lijin Luo, and Chengsen Xia. 2017. "Characterization and corrosion properties of hydroxyapatite/graphene oxide bio-composite coating on magnesium alloy by one-step micro-arc oxidation method." *Surface and Coatings Technology* 317:125-133. doi: https://doi.org/10.1016/j.surfcoat.2017.03.034.

Witte, Frank, Frank Feyerabend, Petra Maier, Jens Fischer, Michael Störmer, Carsten Blawert, Wolfgang Dietzel, and Norbert Hort. 2007. "Biodegradable magnesium–hydroxyapatite metal matrix composites." *Biomaterials* 28 (13):2163-2174. doi: https://doi.org/10.1016/j.biomaterials.2006.12.027.

Xie, Jianhui, and Ben Li Luan. 2008. "Formation of hydroxyapatite coating using novel chemo-biomimetic method." *Journal of Materials Science: Materials in Medicine* 19 (10):3211-3220. doi: https://10.1007/s10856-008-3451-0.

Xiong, Jianyu, Yuncang Li, Peter D. Hodgson, and Cui'e Wen. 2010. "Nanohydroxyapatite coating on a titanium–niobium alloy by a hydrothermal process." *Acta Biomaterialia* 6 (4):1584-1590. doi: https://doi.org/10.1016/j.actbio.2009.10.016.

Xu, Y., H. Li, J. Wu, Q. Yang, D. Jiang, and B. Qiao. 2018. "Polydopamine-induced hydroxyapatite coating facilitates hydroxyapatite/polyamide 66 implant osteogenesis: an in vitro and in vivo evaluation." *Int J Nanomedicine* 13:8179-8193. doi: 10.2147/ijn.S181137.

Yang, Y. C., and Edward Chang. 2001. "Influence of residual stress on bonding strength and fracture of plasma-sprayed hydroxyapatite coatings on Ti–6Al–4V substrate." *Biomaterials* 22 (13):1827-1836. doi: https://doi.org/10.1016/S0142-9612(00)00364-1.

Yang, Y. C., and B. Y. Chou. 2007. "Bonding strength investigation of plasma-sprayed HA coatings on alumina substrate with porcelain intermediate layer." *Materials Chemistry and Physics* 104 (2):312-319. doi: https://doi.org/10.1016/j.matchemphys.2007.03.021.

Yang, Yung-Chin, Chien-Chung Chen, Jhong-Bo Wang, Yen-Ching Wang, and Feng-Huei Lin. 2017. "Flame sprayed zinc doped hydroxyapatite coating with antibacterial and biocompatible properties." *Ceramics International* 43:S829-S835. doi: https://doi.org/10.1016/j.ceramint.2017.05.318.

Yang, Yunzhi, Joo L. Ong, and Jiemo Tian. 2003. "Deposition of highly adhesive ZrO2 coating on Ti and CoCrMo implant materials using plasma spraying." *Biomaterials* 24 (4):619-627. doi: https://doi.org/10.1016/S0142-9612(02)00376-9.

Yao, Hai-Long, Xiao-Zhen Hu, Xiao-Bo Bai, Hong-Tao Wang, Qing-Yu Chen, and Gang-Chang Ji. 2018. "Comparative study of HA/TiO2 and HA/ZrO2 composite coatings deposited by high-velocity suspension flame spray (HVSFS)." *Surface and Coatings Technology* 351:177-187. doi: https://doi.org/10.1016/j.surfcoat.2018.07.082.

Yao, Hai-Long, Hong-Tao Wang, Xiao-Bo Bai, Gang-Chang Ji, and Qing-Yu Chen. 2018. "Improvement in mechanical properties of nano-structured HA/TiO2 multilayer coatings deposited by high velocity suspension flame spraying (HVSFS)." *Surface and Coatings Technology* 342:94-104. doi: https://doi.org/10.1016/j.surfcoat.2018.02.058.

Yin, XinHua, Liang Yan, Ding Jun Hao, Shichang Liu, Ming Yang, BaoRong He, and ZhongKai Liu. 2020. "Calcium alginate template-mineral substituted hydroxyapatite hydrogel coated titanium implant for tibia bone regeneration." *International Journal of Pharmaceutics* 582:119303. doi: https://doi.org/10.1016/j.ijpharm.2020.119303.

Yu, Ji-Min, Hye-Ri Cho, and Han-Cheol Choe. 2022. "Electrochemical characteristics of Sr/Si-doped hydroxyapatite coating on the Ti alloy surface via plasma electrolytic oxidation." *Thin Solid Films* 746:139124. doi: https://doi.org/10.1016/j.tsf.2022.139 124.

Yuan, Bo, Hewei Chen, Rui Zhao, Xuangeng Deng, Guo Chen, Xiao Yang, Zhanwen Xiao, Antoniac Aurora, Bita Ana Iulia, Kai Zhang, Xiangdong Zhu, Antoniac Vasile Iulian, Shen Hai, and Xingdong Zhang. 2022. "Construction of a magnesium hydroxide/graphene oxide/hydroxyapatite composite coating on Mg–Ca–Zn–Ag alloy to inhibit bacterial infection and promote bone regeneration." *Bioactive Materials* 18:354-367. doi: https://doi.org/10.1016/j.bioactmat.2022.02.030.

Yuan, Qiuhua, and Teresa Diane Golden. 2009. "Electrochemical study of hydroxyapatite coatings on stainless steel substrates." *Thin Solid Films* 518 (1):55-60. doi: https://doi.org/10.1016/j.tsf.2009.06.029.

Zandi, Mojgan, Hamid Mirzadeh, Christian Mayer, Henning Urch, Mohamadreza Baghaban Eslaminejad, Fatemeh Bagheri, and Houri Mivehchi. 2010. "Biocompatibility evaluation of nano-rod hydroxyapatite/gelatin coated with nano-HAp as a novel scaffold using mesenchymal stem cells." *Journal of Biomedical Materials Research Part A* 92A (4):1244-1255. doi: https://doi.org/10.1002/jbm.a. 32452.

Zhang, J. X., R. F. Guan, and X. P. Zhang. 2011. "Synthesis and characterization of sol–gel hydroxyapatite coatings deposited on porous NiTi alloys." *Journal of Alloys and Compounds* 509 (13):4643-4648. doi: https://doi.org/10.1016/j.jallcom.2011.01.196.

Zhang, Sam, Zeng Xianting, Wang Yongsheng, Cheng Kui, and Weng Wenjian. 2006. "Adhesion strength of sol–gel derived fluoridated hydroxyapatite coatings." *Surface and Coatings Technology* 200 (22):6350-6354. doi: https://doi.org/10.1016/ j.surfcoat.2005.11.033.

Zhang, Xiao-Juan, Dong-Yang Lin, Xiao-Hui Yan, and Xiao-Xiang Wang. 2011. "Evolution of the magnesium incorporated amorphous calcium phosphate to nano-crystallized hydroxyapatite in alkaline solution." *Journal of Crystal Growth* 336 (1):60-66. doi: https://doi.org/10.1016/j.jcrysgro.2011.09.039.

Zhang, Xiaonan, Hui Yin, Lu Xiao, Zhikun Li, Chuang Ma, Wei Xu, and Yingbo Wang. 2022. "Chitosan regulated electrochemistry for dense hydroxyapatite/MgO nanocomposite coating with antibiosis and osteogenesis on titanium alloy." *Colloid and Interface Science Communications* 48:100616. doi: https://doi.org/10.1016/ j.colcom.2022.100616.

Zhang, Xinxin, Xueqin Lu, You Lv, Lei Yang, Erlin Zhang, and Zehua Dong. 2021. "Enhancement of Corrosion Resistance and Biological Performances of Cu-Incorporated Hydroxyapatite/TiO2 Coating by Adjusting Cu Chemical Configuration and Hydroxyapatite Contents." *ACS Applied Bio Materials* 4 (1):903-917. doi: 10.1021/acsabm.0c01390.

Zhao, Huan, Shu Cai, Shuxin Niu, Ruiyue Zhang, Xiaodong Wu, Guohua Xu, and Zetao Ding. 2015. "The influence of alkali pretreatments of AZ31 magnesium alloys on bonding of bioglass–ceramic coatings and corrosion resistance for biomedical applications." *Ceramics International* 41 (3, Part B):4590-4600. doi: https://doi.org/ 10.1016/j.ceramint.2014.12.002.

Zhao, Yafan, Mingda Song, Chuanzhong Chen, and Jian Liu. 2008. "The role of the pressure in pulsed laser deposition of bioactive glass films." *Journal of Non-Crystalline Solids* 354 (33):4000-4004. doi: https://doi.org/10.1016/j.jnoncrysol.2008.05.019.

Zhao, Zhong, Liqun Du, Yousheng Tao, Qingfeng Li, and Lei Luo. 2016. "Enhancing the adhesion strength of micro electroforming layer by ultrasonic agitation method and the application." *Ultrasonics Sonochemistry* 33:10-17. doi: https://doi.org/10.1016/j.ultsonch.2016.04.021.

Zhong, Zhenyu, Jinli Qin, and Jun Ma. 2015. "Cellulose acetate/hydroxyapatite/chitosan coatings for improved corrosion resistance and bioactivity." *Materials Science and Engineering: C* 49:251-255. doi: https://doi.org/10.1016/j.msec.2015.01.020.

Zhou, Lan, Guo-Hua Lü, Fei-Fei Mao, and Si-Ze Yang. 2014. "Preparation of biomedical Ag incorporated hydroxyapatite/titania coatings on Ti6Al4V alloy by plasma electrolytic oxidation." *Chinese Physics B* 23 (3):035205. doi: 10.1088/1674-1056/23/3/035205.

Zhou, Zhiwei, Bo Zheng, Yuanping Gu, Chong Shen, Jian Wen, Zhengbing Meng, Shuoping Chen, Jun Ou, and Aimiao Qin. 2020. "New approach for improving anticorrosion and biocompatibility of magnesium alloys via polydopamine intermediate layer-induced hydroxyapatite coating." *Surfaces and Interfaces* 19:100501. doi: https://doi.org/10.1016/j.surfin.2020.100501.

Zieliński, A., P. Antoniuk, and K. Krzysztofowicz. 2014. "Nanotubular oxide layers and hydroxyapatite coatings on 'Ti–13Zr–13Nb' alloy." *Surface Engineering* 30 (9):643-649. doi: https://10.1179/1743294414Y.0000000302.

Biographical Sketches

Hossein Eivaz Mohammadloo

Affiliation: Assistant Professor, Colour, Resin and Surface Coatings Department, Iran Polymer and Petrochemical Institute, Tehran, Iran

Education: PhD from the Amirkabir University of Technology in 2016

Business Address: Iran Polymer and Petrochemical Institute (IPPI), Tehran, Iran

Research and Professional Experience: Development of smart coatings such as intelligent coatings for implant applications and biomedical devices, self-healing coatings based on micro/nano carriers, anti-corrosion nanocomposites and surface treatments by conversion coatings.

Publications from the Last 3 Years:

1. Motamedi, M., S. Mohammadkhah, M. Ramezanzadeh, H. Eivaz Mohammadloo, B. Ramezanzadeh, Designing Hybrid Mesoporous Pr/Tannate-Inbuilt ZIF8-Decorated MoS_2 as Novel Nanoreservoirs toward Smart pH-Triggered Anti-corrosion/Robust Thermomechanical Epoxy Nanocoatings. *ACS Applied Materials & Interfaces 2022* 14 (27), 31170-31193.
2. Asemabadi, Z., A. Mardani Korrani, M. Mohammadi Dolatabadi, H. Eivaz Mohammadloo, A.A. Sarabi, Sh. Roshan, Modification of hydroxyapatite coating in the presence of adipic acid for Mg-based implant application, *Progress in Organic Coatings* 172 (2022) 107088, https://doi.org/10.1016/j.porgcoat.2022.107088.
3. Moaref, R., M. H. Shahini, H. Eivaz Mohammadloo, B. Ramezanzadeh, S. Yazdani, Application of sustainable polymers for reinforcing bio-corrosion protection of magnesium implants–a review, *Sustainable Chemistry and Pharmacy* 29 (2022) 100780.
4. Fotovvat, B., M. Behzadnasab, S. M. Mirabedini, H. Eivaz Mohammadloo, Anti-corrosion performance and mechanical properties of epoxy coatings containing microcapsules filled with linseed oil and modified ceria nanoparticles, *Colloids and Surfaces A: Physicochemical and Engineering Aspects* 648 (2022) 129157.
5. AhadiParsa, M., H. Eivaz Mohammadloo, S. M. Mirabedini, Sh. Roshan, Bio-corrosion assessment and surface study of hydroxyapatite-coated AZ31 Mg alloy pre-treated with vinyl tri-ethoxy silane, *Materials Chemistry and Physics* 287 (2022) 126147.
6. Ghamsarizade, R., A. A. Sarabi, Sh. Roshan, H. Eivaz Mohammadloo, Study on release and inhibition action of mixed ZAPP and 8-HQ corrosion inhibitors loaded in pH-sensitive microcapsules for Mg AZ31, *Colloids and Surfaces A: Physicochemical and Engineering Aspects* 644 (2022) 128883.
7. Shahini, M. H., H. Eivaz Mohammadloo, Bahram Ramezanzadeh, Recent approaches to limit the tribocorrosion of biomaterials: A review, *Biomass Conversion and Biorefinery* (2022) 1-21, https://doi.org/10.1007/s13399-022-02719-3.
8. Roshan, Sh., H. Eivaz Mohammadloo, A. A. Sarabi, M. Afshari, Biocompatible hybrid chitosan/hydroxyapatite coating applied on the AZ31 Mg alloy substrate: In-vitro corrosion, surface and structure studies, *Materials Today Communications* 30 (2022) 103153.

9. Shahini, M. H., H. Eivaz Mohammadloo, Mohammad Ramezanzadeh, Bahram Ramezanzadeh, Recent innovations in synthesis/characterization of advanced nano-porous metal-organic frameworks (MOFs); current/future trends with a focus on the smart anti-corrosion features, *Materials Chemistry and Physics* 276 (2022) 125420.
10. Mohammadloo, H. Eivaz, A. A. Sarabi, S. Roshan, A. Eivaz Mohammadloo, 8-Hydroxyquinoline/nanoclay epoxy nanocomposite as a smart coating for early corrosion detection, *Corrosion Engineering, Science and Technology* 56 (8) (2021), 753-766.
11. Afshari, M., H. Eivaz Mohammadloo, A. A. Sarabi, S. Roshan, Modification of hydroxyapatite-based coating in the presence of polyvinylalcohol (PVA) for implant application: Corrosion, structure and surface study, *Corrosion Science* 192 (2021) 109859.
12. Shahini, M. H., N. Taheri, H. Eivaz Mohammadloo, B. Ramezanzadeh, A comprehensive overview of nano and micro carriers aiming at curtailing corrosion progression, *Journal of the Taiwan Institute of Chemical Engineers* 126 (2021) 252-269.
13. Shahini, M. H., H. Eivaz Mohammadloo, B. Ramezanzadeh, Recent advances in steel surface treatment via novel/green conversion coatings for anti-corrosion applications: A review study, *Journal of Coatings Technology and Research* 19 (2022) 159–199.
14. Pouladi, J., S. M. Mirabedini, H. Eivaz Mohammadloo, N. Ghasemi Rad, Synthesis of novel plant oil-based isocyanate-free urethane coatings and study of their anti-corrosion properties, *European Polymer Journal,* (2021) 110502.
15. Shahini, M. H., B. Ramezanzadeh, H. Eivaz Mohammadloo, Recent advances in biopolymers/carbohydrate polymers as effective corrosion inhibitive macro-molecules: A review study from experimental and theoretical views, *Journal of Molecular Liquids,* (2020) 115110.
16. Mohammadloo, H. Eivaz, S. M. Mirabedini, H. Pezeshk-Fallah, Microencapsulation of Quinoline and Cerium based Inhibitors for Smart Coating Application: Anti-Corrosion, Morphology and Adhesion Study, *Progress in Organic Coatings*, 137C (2019) 105339.
17. Nabizadeh, N., A. A. Sarabi, H. Eivaz Mohammadloo, Comparative investigation of Cu ion and adipic acid addition on electrochemical and microstructure characteristics of vanadium conversion coating on AZ31 Mg alloy, *Surface and Coatings Technology* 357 (2019) 1-11.

Shamim Roshan

Affiliation: Lab instructor of Corrosion and Protective Coatings laboratory in faculty of Polymer engineering, Amirkabir University of Technology, Tehran, Iran

Education: BE, ME, and PhD degrees from Amirkabir University of Technology, Tehran, Iran

Business Address: Amirkabir University of Technology, Tehran, Iran

Research and Professional Experience: Smart coatings and microencapsulation, corrosion and bio-corrosion as well as applied surface science, superhydophobicity and nanotechnology. She was recently engaged in the study of the electrochemistry and surface characterization of superhydrophobic coatings and conversion coatings as a visiting researcher at University of Quebec in Chicoutimi, Canada and Instituto Superior Técnico, Lisbon, Portugal.

Publications from the Last 3 Years:

1. One-step fabrication of superhydrophobic nanocomposite with superior anticorrosion performance.
2. Bio-corrosion assessment and surface study of hydroxyapatite-coated AZ31 Mg alloy pre-treated with vinyl tri-ethoxy silane.
3. Study on release and inhibition action of mixed ZAPP and 8-HQ corrosion inhibitors loaded in pH-sensitive microcapsules for Mg AZ31.
4. Biocompatible hybrid chitosan/hydroxyapatite coating applied on the AZ31 Mg alloy substrate: In-vitro corrosion, surface and structure studies.
5. Cerium containing pH-responsive microcapsule for smart coating application: Characterization and corrosion study.
6. Modification of hydroxyapatite-based coating in the presence of polyvinylalcohol (PVA) for implant application: Corrosion, structure and surface study.
7. 8-Hydroxyquinoline/nanoclay epoxy nanocomposite as a smart coating for early corrosion detection.
8. Comprehensive study on the electrochemical, morphological, and adhesion properties of Cr-free thin film: with and without polyurethane coating.

9. Improved performance of Ti-based conversion coating in the presence of Ce/Co ions: Surface characterization, electrochemical and adhesion study.

Reza Ghamsarizade

Affiliation: Department of Polymer Engineering and Color Technology, Amirkabir University of Technology, Tehran, Iran

Education: B.E degrees from Amirkabir University of Technology, Tehran, Iran

Business Address: Amirkabir University of Technology, Tehran, Iran

Research and Professional Experience: protective coatings, which include smart organic coatings reinforced with microcapsules. In addition, he has also conducted studies on biocompatible hydroxyapatite coatings for medical implants. Recently, he has worked on papers in the area of MXene and MOFs. His main focus in coatings is to investigate their anti-corrosion properties.

Publications from the Last 3 Years:

1. Study on release and inhibition action of mixed ZAPP and 8-HQ corrosion inhibitors loaded in pH-sensitive microcapsules for Mg AZ31.

Chapter 2

Removal of Pollutants by Hydroxyapatite Composite

Mojgan Zendehdel*
Department of Chemistry, Faculty of Science, Arak University, Arak, Iran

Abstract

Water is an essential need for human life, which is polluted due to various reasons. The water pollution is a universal subject and the world community is facing problems caused by it. One of the main causes of environmental water pollution is anthropogenic actions and industrial activities in the world. Water pollutants include thousands of organic (pesticides, fertilizers, hydrocarbons, phenols, plasticizers, biphenyls, detergents, oils, greases, pharmaceuticals, etc.), inorganic (heavy metals, nitrate, sulfate, phosphate, fluoride, etc.), and biological pollutants (virus, bacteria, fungi, algae, amoebas, and planktons, etc.). Many studies were performed to remove emerging pollutants by the adsorption process. These processes do not add unfavorable by-products to the environment and have been proven to be superior to other wastewater treatment techniques in terms of the simplicity of design, operation, and insensitivity to toxic compounds. Due to the presence of both cationic and anionic groups in the structure of hydroxyapatite, this compound is a good candidate for absorbing cationic and anionic pollutants from water. Hydroxyapatite is widely used for adsorbing pollutants from water due to its properties such as low cost, ion exchange properties, high biocompatibility, and high adsorption capacity. Although, the use of pure HAP nanoparticles is very useful for removing toxic substances, but its preparation process is expensive. Also, the amount of HAP nanoparticles during filtration shows a significant drop due to their small size. A

* Corresponding Author's Email: m-zendehdel@araku.ac.ir; mojganzendehdel@yahoo.com.

In: What to Know about Hydroxyapatite
Editor: Isaac Lavoie
ISBN: 979-8-88697-523-9
© 2023 Nova Science Publishers, Inc.

literature review shows how to overcome these problems as well as the applications and advantages of HAP nanoparticles. It is essential to prepare adsorbents based on HAP in a usable form by combining them with other inorganic materials such as synthetic and natural zeolite, Fe_3O_4, MnO_2 and Al_2O_3.

Introduction

Nowadays, water contamination with heavy metal ions, anions, organic compounds and biological pollutants has become a burning issue and significant concern due to economic and industrial development. One of the most important problems is increasing water pollutants due to various reasons such as uncontrolled sources [1]. The past years have witnessed several reports on a study about removing different kinds of pollutants from water because of their pivotal roles in biological systems and chemical processes. It is worth noting that much interest has been sparked in using low-cost processes [2, 3]. In last decade, different technologies were reported for purifying water and recycling in different papers. The most important processes to purify water have included ionic exchange, precipitation–coagulation, membrane processing, reverse osmosis, electro-assisted processes, adsorption and biosorption. The adsorption method is one of the most efficient treatment technologies due to the simple operation, cost-effectiveness and no generation of sludge. Hydroxyapatite (HAP) is one of the economical adsorbents which can remove fluoride and toxic cations from water by ion exchange and electrostatic attraction mechanism. The small size of hydroxyapatite regarding higher specific surface area makes it an effectual adsorbent for toxic ions. On the other hand, the process for the preparation of nano-size HAP being very useful for the removal toxic material is expensive and the small particle size HAP during the filtration significantly drops [4-8]. Literature review shows the overcome of this problem and at the same time utilizing the advantages of nano HAP. It is essential to prepare adsorbents based on HAP in a usable form by combining them with other high-surface materials such as zeolite, hydrotalcite, magnetic, polymer and ZIF [9-12]. With this aim, this current chapter offers a comprehensive review of hydroxyapatite composites with inorganic materials to use in the removal of different pollutants from water and wastewater. In addition, some background and basic information such as introducing another phase in the composite, methods for synthesis, and characterization of hydroxyapatite composite are explained. Additionally, removal of toxic materials result by HAP and its composites are discussed (Figure 1).

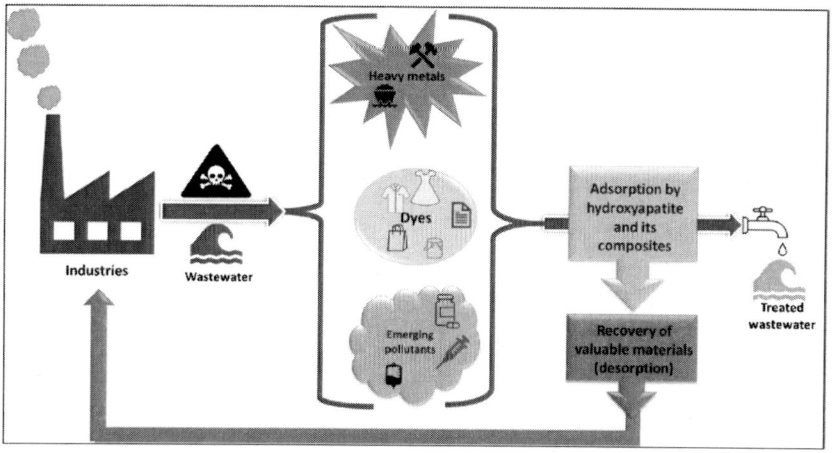

Figure 1. Adsorption of various pollutants by HAP, its composites, and their recovery.

Overview to Hydroxyapatite

Hydroxyapatite (HAP) is a biocompatible, mineral and unique form of calcium apatite with $Ca_5\ (PO_4)_3OH$ formula and is biocompatible. It is crystallized in the hexagonal crystal system in which the separate oxygen atoms are OH^- ions with H lying outside of the cell (Figure 2). The calcium ions are coordinated by phosphate and hydroxide ions. It is the major component of teeth and bone minerals while constitutes about 70% and 90% of bone and enamel weight respectively [13-14]. HAP is usually white to off-white with a molecular weight of 502.31 g/mol [15]. In vertebrates, the key mineral is naturally occurring in the form of calcium phosphate with a Ca/P ratio of less than 1.67 [16-17].

HAP is used in several forms include powders, granules, and porous blocks. In addition, some methods are utilized for the synthesis of HAP, such as sol-gel synthesis, hydrothermal methods, ultrasound and microwave method (U–M), co-precipitation, and sonochemical method [18-19]. These methods include aqueous chemical precipitation and sol-gel process with the use of the different sources of Ca and P. To improve HAP properties, many elaboration methods have been employed in order to control and optimize crystallinity, morphology and particle size. For example, a literature review has shown that nano-sized HAP has high surface energy, which may improve its mechanical properties and allows it for a faster implant surface turnover [20]. Controlling the size, crystallinity and surface properties of HAP crystals is also an important challenge for the design of novel nano-HAP in order to secure the requirements for various applications such as biocompatibility, bioactivity and absorbent [21-24].

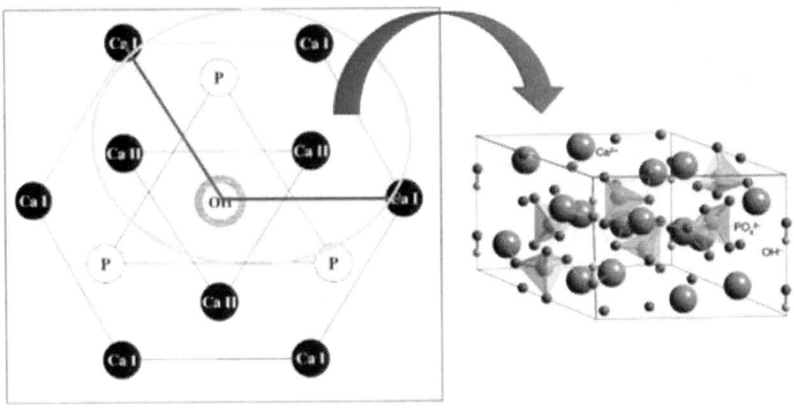

Figure 2. Structure of HAP.

In HAP synthesis, the use of natural resources containing calcium components can be substituted instead of commercial calcium salts is the utmost challenge. Green synthesis involves the use of various waste resources such as eggshells, shells, fish bones and scales, animal bones, and algae or planets that can be used to synthesize HAP. A typical powder XRD pattern of a synthesized nanometer-sized of HAP powder which the main (h k l) indices are (002), (211), (300), (202), (130), (002), (222) and (213) are exhibited in Figure 3(a). The FESEM microscopy technique shows the presence of a spherical like particle morphology with 30 ± 5 nm particle size Figure 3(b).

It is a well-known component with a high application as an adsorbent in the purification of wastewater, fluorescent lamps, fuel cell materials and it has been used in biomedicine, biosensor, catalyst or catalyst support, and also utilized as a bone substitute and ceramic implants [26]. Although, Hydroxyapatite has been studied as a sorbent different in conditions various limitations such as low mechanical reliability in wet environment, thermal and absorption in a narrow range of pH [27]. To prevent the phase impurities in HAP, its crystalline structure and chemical composition should be controlled.

The synthesis has complexities in controlling microstructure, stoichiometry, and agglomeration [28]. Thus, one of the important goals for developing synthetic method is focusing on preparation of HAP composite.

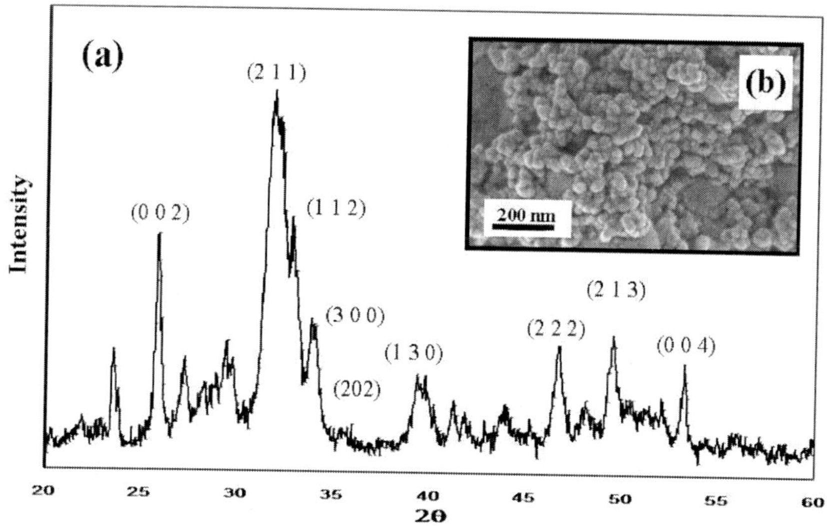

Figure 3. Typical XRD (a) and FESEM micrograph (b) of HAP sample [25].

Fluoride Anion

Fluorine, the most electronegative and extremely reactive chemical element, is found in nature only in combined forms as fluoride ions (F^-), of which fluorite (CaF_2), fluorapatite ($Ca_5(PO_4)_3F$) and cryolite (Na_3AlF_6) represent the most dominant fluoride-containing minerals. Fluoride- bearing rocks are the main sources of fluoride in groundwater. Many factors such as pH, temperature, con-centration of calcium and bicarbonate ions and solubility of fluoride minerals in water have an impact on the amount of fluoride contention the groundwater [29]. Removal of fluoride from water is very important because high levels of fluoride can cause drastic problems for humans such as bone fluorosis, dental caries and lesions of the thyroid, endocrine glands, and brain [30-31]. Therefore, the management of fluoride is necessary for public health. A Low level of fluoride is useful for humans, however, the beneficial fluoride concentration range for human healing this very small. According to the World Health Organization guidelines (WHO), the maximum limit of fluoride concentration in drinking water is 1.5 mg L^{-1}[32]. Thus, it is essential to alleviate the severity of the problem through the provision of drinking water with optimal fluoride concentration. So far, various methods have been applied for the defluorination from aqueous environments, such as ion exchange, adsorption, precipitation, electrodialysis and electrochemical methods [33]. Among these methods, Adsorption has been found as the most effective used method for the removal of fluoride based on low maintenance cost, flexible process and simplicity of design. In recent years, different low-cost adsorbents such as synthetic resin [34], activated alumina [35], carbon nanotubes [36], bone char [37], fly ash and other materials have been used for removal of fluoride ions [38]. Not with standing, the fluoride adsorption capacity of the most conventional adsorbent is not enough for wide applications so, in recent years, many efforts have been made to develop new and effective materials for the removal of this anion. For instance, in many kinds of literatures, using an adsorbent with micro- sized particles for the removal of fluoride was reported, but in recent years, nanomaterials with a high surface-to-volume ratio and short diffusion route have been used as an effective sorbent of it from aqueous solution. Hydroxyapatite can be the main candidate due to its non-toxicity, availability and high defluorination ability [39-41]. In this context, a significant number of nano-HAP adsorbents have been synthesized and studied, showing high defluorination capacities [42]. Although, the potential application of HAP in the field is not easily achievable, due to the cost associated with their production and significant pressure drops during filtration owing to the small particle size. To overcome these problems and at keeping the advantages of n-HAP, it is essential to prepare adsorbents based on HAP in a usable form by combining them with other high-surface materials [43-44], on which surface the HAP nanoparticles are supported. One of the most important supports for HAP is zeolite. These materials are naturally occurring aluminosilicates that can be also obtained and characterized by synthesis and high internal and external surface areas respectively. Several types of moieties such as water and alkaline/alkaline earth metals can be hosted or adsorbed in both natural and synthetic zeolites pores and channels. Literature reviews show that zeolites in general are very effective for the removal/immobilization of toxic and radioactive waste species, heavy metals as well as ammonium from wastewater, seawater potassium extraction, and also, have been applied for the production of environmentally friendly detergent [45].

Removal of Fluoride Anion with the Use of HAP Composite

In this regard, several adsorbents have been reported in the literature from tailoring with natural and synthetic zeolite/HAP composite. Izabel Diaz et al. were prepared a low-cost nano-hydroxyapatite/stilbite (nHAST) composite based on locally available stilbite (STI) zeolite of Ethiopia and its fluoride removal. To prepare nHAST the STI was ground in desired powder form and sieved into an average particle size between 75 - 105 μm and then 2 g of STI was mixed with 30 ml of 1 M $(NH_4)_2HPO_4$ solution; the mixture was kipped under static conditions for 6 days at room temperature until HAP was grown. The adsorbent composite material was separated by filtration, the solid was washed and dried at room temperature overnight [46]. The XRD pattern of the synthesized that composite samples (blue line) and the parent STI zeolite (red line) are very similar, confirming the stability of the STI framework (Figure 4) [47]. However, slight variations are observed in the region between $2\Theta= 23^0$ and 24^0, where the initial stilbite mineral showed three distinguishable peaks characteristic of monoclinic stilbite, while the nHAST composite showed single diffraction (marked by an sterisk), evidencing a change of the space group to orthorhombic; this is caused by the almost-complete exchange of Ca^{2+} and Na^+ of the original stilbite by NH^{4+} ions during the preparation of the composite. The identification of the presence of HAP by XRD was not possible in the case of the nHAST composite, because of the low concentration of HAp in the composite, its nanometric crystal size and the overlapping with intense zeolite diffractions.

Figure 4 shows TGA–DTG curves of the STI (red lines), nHAST (blue lines), in the original STI zeolite, a second strong weight loss, with maximum desorption rates at 175 and 250°C is assigned to desorption of strongly-retained water molecules by coordination with exchangeable Ca^{2+} ions. Interestingly, desorption of poorly-retained water at temperatures below 130 ^{0}C is much more intense for the nHAST composite, while desorption of strongly-retained coordinated water molecules at higher temperatures is much smaller, which is a consequence of the exchange of Ca^{2+} by NH^{4+} ions, which do not coordinate water molecules, and hence the major water desorption occurs at low temperature. In the nHAST composite, an additional broad peak is observed around 460°C, which is attributed to deamination, indicating the release of NH_3 as NH_4-STI transforms into H-STI [48]. SEM micrographs show the original STI zeolite before modification and synthesized nHAST composite in Figure 4 c-d which the nanometric size of HAP grown on the surface of the stilbite crystals cannot be observed clearly at the magnification exhibited in the micrographs. However, some irregularities changes on the surfaces with respect to the parent zeolite can be observed.

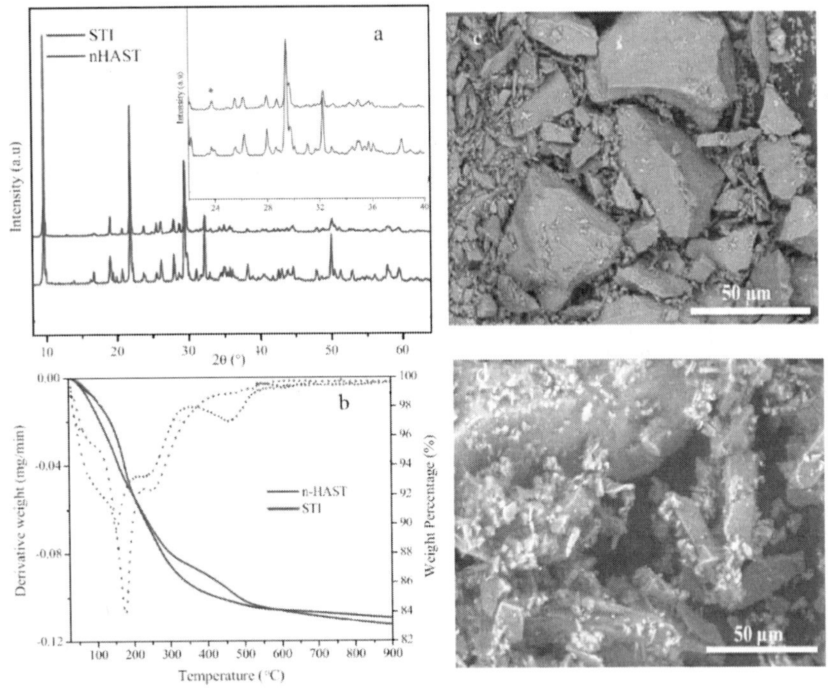

Figure 4. XRD pattern, TGA/DTA and SEM of Stilbite and nano-hydroxyapatite /stilbite.

Adsorption experiments were initially performed for nHAST composite using various doses (4, 6, 8, 10 and 12 g/L) at a constant initial fluoride concentration of 10 mg/L and a contact time of 20 h with continuous agitation at room temperature (Figure 5). As expected, the Fluoride removal efficiency was significantly increased at higher adsorbent doses, and reached about 84.0% of fluoride elimination at a dose of 10 g/L for nHAST. The result compared with bone char (BC) was kindly donated by the Oromo Self-Help Organization (OSHO) Defluoridation Plant located at Modjo, Ethiopia the result shows that both materials can reduce the fluoride concentration below the WHO limit of 1.5 mg/L; worth is noting the high initial fluoride concentration employed (10 mg/L). They observed a much higher capacity for the HAP in the nHAST composite, with an F^- removal capacity (at the dose of 10 g/L) of 9.15 mg F^-/g HAP, significantly higher than that of BC with an intrinsic HAP capacity of 1.08 mg F^-/g HAP, and one of the highest values found in the literature for HAP.

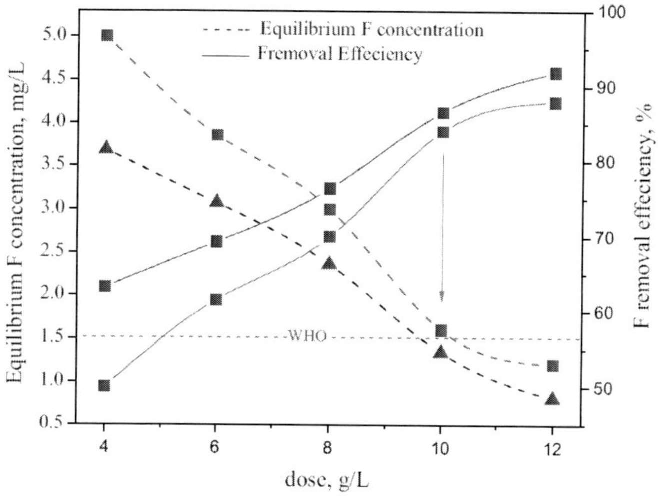

Figure 5. Fluoride equilibrium concentration (dashed lines) and removal efficiency(solid lines) of nHAST (blue) and BC (black) as a function of dose, with initial F^- concentration of 10 mg/L.

The reaction mechanism involving fluoride adsorption is analyzed based on kinetics and isotherm studies. In both nHAST and BC, the kinetic data fitted well to a pseudo-second-order kinetic model of similar characteristics. In contrast, the adsorption isotherm on the nHAST composite fitted best with the Freundlich model, whereas on BC, it correlated well with the Langmuir model, suggesting a different mechanism: adsorption of fluoride on BC was homogeneous, whereas on the composite was heterogeneous, possibly related to the higher loading of fluoride on the hydroxyapatite component in the composite. At low concentrations, both adsorbents behave similarly; however, the maximum adsorption capacity, measured at high concentrations, is higher in the nHAST composite than in BC (Figure 6).

In another work, Zendehdel et al. studied the removal of carcinogenic F^- by a nanocomposite prepared from nano HAP and Iranian clinoptilolite, a low-cost natural zeolite. For modification, this nanocomposite was functionalized by using (3- aminopropyl) trimethoxysilane (APTMS) as the silylation agent. In the first step, to prepare HAP/ clinoptilolite a gel of nano HAP, calcium nitrate and potassium dihydrogen phosphate were completely dissolved in deionized water at ambient temperature. Then, different amounts of clinoptilolite (0.5, 1,2,3,4 and 5 g) were added to it. In the second step, the pH value was adjusted to 11 by Ammoniac solution. The reaction mixture was stirred for 24 h to complete the reaction. The powder was filtered and washed two times with hot water for the removal of residual potassium, ammonium and nitrate ions. After drying the nanocomposite at 40ºC for 24 h, the powder was calcined at 600ºC for 1 h and cooled in the furnace until room temperature in atmospheric pressure.

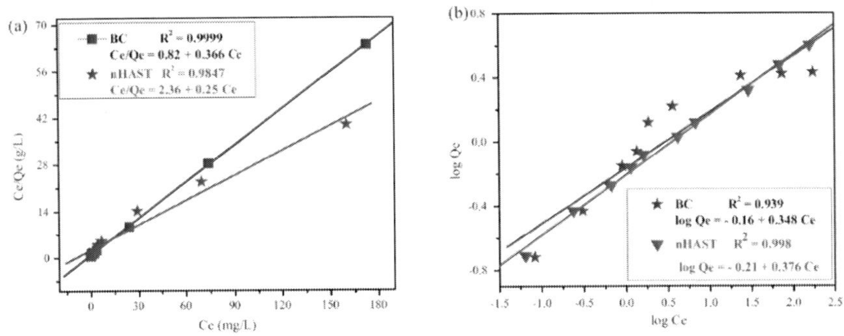

Figure 6. (a) Linearized Langmuir adsorption isotherm for fluoride adsorption, and (b) Linearized Freundlich adsorption isotherm for fluoride adsorption on nHAST (blue)and BC (black) samples.

To prepare Amino-functionalized HAp/ clinoptilolite (HAP/Clin/NH_2), HAp/ clinoptilolite and 3-aminopropyltriethoxysilane (3-APTES) in dry toluene were mixed and refluxed overnight under N_2 atmosphere. The prepared solid was filtered and washed with ethanol and acetone several times, and air-dried at room temperature [49].

Figure 7A (a-d) shows the FT-IR for pure HAP, clinoptilolite, HAP/Clin, and HAP/Clin/NH_2 with an optimum amount of zeolite (2 g). In the case of HAP/Clin and HAP/Clin/NH_2, the presence of the characteristic band of Hydroxyapatite and clinoptilolite in Figure 7c and d, confirms the creation of two HAP as well as zeolite phases in nanocomposite [50]. After functionalizing HAP/Clin with amine groups the C-H and N–H vibration bands in the 2800–3500 and 1500–1650 cm^{-1} regions can be observed. In Figure 7b, a broad and sharp band related to hydroxyl groups stretching vibration and H–O-H bending vibration can be seen at 3466 and 1653 cm^{-1} [51] considering that the intensity of these bands after creating HAP/Clin, and HAP/Clin/NH_2 nanocomposite decreased that may be due to hydrogen bond interaction between HAP and zeolite.

Figure 7B (a and b) illustrated the XRD patterns for HAP/Clin, and HAP/Clin/NH_2 with 2 g of clinoptilolite. The peaks at 2θ = 25.8°, 27.7°,34.1° and a broader diffraction peak around 29° confirms the presence of nano-hydroxyapatite which matches the standard diffraction pattern for HAP (JCPDS 9–432) [52]. In addition, the peaks at 2θ = 9.9°,11.2°, 22.4°,31° and 34.2° match with used to identify the phases clinoptilolite with the highest intensity at 2θ = 31° [25–1349 card numbers of clinoptilolite [$(NaKCa)_6$ (Si_6Al_{30}) O_{72}] as principal crystalline phase about 80 percent. A minor amount of natural impurities such as heulandite [$CaAl_2Si_7O_{18}{\cdot}6H_2O$] and quartz [$SiO_2$] in these samples is normal for this type of materials [53]. The result shows after functionalizing with amine groups the intensity of peaks related to zeolite was reduced and a small amount of amorphous phase can be seen which may be due to creation of a covalent bond between zeolite and 3-APTES groups that destroy some parts of zeolite frameworks.

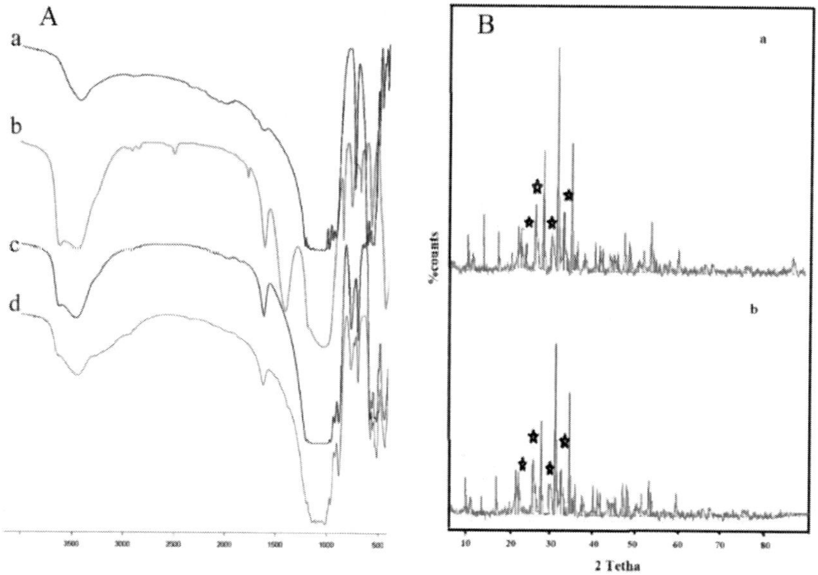

Figure 7. A, The FT-IR for (**a**) pure HAP, (**b**) clinoptilolite, (**c**) HAP/Clin, and (**d**) HAP/Clin/NH_2 B, XRD patterns (**a**) HAP/Clin and (**b**) HAP/Clin/NH_2.

Figure 8a-d shows the morphology of clinoptilolite, HAP, HAP/Clin and HAP/Clin/NH_2, respectively. The morphology of HAP is spherical, rod-like and it has the particle size of about 40–70 nm (Figure 8b). The SEM image of HAP/Clin and HAP/Clin/NH_2 composite (Figure 8c-d) clearly indicates the presence of both zeolite and HAP phase which show with yellow circle. In addition, some irregularities can be observed on the surfaces as compared to the parent cliniptilolite. N_2 absorption–desorption of clinoptilolite, HAP/Clin and HAP/Clin/NH_2 were shown in Figure 6g-i. As can be seen, all samples have shown the isotherm profiles of type IV in which the hysteresis loop started from around $P/P_0 = 0.5$ and demonstrates the adsorption of N_2 at a higher pressure and consequently the existence of mesopores, especially in the parent clinoptilolite case. In addition, the type I curve related to the micropores phase can be seen before $P/P_0 = 0.5$ for all samples. The BJH distribution curves further prove the isotherm profiles by showing the presence of mesopores with a pore size in the range of 5–50 nm, which decreased after introducing HAP to zeolite (Figure 6j-l) and describe the presence of higher micropores in the HAP/Clin and HAP/Clin/ NH_2. The results for BET surface area are 56.99, 39.66, 23.91 m^2/g and total volumes are 0.19, 0.15 and 0.13 cm^3/g for clinoptilolite, HAP/Clin and HAP/Clin/NH_2 respectively. This would support this hypothesis, in which nano HAP particles and functional groups are introduced onto clinoptilolite pores [54].

Figure 8. SEM images for (a) cliniptilolite (b) hydroxyapatite (c) HAP/Clin and (d) HAP/Clin/NH_2, N_2 absorption–desorption of (g) clinoptilolite, (h) HAP/Clin and (i) HAP/Clin/NH_2.

Subsequently, batch adsorption studies were implemented to consider the effect of parameters such as the contact time, pH, temperature, and initial concentration of pollutants, types of adsorbent and adsorbent dosage considered to remove F^- by HAP/Clin and HAP/Clin/NH_2. The optimum condition for F^-,60 min, pH = 7, initial concentration 20 ppm,0.05 g of absorbent and 25°C are the best conditions for removal. The results for

HAP/Clin and HAP/Clin/NH_2 show higher than removal with respect to pure HAP and clinoptilolite. In addition, NH_2 had better performance for absorption of F^- based on scheme1. As, we know, HAP isn't regenerated and it is not reusable [55] additionally very likely combined with the relative merits of natural zeolite adsorbents with large surface area and low cost.

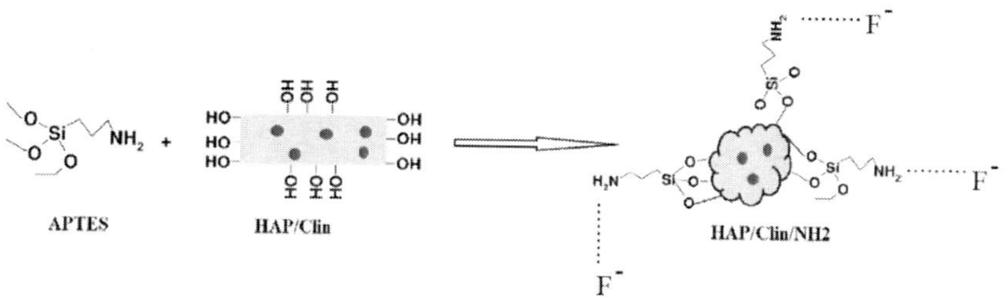

Scheme 1. The interaction of F^- with HAP/Clin/NH_2.

The consideration of solution pH is a very important factor in the adsorption process [56]. The influence of pH was investigated at various pH values from 2.0 to 11.0 for absorption of and F^- ions. The maximum removal of F^- was found in the pH of about 7 suitable for practical application in the household system. The Literature review confirmed the electrostatic attraction between the positively charged metal in zeolite, hydroxyl apatite surface and fluoride anions which can be useful for the removal of F^- [57]. The fluoride removal rate dramatically decreased in the acidic pH and alkaline pH. The following mechanism for fluoride sorption on HAP/Clin and HAP/Clin/NH_2 surface may be given as [58, 59]:

$$\mathrm{M-OH+F^- \leftrightarrow M-F+OH^-} \tag{1}$$

$$\mathrm{T\ OH+F^- \leftrightarrow T\ OH.....F^-} \tag{2}$$

$$\mathrm{HAP/Clin/NH2+F^- \leftrightarrow HAP/Clin/NH_2.....}F^- \tag{3}$$

The two Langmuir and Freundlich adsorption isotherm models are usually used to describe the relationship between the adsorbent and adsorbate. These two isotherms parameters are presented in Table 1 for the removal of F^- by HAP/Clin/NH_2 in a temperature range of 25- 45°C. In the case of F^-, the correlation coefficients (R^2) of Langmuir and Freundlich isotherm models are larger than 0.99, which indicated that both models could well fit the adsorption process of fluoride on HAP/Clin/NH_2. Literature review [60] showed that the maximum adsorption capacity q_{max} from the Langmuir model is above 11 mg/g, which is higher than reported nano-size HAP (below 1.5 mg/g). This result demonstrates that the presence of zeolite and NH_2 plays an important role in F^- adsorption [61].

Table 1. Isotherm parameters for sorption of F− by HAP/Clin/NH_2

Isotherm		Langmuir				Freundlich		
Metal ion	Temperature °C [a]	q_{exp} (mg/g)	q_{max} (mg/g)	K_L (L/mg)	R^2	K_f (mg/g)	N	R^2
F^-	25	9.8	11.9	3.61	0.995	1.90	1.33	0.997
	35	10.2	11.4	4.37	0.996	2.5	1.34	0.987
	45	11.3	12.3	4.85	0.996	2.9	1.37	0.990

The new nanocomposite of NaP zeolite-Hydroxyapatite (NaP:HAP) has been developed for the removal of fluoride from an aqueous solution. NaP zeolite with a gismondine (GIS) framework topology and interesting channels of 0.31 nm × 0.44 nm in [100] and 0.26 nm × 0.44 nm in [010] has a good potential for adsorption of small gases, liquid molecules, and hazardous materials. According to the above introduction, zeolite: HAP nanocomposite may be a potential adsorbent for the simultaneous removal of fluoride from an aqueous solution because of joint characteristics of HAP and zeolite in zeolite: HPA [9]. For the preparation of NaP: HAP nanocomposite different amounts of HAP were added to zeolite gel (the molar composition 16Na2O: Al2O3: 15SiO2: 320H2O and aged 5 days) and the mixture was stirred for 15 min continuously. In order to prepare of the nanocomposite, the mixture was then placed in an autoclave at 100°C for 26 h. Next, the precipitated sample was filtrated and washed with deionized water repeatedly until it reaches a pH value of 7. The resulting product was then dried at room temperature for 24h. Figure 9 a-d shows the XRD pattern of NaP: HAP nanocomposite with different ratio (NaP: HAP ratios: 1:1, 1:2, 1:4, 1:6) and shows the presence of gismondine-like zeolite in the synthesized sample for 1:1 ratio (Figure 9a) with reflections on 2Θ = 12.48°, 17.92°,21.79°, 28.1° and 33.4° [62] while the minor, broad peak about 2Θ = 29° corresponding to Hydroxyapatite. This suggests that the zeolite pores could also be incorporated by nanoparticles of HAP due to their very small nano size.

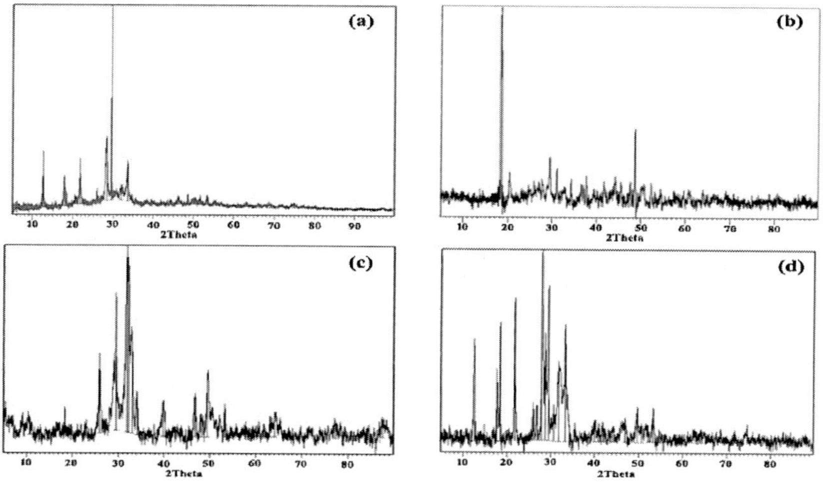

Figure 9. (a–d) XRD pattern of NaP:HAP nanocomposite with different ratio (NaP:HAP ratio: 1:1, 1:2, 1:4, 1:6).

Figure 10a–b is SEM micrograph for NaP:4HAP, (d) NaP:6HAP nanocomposite. Similar to other observations found in the literature, both cactus-like and diamond-like morphology with 1–2 μm size of each diamond-shaped particle appeared in the SEM image of NaP: HAP nanocomposite [63]. Also, we can see with increasing the percentage of HAP diamond-like morphology for NaP zeolite has become visible.

In order to investigate the efficiency of Nap: HAP nanocomposite on the removal of fluoride ion, Box–Behnken design with three-level and four-factor has been employed for determination of effective process parameters such as solution pH (4–11), temperature (25–55°C), initial adsorbent dose (1–3 g) and initial fluoride concentration (5–25 mg/L) on the removal of fluoride from aqueous solution. After 29 batch runs, the Quadratic model was established by regression analysis of the experimental data obtained from 29 batch runs (Figure 11). The results showed that the second-order polynomial regression model was capable of accurately interpreting the experimental data with a coefficient of determination (R^2) value of 0.9327 and an F-value of 13.87. The optimum pH value was found to be 7 at 60 min with 5 mg/L of fluoride concentration. It is evident from the dosage study that there was an increase in removal percentage because of rising in dosage amount from 1 g/L to 3 g/L.

Figure 10. a–b is SEM micrograph for NaP:4HAP, (d) NaP:6HAP nanocomposite.

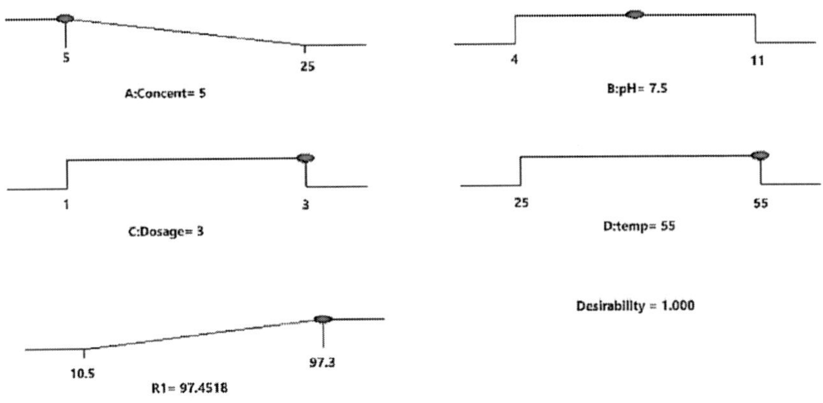

Figure 11. Desirability ramp for optimization of four parameters.

Quantity uptake of fluoride was evaluated using the Langmuir, Freundlich and Dubinin Radushkevich (DR) models (Table 2) which the result confirms it was correlated well with the Langmuir adsorption model.

Table 2. Langmuir and Freundlich sorption isotherm constant and correlation coefficient (R^2) for fluoride ion on NaP:HAP nanocomposite

Temperature (°C)	Langmuir constant				Freundlich constant			D–R constant		
	q_m	K_L	R^2	K_f	1/n	R^2	K_D	Lnq_0	E (Kj/mol)	R^2
25	11.947	3.612	0.9955	1.901	0.7472	0.9974	–	–	–	–
35	11.428	4.376	0.9952	2.553	0.7462	0.9826	–	–	–	–
45	12.269	4.859	0.9959	2.895	0.7388	0.9788	–	–	–	–
55	12.69	6.27	0.9933	3.607	0.7294	0.981	7×10^{-8}	2.0506	2.672	0.978

In order to study the temperature effect on the removal efficiency, the adsorption experiments were conducted in the temperature range of 25°–55°C. Figure. 12, show the effect of temperature on adsorption while keeping two parameters constant. It was found that the percentage of removed fluoride ions, leads to increase slowly with rising in solution temperature from 25°to 55°C, indicating that the adsorption is an endothermic process. This was further supported by calculating of thermodynamic parameters such as Gibbs free energy(ΔG°), entropy (ΔS°) and enthalpy (ΔH°). The changes in these parameters of adsorption can be determined by the following equations:

$$\mathrm{Log}K_c = \frac{\Delta S}{2.303R} - \frac{\Delta H}{2.303RT} \tag{4}$$

$$\Delta G = \Delta H - T\Delta S \tag{5}$$

where, ΔH and ΔS are the changes in enthalpy and entropy of adsorption, respectively and R refers to the universal gas constant (8.314 KJ/mol K). K_c value is the equilibrium constant and calculated from the Langmuir isotherm. Values of ΔH and ΔS were calculated from the slope and intercept of Van't Hoff plot (Log K_c versus 1/T) and represented in Table 3. The Gibbs free energy change values in each temperature were found as negative which indicates the spontaneity and feasibility of the positive value of enthalpy indicating the endothermic process of adsorption. The positive value of ΔS suggests that the entropy is responsible for making the ΔG negative value. Therefore, the adsorption process was spontaneous.

Desorption study gives knowledge about the nature of adsorption. In order to understand the desorption, after the first adsorption of fluoride ions by Nap: HAP nanocomposite, the waste of adsorbent put in the 25 mL of 0.1 N hydrochloric acid solution for1 h and then, the filtrate was measured for fluoride content. After that, this adsorbent was floated in the desired solution (pH of 7 at 55°C and 3 g/L of adsorbent with 5 mg/L of fluoride solution) for 1 h and the concentration of residue ions was determined as well and this cycle was repeated 10 times. Also, the same condition was carried out for sodium hydroxide solution (0.1 N of NaOH) to recognize the reuse-ability of NaP: HAP nanocomposite in acidic and alkaline solutions. These results were compared and demonstrated in Figure 13, desorption of adsorbed fluoride in acidic solution and alkaline solution resulted about 65.2% and 51% respectively. It is evident from high desorption values in both solutions that the adsorption of fluoride onto NaP: HAP nanocomposite was physical in nature.

Table 3. Thermodynamic parameters of fluoride ion on NaP: HAP nanocomposite

Thermodynamic parameters				
Temperature (°C)	ΔG (KJ/mol)	ΔH (KJ/mol)	ΔS (J/mol K)	R^2
25	−3.181			
35	−3.779			
45	−4.177			
55	−5.004	10.81	47.173	0.966

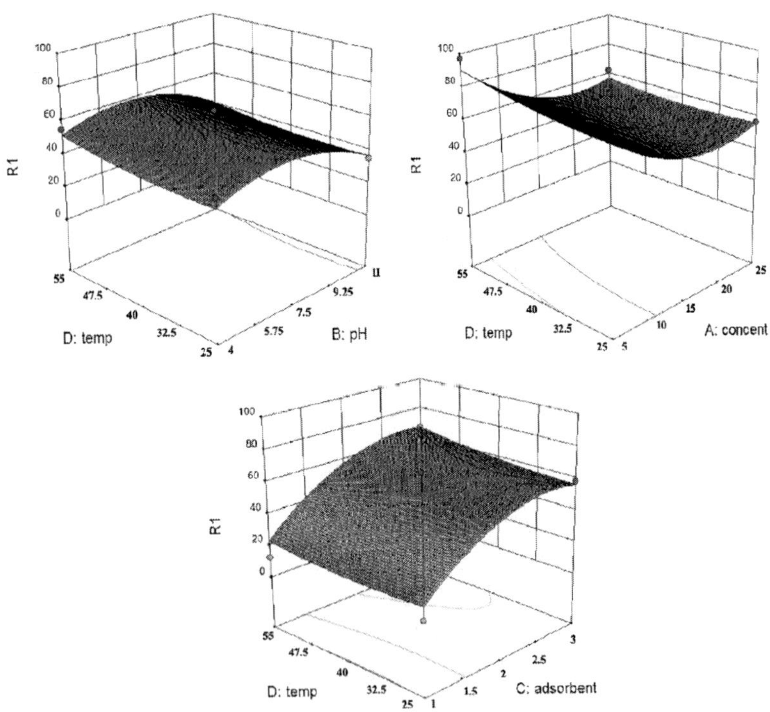

Figure 12. Response surface plot showing the effect of temperature on pH, concentration and adsorbent dosage on F removal in pH and initial concentration: 7 and 5 mg/L, respectively.

In other reports, Isabel Diaz et al. studied the potential of a newly developed nano-hydrotalcite/hydroxyapatite (n-HT/HAP) composite, and its constituent materials, in fluoride removal. In this manuscript, they studied the effect of the type and number of divalent cations (nickel and magnesium) of hydrotalcite in their defluorination capacity via the memory effect [64]. Hydrotalcites (Figure 13), also known as layered double hydroxides (LDH), are anionic clays with positively charged octahedral hydroxide layers, which are neutralized by interlayer anions and water molecules. They are generally formulated as $[M^{2+}_{1-x}M^{3+}_{x}(OH)_2]\,(A^{m-})_{x/m}\cdot nH_2O$, where M^{2+} and M^{3+} are di- and trivalent cations, respectively, A^{m-} is an anion whose charge is m-. Magnesium-aluminum hydrotalcite is stable up to 400°C. It decomposes into a NaCl-type magnesium-aluminum mixed oxide above 400°C and into MgO and $MgAl_2O_4$ above 800°C [65]. The Mg/Al and Ni/Mg hydrotalcite samples were synthesized from water solutions, of Mg $(NO_3)_2\cdot 9H_2O$, Ni $(NO_3)_2\cdot 6H_2O$, and Al $(NO_3)_3\cdot 9H_2O$. A second aqueous solution was prepared with NaOH. The precipitation was carried out by adjusting the flow of each solution to maintain a pH constant at 9. The solution amounts were adjusted to correspond to Mg/Al or Ni/Al molar ratios of 3:1. The resulting gels were treated in a microwave autoclave for 10 min operating at 2.45 GHz. The microwave irradiation power was 200 W and the temperature was fixed at 80°C. The solids were recovered by decantation and washed several times with distilled water until the residual solution reached a pH value of 8. The solids were, then, dried in an oven at 70°C for 24 h. Part of these samples was calcined in air at 500°C for 8 h to be used in the fluoride removal experiments. To prepare the Nano-hydrotalcite/Hydroxyapatite composite, the same solutions used to synthesize an Mg/Al or Ni/Al 3:1 hydrotalcite sample were used to synthesize a composite by dropping to solutions which ready to prepare HAP from H_3PO_4 solution dropped in $Ca(OH)_2$ solution adjusting the flow in such a way that the pH was equal to 11.5. The resulting mixture was submitted to microwave irradiation for 10 min as previously. The solids were washed up to pH 10, recovered by decantation, and dried at 70°C.

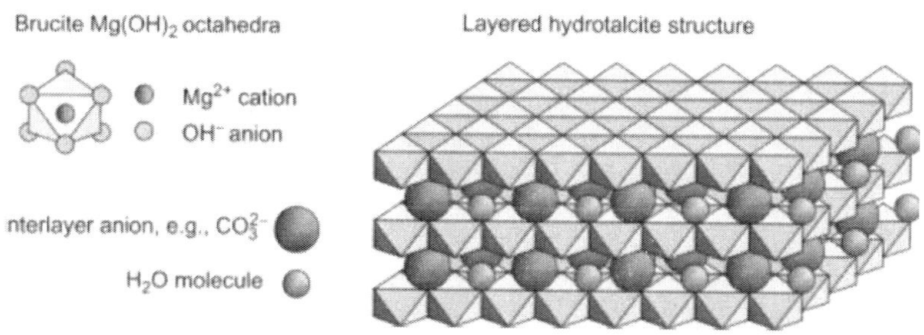

Figure 13. Structure of hydrotalcites.

Figure 14 displays the fluoride retention (%) and defluorination capacity (DC) of the cHT samples with different composition. Figure 14a corresponds to an initial fluoride concentration of 20 mg/L and Figure 14b to 5 mg/L, both at an adsorbent dose of 10 g/L. If the initial fluoride concentration is 20 mg/L, the performance of the cHT samples Ni-Al (2:1), Ni-Al (3:1), Mg-Al (3:1), and Mg-Al (4:1) reaches 58, 82, 95, and 97% of fluoride retention, respectively.

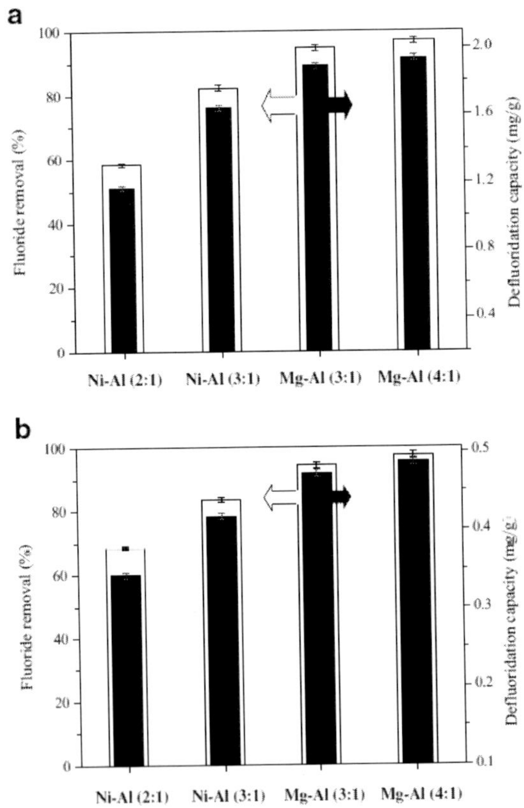

Figure 14. Fluoride removal percentage retention (%) and defluorination capacities (DC) of calcined hydrotalcites (cHT) with different composition at an initial fluoride concentration of a 20 mg/L and b 5 mg/L at an adsorbent dose of 10 g/L.

Figure 15 describes the X-ray diffraction patterns of n-HT/HAP composite before and after F removal, and HAP for comparison purposes. The profile of the composite is similar to the one of the hydroxyapatites, and the presence of hydrotalcite is envisaged as peaks of low intensity, found at 12°

to 13° and 22° (2θ) whose broadness has to be attributed to very small particle sizes. Therefore, the hydrotalcite crystallites covering the large particles of hydroxyapatite must be nano-metrical. The diffusion of fluoride ions to the center of the hydrotalcite particles is then easier and faster. Such a scenario is based also on the fact that the sample n-HT/HAP was not calcined previously; therefore, hydrotalcite on the hydroxyapatite support does not retain fluoride through a memory effect; the exchange is a simple ionic exchange.

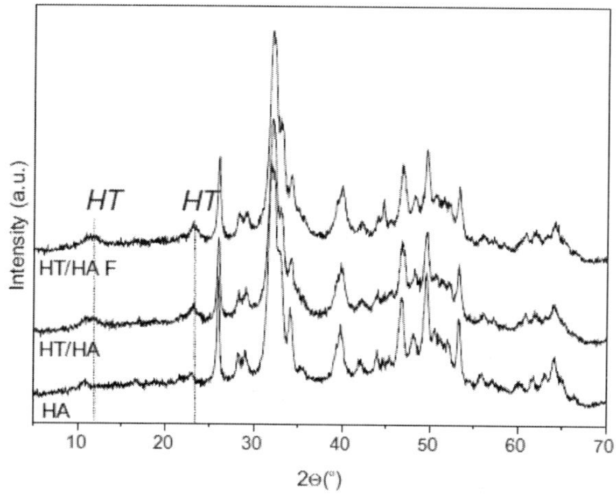

Figure 15. Comparison of thc X-ray diffraction patterns of the pure hydroxyapatite (HAP), composite n-HT/HAP before and after F^- removal experiment.

Mechanism of F^-Removal

The fluoride removal by NaP: inorganic material can be controlled by both adsorption and ion exchange mechanisms. Theoretically, three essential steps can be taken during the removal process of fluoride onto solid particles [66]:

a) External mass transfer has occurred by transfer or diffusion of fluoride ions to the external surface adsorbent from bulk solution;
b) Particle surfaces of adsorbed fluoride ions;
c) The adsorbed fluoride ions may exchange with the structural elements inside the adsorbent particle; whereas porous material: HAP nanocomposite contains two parts of porous material and Hydroxyapatite so the mechanism of fluoride removal can be

described by both reactions in porous materials and Hydroxyapatite. As described above, some percent of HAP is distributed onto the external surface of porous material, so fluoride ions get adsorbed onto Hydroxyapatite and the OH group present in the Hydroxyapatite can be exchanged with fluoride ions and caused the formation of fluorapatite or hydroxy fluorapatite that indicates according to the reaction [67]:

$$Ca_{10}(PO_4)_6(OH)_2 + 2F^- \rightarrow Ca_{10}(PO_4)_6(F)_2 + 2OH^- \quad (6)$$

or

$$Ca_{10}(PO_4)_6(OH)_2 + F \rightarrow Ca_{10}(PO_4)_6\,FOH + OH^- \quad (7)$$

In addition, porous material such as zeolite and hydrotalcite in nanocomposite has an efficient role in removal of fluoride. The hydrolysis process of zeolite is illustrated according to the reaction [68]:

$$Me - Z\,(s) + H_2O\,(l) \leftrightarrow H - Z + Me^{n+} + nOH^- \quad (8)$$

where Z is zeolite or hydrotalcite, Me is an exchangeable cation (Na^+, K^+, Ca^{2+}, Mg^{2+}) and n is cation charge. A reaction of metallic ions occurs at the same time:

$$Me^{n+}\,(l) + H_2O\,(l) \leftrightarrow [Me{-}OH]^+ + H^+ \quad (9)$$

According to the latest reaction, adsorbed fluoride ions onto the zeolite surface can be exchanged with OH group that adsorbed onto the surface zeolite or hydrotalcites and fluoride anions on the zeolite surface or hydrotalcites may form with exchangeable cations more or less stable complexes:

$$Me^{n+}\,(l) + mA^{y-}\,(l) \leftrightarrow (MeA_m)^{n-my}\,(l) \quad (10)$$

Hence, it can be concluded that the fluoride removal by NaP: HAP nanocomposite is carried out by both adsorption and ion exchange mechanisms and thus it can act as an effective adsorbent.

Heavy Metals

The contamination of surface and ground waters caused by toxic heavy metal ions such as Pb(II), Cd(II), Cu(II), Cr(VI) and Zn(II) has become a serious environmental issue because these ions are virtually non-biodegradable and tend to accumulate in living organisms, causing severe disorders and diseases. Utilization of water containing very low concentrations of Pb (II) ions in the long term causes several health problems such as coma, nausea, renal failure, and cancer [69]. It can be suggested that Pb (II) replaces calcium in the bones forming repository sites for long term-release [70]. Cadmium has severe toxicity and causes a wide range of health problems such as kidney disease, blood pressure, and reduction in red blood cells [71]. Chromium (Cr) exists in two valence states Cr (III) and Cr (VI) in wastewater and is also widely utilized in various industries. Although, Cr (III) is an essential element for living beings, Cr (VI) exhibits high toxicity. Literature review shows the carcinogenicity and mutagenicity effect, which may produce severe health problems, such as genetic defects, liver damage and cancer [72]. The acceptable level of chromium in drinking water according to the World Health Organization (WHO) standards is 0.05 mg/L for Cr (VI) and 0.1 mg/L for total Cr [73]. Despite being an essential micronutrient for life, zinc can also lead to depression, lethargy, increased thirst and neurologic symptoms beyond the maximum acceptable concentration in the living organisms [74]. Therefore, it is very important to remove heavy metal ions from contaminated water in terms of protecting human health and the environment. There are several physicochemical methods available for heavy metal ions removal from an aqueous solution, including chemical precipitation, reverse osmosis, filtration, solvent extraction, electrodeposition, electrodialysis, ion exchange and adsorption [75]. The researchers are interested in adsorption as an attractive technology due to its high efficiency, simple process, and reduced cost [76]. A lot of adsorbents have been reported for the removal of heavy metals from polluted water. These include activated carbon [77], chitosan [78], and mineral materials such as metal oxides [79], activated clay [80], natural zeolite, and bentonite [81]. In recent years, apatite-group minerals, which have crystal chemistry with special characteristics, have come to be considered an environmentally benign material suitable for water purification. Therefore, it seems to prepare an adsorbent based on n-HAP in a usable form by combining them with other high-surface materials such as zeolite, magnetic and manganese oxide. The adsorption capacity of heavy metal ions (mg/L) onto the adsorbent was calculated according to the following equation:

$$q_e = \frac{(C_0 - C_e)V}{W} \tag{11}$$

where C_0 and C_e are the initial and equilibrium concentration (mg/L) of heavy metal ions, respectively. Also, q_e (mg/g) is the amount of adsorbate adsorbed per unit mass of adsorbent, V is the volume solution (in liter), and W is the weight of adsorbent (in grams). The removal efficiency of metal ions from aqueous solution (in percent) was calculated by the following equation:

$$\%\text{Removal} = \frac{C_0 - C_e}{C_0} \times 100 \tag{12}$$

Removal of Heavy Metal with the Use of Hydroxyapatite Composite

Zhiliang Zhu et al. were prepared a novel hydroxyapatite/magnetite (HAP/Fe_3O_4) composite adsorbent to investigate the removal capability of Pb^{2+} from aqueous solutions [82]. Magnetic Fe_3O_4 has the advantage of the being used as a support material of composite adsorbent because it can be easily manipulated by an external magnetic field [83]. Despite up to date several studies carried out on surface modification of Fe_3O_4 nanoparticles for the adsorption of heavy metals [84], it has not been documented that HAP is combined with Fe_3O_4 to produce composite adsorbent for the removal of heavy metals from aqueous solutions. Magnetite (Fe_3O_4) was prepared using a modified coprecipitation method Iron salts of Fe (III) and Fe (II) (in a 2:1 molar ratio) were dissolved in distilled water. A solution of NaOH (2 mol/L) was added to the ferrous solution under N_2, producing iron oxide solids immediately [85]. After stirring for 0.5 h, the magnetite (Fe_3O_4) particles were washed repeatedly with distilled water. HAP was combined with Fe_3O_4 by in situ precipitation. A suspension of Fe_3O_4 in CH_3COOH solution (2.1 wt.%) was added to the mixture solution of Ca $(NO_3)_2$ with KH_2PO_4 in a molar ratio of 5:3. The molar ratio of Fe_3O_4 to HAP was 1:1. After stirring for 0.5 h, a solution of NaOH (2 mol/L) was added to the mixture while the pH was remained at 13. The mixture was kept at 37°C for 6 h, and then the obtained HAP/Fe_3O_4 composite adsorbent was washed repeatedly with distilled water and dried before using.

Figure 16 a and d shows the XRD patterns for the composite adsorbents HAP/Fe_3O_4 before and (d, e, f) after Pb^{2+} adsorption. All the peaks are consistent with those of JCPDS International Center for Diffraction Data (ICDD) Powder Diffraction File for this phase (filenumber72-1243, 75-1372). There are five characteristic peaks for HAP (2Θ = 25.87°, 31.74°, 39.75°, 46.61°, 49.47°) and three for Fe_3O_4 (2Θ = 35.43°, 56.94°, 62.53°). Besides characteristic peaks for HAP and two for Fe_3O_4, new peaks (2Θ = 27.68°, 31.48°, 40.85° 49.32°, 57.53°, 61.84°, 69.80°) are found, which belong to $Pb_{10}(PO_4)_6(OH)_2$ (File number 87-2478). This result confirms that the dissolution precipitation mechanism plays an important role for Pb^{2+} adsorption on HAP/Fe_3O_4.

For comparison of HAP/Fe_3O_4 samples before and after the adsorption of lead ions, were analyzed for their surface appearance. The SEM micrographs are shown in Figure 16 b and e. It is found that before adsorption, the HAP/Fe_3O_4 material was composed of irregularly shaped particles, and large flake-like species were formed on the surface of the adsorbent after the adsorption of Pb^{2+}.

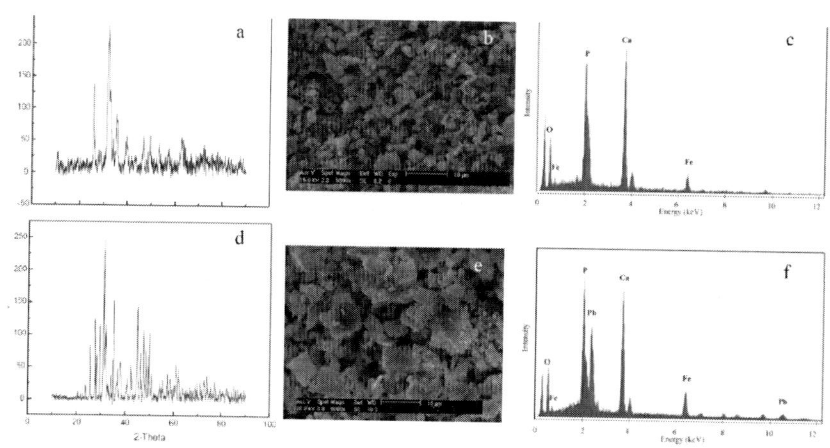

Figure 16. The XRD pattern, SEM and EDAX spectra of HAP/Fe_3O_4: (a, b, c) before and (d, e, f) after Pb^{2+} adsorption.

EDAX spectra for the HAP/Fe_3O_4 samples before and after Pb^{2+} absorption is shown in Figure 16 c and f, respectively. The EDAX spectra of HAP/Fe_3O_4 exhibit peaks for Ca, P, Fe, O, which are the major constituents of HAP and magnetite. For the Pb^{2+} absorbed sample, there are new peaks appearing in the EDAX spectrum, which is attributed to Pb element. Therefore, the strong peaks at 2.34, 10.5 keV in the EDAX spectra provided important evidence for Pb^{2+} adsorption on HAP/Fe_3O_4. These results confirm that Pb^{2+} was efficiently adsorbed on the surface of HAP/Fe_3O_4. HAP/Fe_3O_4 composite as the novel absorbent has been developed and its adsorption behavior on Pb^{2+} has been investigated. The experimental results confirm that this adsorbent has the potential application for removal of Pb^{2+} from aqueous solution (99%). Efficient absorption of Pb^{2+} can be realized in a wide pH range when an adsorbent dosage of more than 0.1 g/L is used. The effect of coexistent substances such as competing cations (K^+, Na^+ and Ca^{2+}, Mg^{2+}, which exist widely in natural water sources) and humic acid was considered. Humic substances (HSs) are the major components of the natural organic materials found in soil, water and geological organic deposits [86]. Because of the capability of HSs in complexation with metal ions, compounds in this catalog have recently been recognized as the important materials that affect the redistribution of heavy metal ions in soil and the hydrosphere [87]. In their experiments, coexistent of cations, e.g., Ca^{2+}, Mg^{2+}, K^+ and Na^+ cations, do not interfere the elimination of Pb^{2+}, while humic acid does. the presence of HA introduced a significant effect on the removal of Pb^{2+}. It is obvious that the removal efficiency of Pb^{2+} decreased as HA increased from 1 to 20mg/L. It is worthwhile to note that the Pb^{2+} removal efficiency increased first as the HA concentration was up to 50 mg/L, and then, kept almost constant within the HA concentration range of 50–100 mg/L.

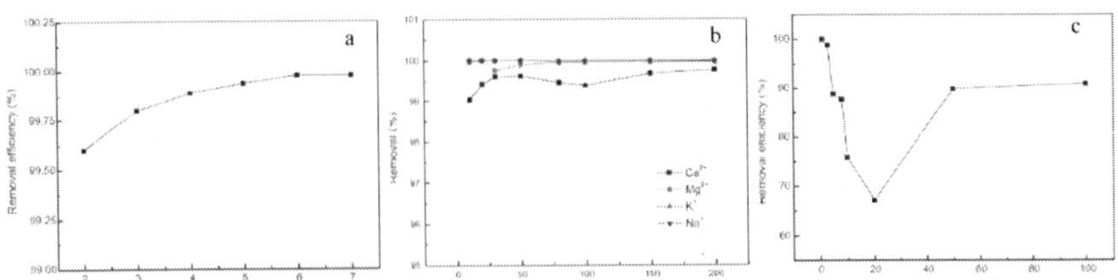

Figure 17. (a)Effect of pH on the removal of Pb^{2+} ions by HAP/Fe_3O_4, b) Effect of competing cations and, c)Effect of HA on the removal of Pb^{2+} ions by HAP/Fe_3O_4.

Mechanistic studies show that the adsorption data fits the Langmuir and Freundlich isotherm models reasonably well and the adsorption processes obey the pseudo-second-order kinetics model. Based on the experiment results, it is concluded that dissolution/precipitation and the surface complexation are mutually responsible for lead adsorption on HAP/Fe_3O_4, which should be attributed to the multiple functionalities of the material (Figure 18 a-c).

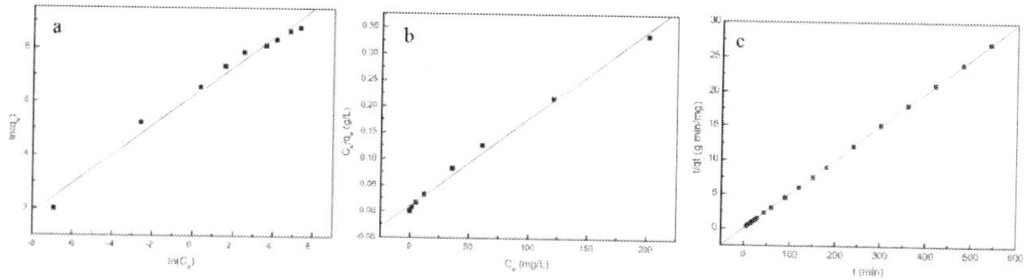

Figure 18. The Freundlich isotherm plot (a) and The Langmuir isotherm plot (b) The pseudo-second-order kinetics of Pb^{2+} adsorption by HAP/Fe_3O_4 (c).

In another work, the effect of different percent of Fe_3O_4 in the HAP/Fe_3O_4 on the structure and applicability of adsorbent was considered [88]. The XRD patterns of the FeHx composites together with the pure HAP and Fe_3O_4 powders are shown in Figure 19a. The well-resolved diffraction peaks of pure HAP correspond to the reference hydroxyapatite planes can be clearly seen in all the FeHx composites. As the Fe_3O_4: HAP ratio increases, the intensity of the diffraction peaks corresponding to the (311), (511) and (440) planes of the magnetite Fe_3O_4 structure increases. The increase of other peaks, especially the (220) diffraction is more difficult to follow because of overlapping exists with HAP diffraction peaks. It is worth noting that the presence of γ- Fe_2O_3 particles (JCPDS-39-1346), which can be result from powder drying, cannot be fully put aside as the XRD patterns of two iron oxide phases differ mainly by small intensity peaks. Noticeably, the HAP diffraction peaks are still visible for the FeH4 sample and no additional diffraction peaks are apparent corresponding with another calcium phosphate phase. This suggests that the presence of Fe_3O_4 does not interfere with the precipitation of the HAP phase. Figure 19b shows the nitrogen sorption isotherms of the FeHx composites, together with the HAP and Fe_3O_4 pristine powders. The Fe_3O_4 powder exhibits a type IV isotherm with a H_1 hysteresis according to IUPAC [89] which reflects inter-particular porosity due to the packing of spherical particles of size of 8–15 nm. FeHx isotherms have a similar shape to that of HAP materials and can be classified as type IV isotherms with a type H_3 hysteresis [89]. This can be attributed to the platelet-like shape of HAP nanoparticles [90]. In the high-p/p_0 pressure domain, the HAP and FeHx isotherms do not exhibit a well-defined plateau, reflecting the absence of a closed mesoporosity.

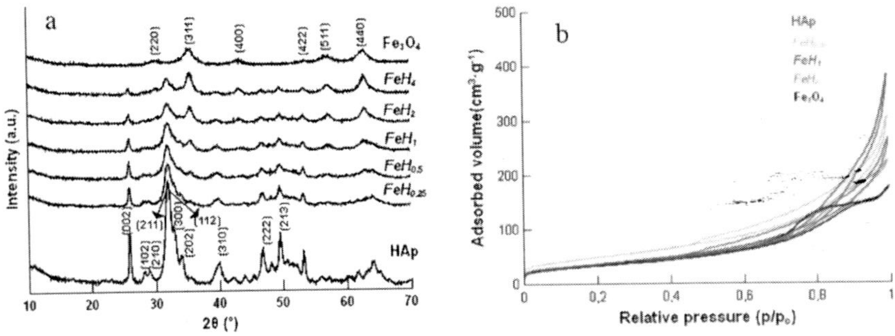

Figure 19. XRD patterns (a) and nitrogen sorption(b) of Fe_3O_4, HAP and selected nanocomposites.

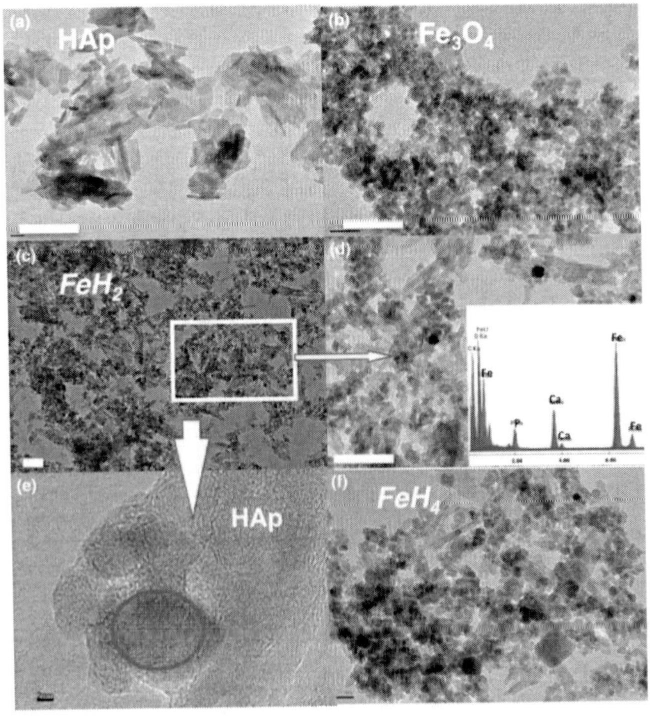

Figure 20. TEM images of the nanoparticles: a) HAP, b) Fe_3O_4, c–e) FeH_2 (insert: EDX spectra) and f FeH_4 (scale bar = 200 nm, except (e) 2 nm).

TEM images (Figure 20) provide more details about the organization of the composite powders. The HAP (Figure 20a) and iron oxide particles (Figure 20b) have distinct size, morphology and contrast so that they can be easily distinguished in the different FeHx composites (Figure 20c–f and ESM-2). In all situations, Fe_3O_4 colloids uniformly surround HAP particles. The energy-dispersive X-ray (EDX) spectra, inset of Figure 20d, provide evidence that the Ca/P molar ratio of the composite phase is close to 1.67, which corresponds to the stoichiometric composition of HAP. Figure 20d is an enlarged TEM picture of FeH_2 composite at high magnification, which clearly shows the good dispersion of Fe_3O_4 colloids within the HAP aggregates. In addition, the observation of continuous lattice fringes across individual FeH_x nanorods in HRTEM images (Figure 20e) indicates the possible coexistence of the well crystallized phases of HAP and Fe_3O_4 within a same grain.

Figure 21 presents the lead removal curves obtained at pH 5. First, it is important to point out that Pb (II) removal was almost total (99%) at low initial lead content, so that several sorption data overlap in the first part of the plotted isotherms. At high initial lead content, the removal capacity of composites is significantly enhanced compared to pure HAP, especially at low magnetite content. Noticeably, the pure Fe_3O_4 nanoparticles show no Pb(II) removal capability (data not shown). The maximum removal capacity q_{max} and Langmuir constant k were calculated for each sample, and the results are presented in Table 4. In all situations, the correlation coefficient R^2 for the nonlinear fitting was found C 0.978, suggesting that Langmuir model is well adapted to describe the sorption process and therefore that lead sorption occurs via monolayer coverage on the adsorbent surface.

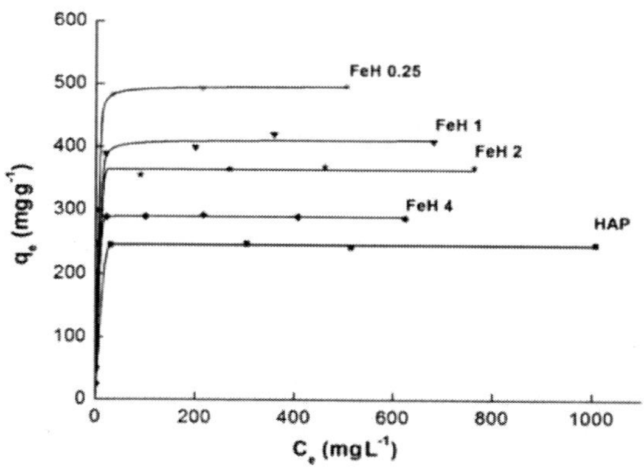

Figure 21. Pb (II) sorption isotherms on selected materials. Plain lines indicate the result of data fitting using the Langmuir equation.

Table 4. Fe_3O_4 content and specific surface area (S_{BET}) of pure phases and composite

	HAp:Fe_3O_4 molar ratio	Fe_3O_4 (wt%)	S_{BET} ($m^2 * g^{-1}$)	q^a_{max} [$q_{max-HAp}$]^b (mg g^{-1})	k [k_{HAp}] (L g^{-1})	R^2
HAP	–	0	125	246 [246]	4.5 [4.5]	0.984
$FeH_{0.25}$	4	5.48	175	496 [524]	1.0 [1.0]	0.996
$FeH_{0.5}$	2	10.39	170	–	–	–
FeH_1	1	18.83	160	412 [507]	0.8 [1.1]	0.993
FeH_2	0.5	31.69	140	372 [541]	0.8 [1.1]	0.978
FeH_4	0.25	48.13	125	294 [566]	0.8 [1.5]	0.986
Fe_3O_4	–	100	105	–	–	–

Maximum removal capacity q_{max} and Langmuir constant k. R^2 corresponds to the correlation coefficient obtained from the nonlinear fitting procedure. [$q_{max-HAp}$] and [k_{HAp}] calculated per g of HAp

^{a} ±15 mg g^{-1}, ^{b} ±20 mg g^{-1}

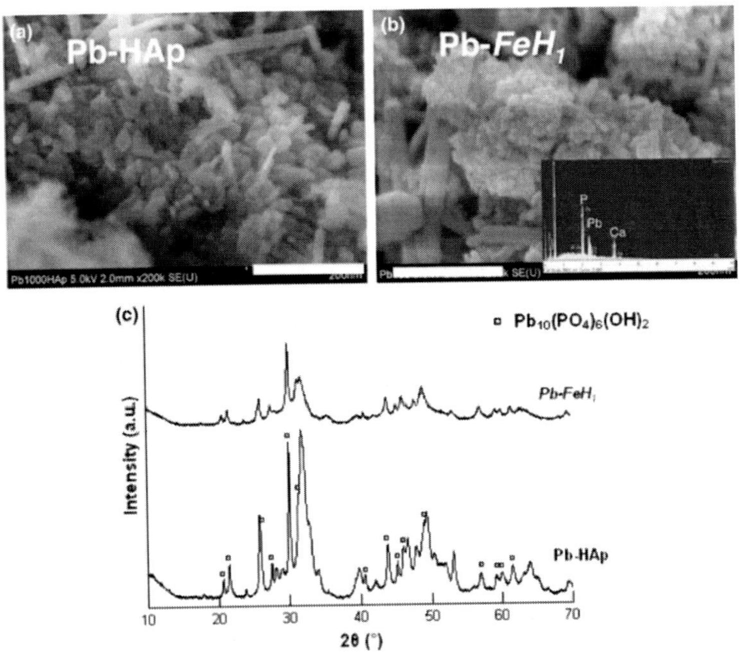

Figure 22. a, b FEG-SEM images and c XRD patterns of HAP and FeH1 after Pb(II) sorption. Insert EDX spectra of Pb–FeH_1.

The FEG-SEM images of the Pb–HAP and Pb–FeH_x particles are presented in Figure 22 a, b, respectively. Large needle-like particles can be seen on the particle surfaces, whose size is 50–300 nm in length and 20–50 nm in width. The related EDX spectrum clearly shows the presence of Pb in all the samples. The XRD diffractograms show distinct peaks that can be attributed to hydroxy pyromorphite with a formula of $Pb_{10}(PO_4)_6(OH)_2$ (JCPDS 02-0700) (Figure 22c). This agrees with previous reports from the literature, indicating that HAP is transformed into hydroxy pyromorphite via the sorption of Pb (II) followed by the cation exchange reaction between Pb (II) ions in aqueous solution and Ca (II) ions of HAP in the composites [91].

Manganese dioxide (MnO_2), an important adsorbing material with a relatively high surface area, microporous structure, and -OH functional groups, can increase the adsorption capacity. It has been exploited as a useful adsorbent for heavy metals such as Cd^{2+} [92], Co $^{2+}$, Zn $^{2+}$,and Cu $^{2+}$ [93]. Despite several studies carried out on the use of novel adsorbents made by loading MnO_2 on another solid such as hydroxyapatite [94]. To preparation of HAp/MnO_2 composite, a suspension of MnO_2 in CH_3COOH solution was added to the solution of Ca $(NO_3)_2$ with KH_2PO_4 at a molar ratio of 5:3. The molar ratio of MnO_2 to HAP was 3:1. After stirring, a solution of NaOH was added to the mixture to by reaching pH value of 13. The mixture was kept for 6 h at 37°C, and then the resulting HAP/MnO_2 composite was washed repeatedly with distilled water and dried before usage. Figure23a shows the XRD patterns of the composite adsorbents HAP/MnO_2. There are five characteristic peaks for HAp (2θ = 25.78°, 32.05°, 34.14°, 39.82°,46.66°) and four for MnO_2 (2θ = 28.71°, 36.89°, 42.12°,49.89°). SEM image and EDAX spectra of the HAP/MnO_2 samples are shown in Figure 23b and c. This adsorbent has been found to be composed of particles with irregular shapes and different size. The EDX spectrum of HAP/MnO_2 exhibited peaks for Ca, P, Mn, and O, which are the major constituents of HAP and manganese dioxide.

Figure 23. XRD pattern, SEM and EDX of composite adsorbent HAP/MnO_2.

In the case of Pb (II) removal, the effect of various factors that may affect the adsorption efficiency have been investigated in optimum solution pH (5), initial solute concentration (5-500mg/L), and the duration of the reaction (24h) and dosage of adsorbent (0.25g/L). The adsorption equilibrium was described well by the Langmuir isotherm model, and the calculated maximum adsorption capacity was 769 mg/g. The effect of coexistent substances such as humic acid (HA) and competing cations (Ca^{2+}, Mg^{2+}) was considered. The presence of HA showed a significant effect on adsorption. It is worthwhile to note that the removal efficiency of Pb^{2+} decreased as HA concentration increased from 1 to 8 mg/L. It also increased within an HA concentration range of 10–50 mg/L. Within an HA concentration range of 50–100 mg/L, the removal efficiency remained almost constant. Similar phenomena were also found in some previous reports [95]. These were attributed to the competitive adsorption of HA and heavy metals. In this Pb^{2+}-HA-HAP/MnO_2 system, the results showed that the influence of HA on the Pb^{2+} adsorption was dependent on HA concentration. At low HA concentrations, the reduction of adsorption efficiency can be explained by the formation of soluble HA-Pb(II) complexes in solution, which stabilized Pb^{2+} ions in aqueous solution. However, at relatively high HA concentrations, the adsorption of HA on the HAP/MnO_2 surface followed by the interaction of Pb^{2+} ions with surface adsorbed HA might increase the adsorption of Pb^{2+}. The effects of Ca^{2+} and Mg^{2+} on Pb^{2+} sorption on HAP/MnO_2 were also studied as the functions of competing cations. As shown in Figure 24, the influence of Ca^{2+} and Mg^{2+} on Pb^{2+} adsorption (more than 99% of Pb^{2+} adsorbed) under experimental conditions could be neglected. These results confirmed that the HAP/MnO_2 could selectively and efficiently remove Pb^{2+} ions from slightly polluted natural water containing the coexistent of Ca2+ and Mg2+ mineral cations.

S.A. Hassanzadeh-Tabrizi et al. have been synthesized a magnetic $CoFe_2O_4$–hydroxyapatite core–shell nanoparticles by reverse microemulsion method (Figure 25) and used as adsorbent to remove Zn (II) from aqueous solution [96]. X-ray diffraction analysis of the sample is shown in Figure 26 (a). The clear and well-defined peaks are related to HAP (JCPDS No.073-0293), the peaks at 2Θ=30.13, 35.76,37.71, 54.01, 55.13, 57.08, 62.63 are related to $CoFe_2O_4$ (JCPDS No.01-1121) [97]. Figure 26b shows the TEM images of $CoFe_2O_4$– HAP nanoparticles. As can be seen, the particles are relatively spherical in shape and appear to be mono dispersive. As known, in the microemulsion method, particles are formed in the internal structure of the nano reactors, which is determined by water to surfactant ratio. The results clearly reveal that a hydroxyapatite nanoshell was formed on the $CoFe_2O_4$ nanoparticles.

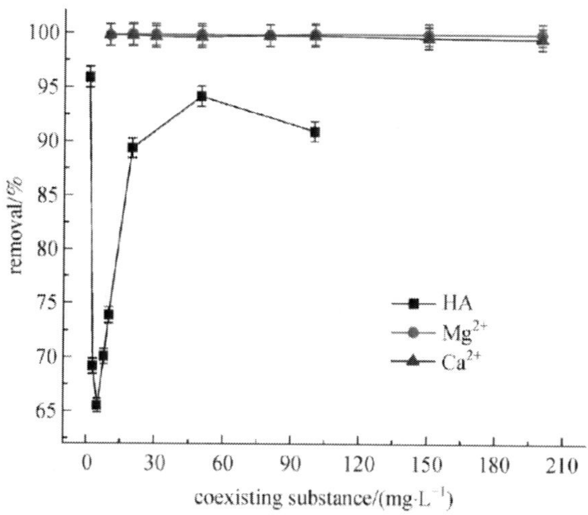

Figure 24. Coexisting substances and removal of Pb^{2+} ions by HAP/MnO_2.

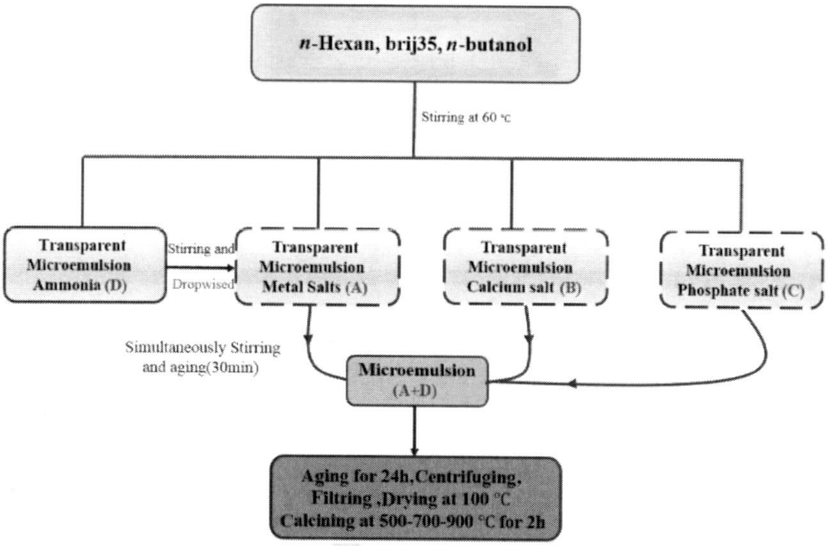

Figure 25. The flowchart of the processing of CoFe2O4– hydroxyapatite core-shell nanoparticles.

The adsorbent showed high adsorption capacity for Zn (II) and could be easily separated from aqueous solution by the application of an external magnetic field. The isotherm adsorption data fit well with the Freundlich model, and the kinetic data fit well with the pseudo-first-order model in the present of 1mg of adsorbent, initial concentration of Zn (II) 32.69 mg/L, after 120 min, in natural pH and room temperature.

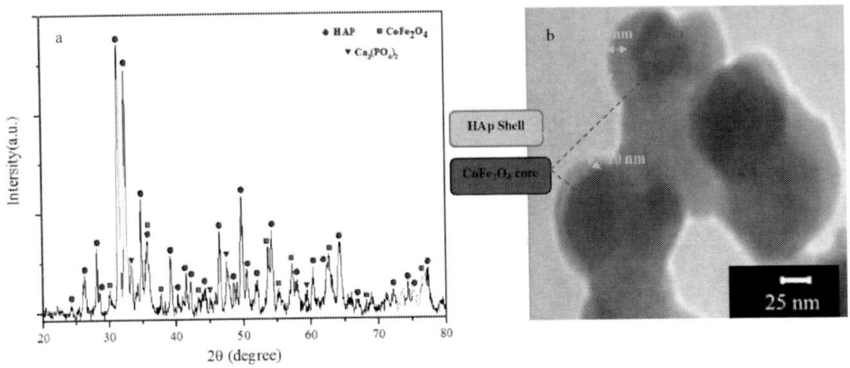

Figure 26. (a) X-Ray diffraction pattern and (b) TEM micrographs of nanoparticles calcined at 700°C.

Another composite material based on inorganic compound and HAP being useful for removal of heavy metals are Zeolite/HAP composite. Zendehdel et al. reported the synthesis and characterization of hydroxyapatite/NaP nanocomposite. It explored the possibility of using this composite as an adsorbent for efficient removal of Pb (II) and Cd (II) from aqueous solution. The method of preparation and characterization were considered in pervious section [7, 9]. The HAP/NaP nanocomposite showed a good efficiency in Pb(II) and Cd(II) removal from aqueous solution. Optimum dose of adsorbent was found to be 2 g/L, and the adsorption capacity of nanocomposite decreased with increasing pH for the removal of Pb(II), while removal of Cd(II) ions increased with increasing pH from 5 to 8. It was found that the maximum adsorption with $\% R = 97.8$ and $\% R = 98.8$ is achieved at pH of 5 for Pb (II) and pH of 8 for Cd (II), respectively. The effect of temperature on the adsorption of Pb (II) and Cd (II) ions was studied using optimum conditions such as adsorbent dose, pH, contact time, and initial concentration of aqueous solution. It was observed that there was a small decrease in adsorption by increasing temperature. The effect of initial concentration was studied. It can be seen that the removal percentage of both ions decreases with an increase in initial ion concentration. The optimum initial concentration was found to be 25 mg/L for Pb (II) ion and 5 mg/L for Cd (II) ion solution (Figure 27). As shown in Figure 27c, the removal of both ions with HAP/NaP nanocomposite was higher than hydroxyapatite and zeolite NaP adsorbents.

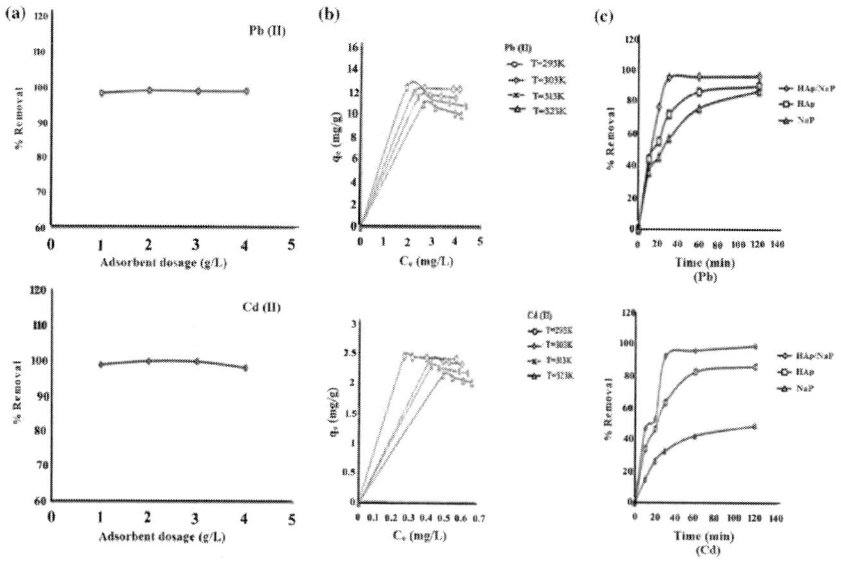

Figure 27. a) Effect of adsorbent dosage on the removal of Pb(II) and Cd(II), b) Effect of temperature on the adsorption of Pb(II) and Cd(II) to HAP/NaP. c) The comparison of HAP/NaP with HAP and zeolite as an adsorbent for the removal of Pb (II) and Cd (II).

Table 5. Isotherm parameters for sorption of Pb(II) and Cd(II) by HAP/NaP

Isotherm	Pb(II)	Cd(II)
Freundlich	$n_f = 1.629$ $K_f = 18.5079$ (mg g^{-1}) $R^2 = 0.8941$	$n_f = 2.047$ $K_f = 7.530$ (mg g^{-1}) $R^2 = 0.9923$
Langmuir	$q_m = 55.55$ (mg g^{-1}) $K_L = 0.747$ (L g^{-1}) $R^2 = 0.9762$	$q_m = 40.16$ (mg g^{-1}) $K_L = 0.229$ (L g^{-1}) $R^2 = 0.9638$
Temkin	$b_T = 247.37$ $A_T = 14.253$ (L mol^{-1}) $R^2 = 0.9932$	$b_T = 410.44$ $A_T = 7.4939$ (L mol^{-1}) $R^2 = 0.8965$
Dubinin–Radushk-evich	$q_D = 34.9808$ (mg g^{-1}) $K_D = 0.0676$ (mol^2 Kj^{-2}) $E = 2.72$ (Kj mol^{-1}) $R^2 = 0.979$	$q_D = 19.3791$ (mg g^{-1}) $K_D = 0.0632$ (mol^2 Kj^{-2}) $E = 2.84$ (Kj mol^{-1}) $R^2 = 0.7134$

The obtained parameters of all isotherm models for adsorbed Pb (II) and Cd (II) on the HAP/NaP nanocomposite along with coefficient correlation (R^2) are shown in Table 5. The equilibrium adsorption data of Pb (II) were fitted to the Temkin isotherm rather than the other isotherm models, and in case of Cd (II), the equilibrium data were fitted to the Freundlich isotherm model. The D-R isotherm model was also applied to analyze adsorption data. The value of obtained E indicates a physical adsorption mechanism for Pb (II) and Cd (II) on HAP/NaP. The plot of kinetic model for adsorption of Pb (II) and Cd (II) on HAP/NaP nanocomposite show the high correlation coefficient (R^2 = 0.999) for both Pb (II) and Cd (II) indicated the pseudo-second model which perfectly fitted the experimental data.

Natural zeolites are aluminosilicates having a negatively charged in the lattice. Importantly, they have been used as a balancer for cations with their merits including abundance, easy availability, low-cost, high ion exchange capacity. Zeolites possess net permanent negative charges on their framework surfaces resulting from isomorphic substitution of Si^{4+} by Al^{3+}, making them suitable for surface modification with different groups such as cationic surfactants [98-100] amino-functionalize increases the ability to remove a variety of heavy metal due to formation of covalent bonds between surface of absorbent and metals [101-103].In this aim Hydroxyapatite/zeolite composite (HZC) and surfactant-modified HZC (SMHZC) were prepared by J. Lin et al. and was used as an adsorbent to remove humic acid (HA) and copper(II) from aqueous solution [104]. For preparation of SMHZC, 10 g of the HAP/Zeolite composite was added into150 mL of 25 mmol/L Cationic surfactant CPB solution. The mixture was then placed in a thermostatic water bath shaker at a constant shaking speed of 150 rpm for 24 h, with the temperature maintained at 313 K. The XRD patterns of natural zeolite, HZC, and SMHZC are shown in Figure 28a. The XRD pattern of the natural zeolite showed the characteristic peaks of clinoptilolite, mordenite, and quartz, indicating the presence of these minerals in the natural zeolite. The main XRD peaks at 2θ=9.94°, 20.9°,22.4°, 25.7°, 26.7°, 27.8°, and 50.2°, which were in agreement with the characteristic XRD peaks of the natural zeolite, were observed in the HZC. This indicates the presence of the natural zeolite in the HZC. In addition, a comparatively broad diffraction peak at about 2θ=31.8°, being the characteristic of HAP, was observed in the HZC. This indicates the presence of HAP in the HZC. The positions of the main XRD peaks of SMHZC were similar to those of HZC.

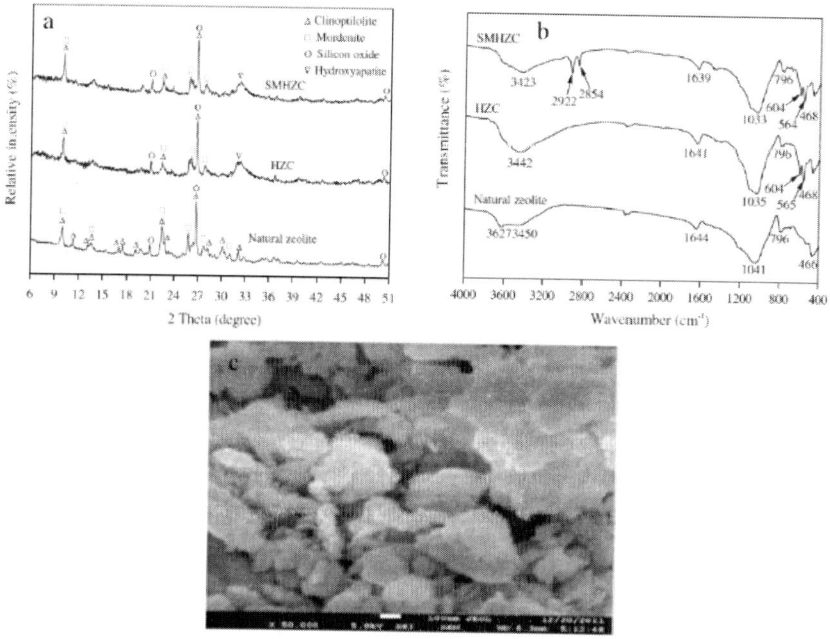

Figure 28. XRD patterns and FT-IR of natural zeolite, HZC, and SMHZC (a,b) and SEM for SMHZC.

This indicates that the modification of HZC with CPB does not lead to the significant structural change of HZC. The FT-IR spectra of natural zeolite, HZC, and SMHZC are shown in Figure 28b. The result shows that all main adsorption bands for the natural zeolite and HAP were listed as follows: 3,627, 3,450, 1,644, 1,041, 796, and 466 cm^{-1} and 604 and 565 cm^{-1} for HZC, and SMHZC, respectively. The FTIR spectrum of SMHZC showed two pronounced adsorption bands at 2,854 and 2,922 cm^{-1}, attributed to the symmetric and asymmetric $-CH_2$ stretching vibrations, respectively that indicates that CPB molecules have been loaded onto the surface of HZC after the contact of HZC with CPB solution. The surface morphologies of SMHZC were determined by using a FE-SEM as shown in Figure 28c. The surface morphologies of SMHZC were significantly different to that of the natural zeolite and surface of the natural zeolite was covered by agglomerated HAP particles, the zeolite crystals could not be seen clearly. When the surface of the natural zeolite was covered by agglomerated HAP particles and CPB molecules, the zeolite crystals also could not be seen clearly (Figure 28c).

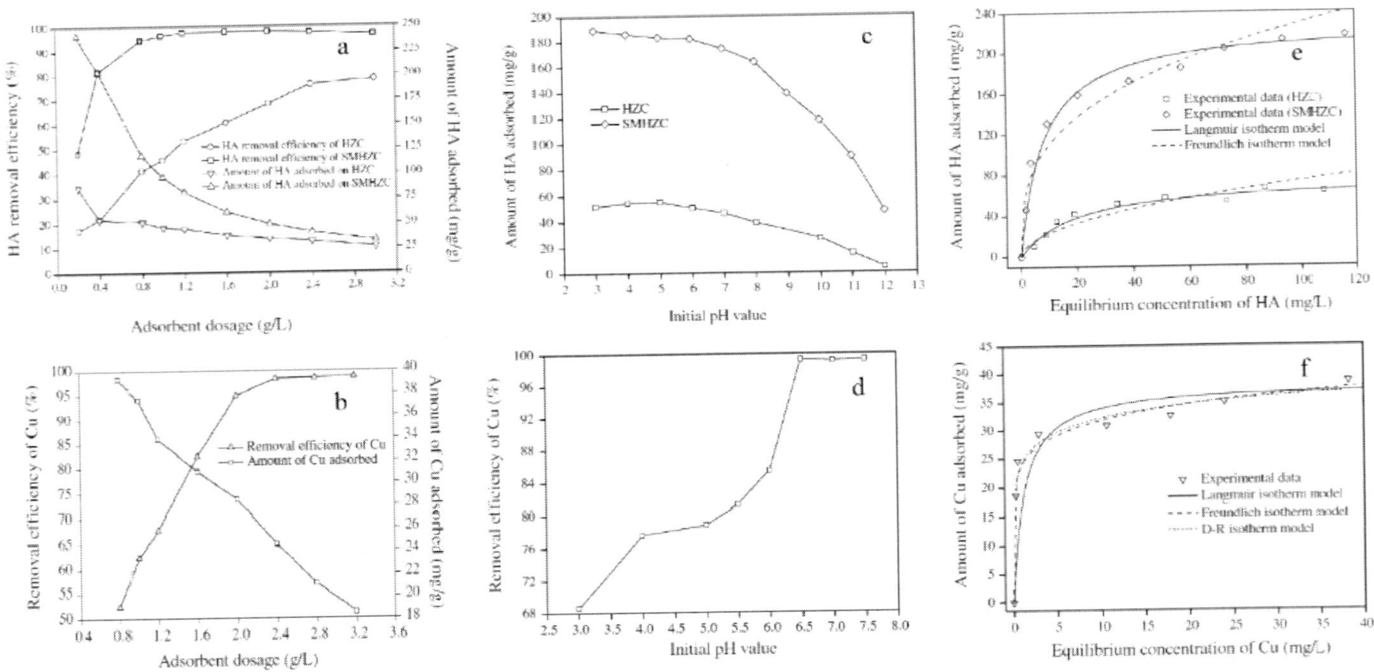

Figure 29. Effect of adsorbent dosage (a, b), Effect of solution pH(c,d) and Adsorption isotherm(e,f) on HA and copper (II) adsorption onto HZC and SMHZC (solution volume 50,25 mL; initial HA concentration 100, 60 mg/L; pH 7,6; contact time 24 h; temperature 303 K, respectively).

The adsorption of HA and copper (II) on SMHZC was investigated. For comparison purposes, HA adsorption onto HZC was also investigated. SMHZC exhibited much higher HA adsorption capacity than HZC. The HA adsorption capacity for SMHZC decreased slightly with increasing pH from 3 to 8 but decreased significantly with increasing pH from 8 to 12. The copper (II) adsorption capacity for SMHZC increased with increasing pH from 3 to 6.5. The adsorption isotherms of HA on HZC and SMHZC are shown that SMHZC exhibited a much higher adsorption capacity for HA than HZC. This indicates that the surface modification of HZC with CPB significantly enhances its HA adsorption capacity and thus SMHZC is more suitable as an efficient adsorbent for removing HA from aqueous solution than HZC. The Langmuir and Freundlich isotherm models were applied to analyze the equilibrium adsorption data. The equilibrium adsorption data of HA on SMHZC better fitted to the Langmuir isotherm model than the Freundlich isotherm model. The equilibrium adsorption data of copper (II) on SMHZC could be described by the Langmuir, Freundlich, and Dubinin– Radushkevich isotherm models (Figure 29). The presence of copper (II) in solution enhanced HA adsorption onto SMHZC. The presence of HA in solution enhanced copper (II) adsorption onto SMHZC. The mechanisms for the adsorption of HA on SMHZC at pH value of 7 may include electrostatic attraction, organic partitioning, hydrogen bonding, and Lewis's acid–base interaction. The mechanisms for the adsorption of copper (II) on SMHZC at pH value 6 may include surface complexation, ion exchange, and dissolution–precipitation. The obtained results indicate that SMHZC can be used as an effective adsorbent to simultaneously remove HA and copper (II) from water.

Zendehdel et al. focused on the removal of toxic and carcinogenic Cr (VI) by a nanocomposite which was prepared from nano HAP and Iranian clinoptilolite as a low-cost natural zeolite. For modification, this nanocomposite is functionalized by (3- aminopropyl) trimethoxysilane (APTMS) as the silylation agent [49]. Surprisingly, the synthesized HAP/Clin/NH_2 not only detected Cr(VI) but also recognized fluoride anions, remarkably was considered in previous part. Expectedly, in all cases with functionalized amino group, the percentage of removal of pollutants was increased about 97.5% for F^- and 95.5% for Cr (VI). Subsequently, batch adsorption studies were implemented to consider the effect of parameters such as the contact time (120 min), pH (pH < 7), temperature (room temperature), and initial concentration of pollutants (30 ppm), types of adsorbent and adsorbent dosage (0.025g). The consideration of solution pH is very important factor for the adsorption process. The removal of Cr(VI) at pH < 7 was significantly higher respect to pH > 7 which indicates the reaction was facilitated under acidic and neutral conditions. According to Eqs. 13–15, chromate ions mainly exist in the form of $HCrO4^-$ under acidic conditions which are considered as suitable ions for interaction by positive charge due to presence of metals and absorption of H^+ on the surface of HAP and zeolite composite [105].

$$Cr_2O_7^{2-} + H_2O \rightleftharpoons 2HCrO_4^-, pK\alpha = 2.2 \quad (13)$$

$$H_2CrO_4 \rightleftharpoons HCrO_4^- + H^+, pK\alpha = 4.1 \quad (14)$$

$$HCrO_4 \rightleftharpoons CrO_4^{2-} + H^+, pK\alpha = 5.9 \quad (15)$$

In the pH > 7, the excess OH^- ions competed with chromate ions for adsorption sites, which caused the removal of Cr (VI) decreased at higher pH [106]. Langmuir and Freundlich's models were adopted as the main sorption isotherms (Table 6), while the pseudo second-order kinetic model was the most fitted Cr (VI) kinetic model in aqueous systems.

Table 6. Isotherm parameters for sorption of Cr (VI)

Isotherm		Langmuir				Freundlich		
Metal ion	Temperature °C [a]	q_{exp} (mg/g)	q_{max} (mg/g)	K_L (L/mg)	R^2	K_f (mg/g)	N	R^2
Cr(VI)	25	159.2	188.7	0.05	0.995	17.6	1.9	0.931
	35	156.7	181.8	0.07	0.996	21.7	2.1	0.945
	45	149.6	175.4	0.10	0.992	0.60	2	0.934

Mechanism Discussion to Removal Heavy Metals

For the composite adsorbent HAP/Inorganic materials, HAP was immobilized on the surface of Inorganic materials during the preparation, being the major contributor for Pb^{2+} adsorption. To elucidate the adsorption of heavy metal ions on different HAP-containing adsorbents, three possible mechanisms, including ion exchange, dissolution/precipitation and surface complexation, have been proposed [107]. In acidic and natural aqueous solutions, HAP can dissolve partly and hydroxypyromorphite [$Pb_{10}(PO_4)_6(OH)_2$, HPy] may precipitate. The dissolution/precipitation mechanism is given as follows:

$$\text{Dissolution: } Ca_{10}(PO_4)_6(OH)_2 + 14H^+ \rightarrow 10Ca^{2+} + 6H_2PO_4^- + 2H_2O \quad (16)$$

$$\text{Precipitation: } 10Pb^{2+} + 6H_2PO_4^- + 2H_2O \rightarrow 14H^+ + Pb_{10}(PO_4)_6(OH)_2 \quad (17)$$

Calcium cation in HAP/ Inorganic materials has stronger affinity to the HAP/ Inorganic materials surface sites than the co-existing Mg^{2+}, K^+ and Na^+ cations. But the solubility product constants of Pb(II)- phosphate are much smaller than Ca(II)-phosphate (for K_{sp} [$Pb_3(PO_4)_2$] = 8.0×10^{-43}; Ksp[$Ca_3(PO_4)_2$]=1×10^{-25}). Consequently, accompanied by dissolution of HAP, more stable Pb (II)-phosphate is formed. This explains the fact that the presence of Ca^{2+} did not influence the removal of Pb^{2+} significantly. To gain additional understanding for the mechanism of Pb^{2+}adsorption on HAP/ Inorganic materials, the concentrations of $Pb^{2+}$and Ca^{2+} ions as well as the pH variation before and after adsorption were monitored in the adsorption isotherm experiments. The amount of adsorbed Pb^{2+} and released Ca^{2+} in the solution at different initial concentrations of Pb^{2+} are presented in Figure 30.

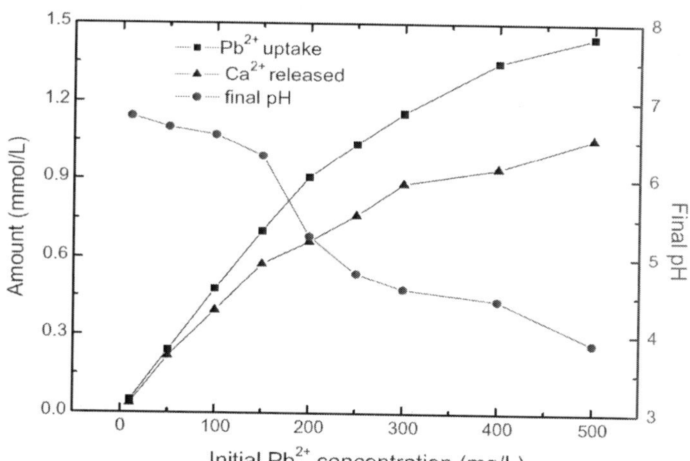

Figure 30. The amounts of Pb^{2+} adsorbed, Ca^{2+} released and final pH of the solution at varying initial concentrations of Pb^{2+}.

The difference between the amounts of Pb^{2+} adsorbed and Ca^{2+} released suggests that in addition to dissolution-precipitation, another adsorption mechanism may exist for the Pb^{2+} uptake process and also shows that the sorption of Pb^{2+} onto adsorbent caused a decrease of the final pH values. It is observed that the final pH value is reduced as the amount of adsorbed Pb^{2+} ions increased.

The pH value drop in equilibrium suggests that H^+ ions are liberated from the solid surface into the solution as a result of exchange with Pb^{2+}, which confirmed formation of surface complexes takes effect to the removal of Pb^{2+} [108].

$$HAP - OH + Pb^{2+} \Leftrightarrow HAp - O - Pb^+ + H^+ \quad (18)$$

$$2HAP - OH + Pb^{2+} \Leftrightarrow (HAP - O)_2 - Pb + 2H^+ \quad (19)$$

Dyes as an Environmental Pollutant

As is well known, dyes are giving colors and life to our nowadays world, but unfortunately it is becoming one of the major problems of environmental pollution that requires being controlled to prevent further issues for posterity. Complex structures, toxicity, high molecular weight, carcinogen, nonbiodegradability, and persistency in the environment are the main cause of dye removal from environment. For instance, the hydrolyzed form of Reactive Blue 19 might remain for 46 years in the environment at neutral pH and room temperature [109]. Furthermore, low density of dyes leads to their congregation on the surface of water bodies and consequently impediment of light penetration and then a significant reduction of dissolved oxygen will be occurred and of course development of anaerobic zones is not unavoidable especially in depth. In case of human, dyes engender dysfunctional of liver, kidney, and reproduction system. Dyes are mainly classified with various group such as acid, basic, disperse, reactive, and direct groups and extensive use of these dyes in distinctive industries such as textile, food, paper, leather, rubber, cosmetic and printing have enabled them to enter the environment [110-111]. It is considered that textile industry is the main responsible for dye contamination and also the World Bank estimated that 17-20% of industrial water contamination is caused by textile dyeing [112]. According to statics, yearly 7 × 105 tones dyes are generated in which 100 tones/year is released into the water bodies [113]. Considering the effluent of a typical textile company, color has been detected in the range of 50-2500 based on the Pt-Co standard color scale [114]. Physical, chemical and biological methods have been employed for dyes removal, including filtration, reverses osmosis, aerobic and anaerobic reactors, advanced oxidation processes, adsorption, precipitation, and coagulation/flocculation. While it is true to say that almost all of these methods have shown suitable effectiveness and efficiency for dye removal at laboratory scales, in fact, they all might not be utilized in industries particularly because are not cost-effective and hard operation. Adsorption, a physical treatment process, involves the movement of dye molecules from the bulk solution to the regions surrounding a solid phase called adsorbent. Adsorption process is considered to be a promising alternative for dye removal in terms of initial cost, design simplicity, high selectivity, insensitivity to toxic substances, and ease of operation [115]. Efficiency of the adsorption process is highly dependent on the adsorbent-adsorbate interaction. Since now, various inorganic materials, such as clay, alunite, dolomite, perlite, zeolite, and metal oxide have extensively used as low-cost adsorbents for dye removal. After complete saturation of the adsorbent surface, an issue regarding separation of the adsorbent particles will arise.

Enhanced Adsorption of Dyes on Hydroxyapatite Nanocomposite

A. Mollahosseini et al. magnetic zeolite/hydroxyapatite (MZeo-HAP) nanocomposite was successfully synthesized using microwave-assisted method (Figure 31) for the removal of Reactive Orange 5 (RO5), Reactive orange 16 (RO16), and Congo red (CR) anionic dyes (Figure 32) from aqueous solutions through batch adsorption [116].

SEM images were taken in order to identify the morphology and structure of the synthesized nano adsorbent. Figure 33 (a, b) shows the SEM image of the synthesized MZeo-HAP nanocomposite. Evidently, the figure is constituted of different parts that small particles are gather around a large sphere. EDAX analysis had proven that light regions are HAP particles. The central part of the nano composite with an unregularly spherical structure is indication of the magnetic zeolite particles. As a matter of fact, formation of HAP layer around Fe_3O_4 nanoparticles can be observed [117-118]. According to the SEM images, porous structure of the MZeo-HAP is beneficial in terms of more dye removal and high adsorption capacity. Figure 33c shows the EDAX spectra of MZeo-HAP nanocomposite. The figure shows the existence of constituent elements Fe, Si, O, Ca, Al, and P, owning various energy values. The presence of Fe and O in the result was due to the magnetic nanoparticles. Considering the structure of zeolite, elements Al and Si are dedication of zeolite presence in the final nanocomposite. Clearly, the peaks C and P were specifically related to HAP. EDAX spectra confirmed that the synthesized nanocomposite was formed from three different components. TEM analysis was carried out to find the shape, and the morphology of the synthesized adsorbent.

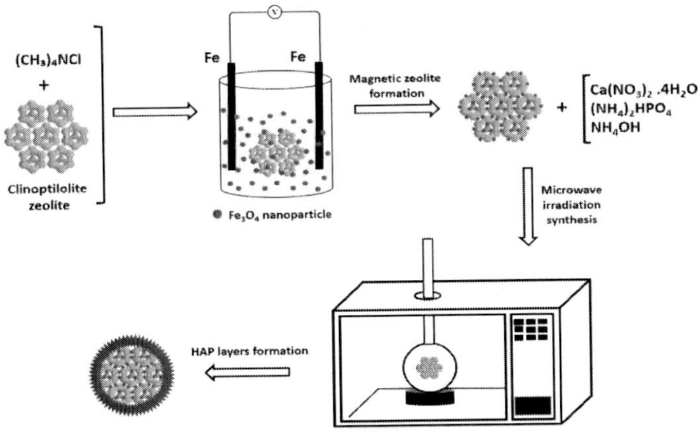

Figure 31. Synthesis scheme for MZeo-HAP nanocomposite preparation.

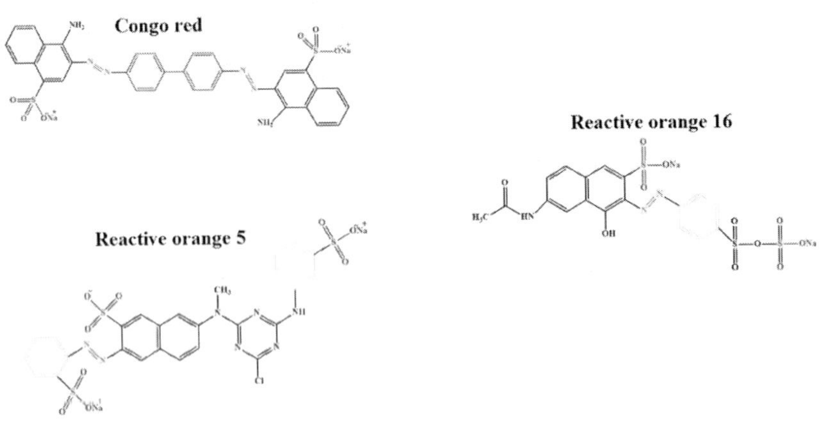

Figure 32. Chemical structure of selected dyes.

Figure 33(d, e) exhibits the TEM image of MZeo-HAP being synthesized by microwave-assisted method. As shown in SEM and EDAX images, formation of the HAP crystals around the magnetic zeolite is expected. It can be observed that HAP crystals has coated around the MZeo, having a lighter part and agglomeration particles. Darker color is specifically assigned to MZeo owning almost a spherical shape.

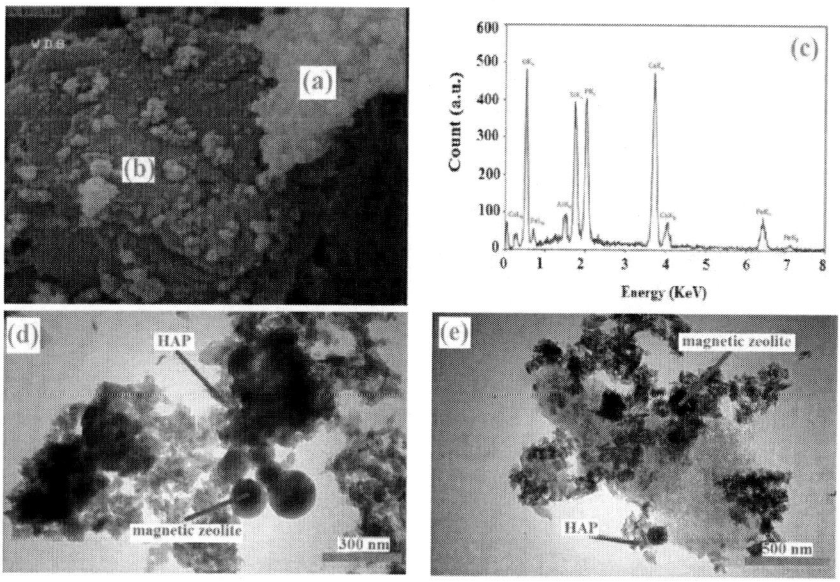

Figure 33. (a, b) SEM image, (c) EDAX of MZeo-HAP, (d, e) TEM images of MZeo-HAP.

XRD pattern of MZeo-HAP nanocomposite ranging (2θ) from 5° to 60° is depicted in Figure 34 a having numerous detectable peaks. All the peaks of synthesized nanocomposite are consistent with data base of HAP (JCPDS No. 09-0432), clinoptilolite (JCPDS No. 39-1383) and Fe_3O_4 (JCPDSNo. 19-0629) in the JCPDS-International Center for Diffraction Data file. These peaks confirmed that the synthesized adsorbent was compromised of magnetic nanoparticles, zeolite, and HAP. The magnetic properties of pure Fe_3O_4 nanoparticles, MZeo, and MZeo-HAP were studied using a VSM and the results are shown in Figure 34b. The bare Fe_3O_4 had the greatest saturated magnetization value of 68 emu/g. The magnetization value of the MZeo, and MZeo-HAP were found to be 1.70 and 0.85 emu/g, respectively. The reduction of saturation magnetization of MZeo, and MZeo-HAP was mainly attributed to the presence of non-magnetic layer to the magnetic nanoparticles. N_2 adsorption-desorption analysis was carried out to figure out the textural properties of MZeo- HAP. It can be seen that S_{BET} and pore volume of MZeo-HAP could attain 103.56 m^2/g and 0.34 cm^3/g which are more than that of magnetic zeolite and zeolite. The N_2 adsorption-desorption isotherms exhibit typical type IV isotherm and clear H_3 hysteresis loop was observed with MZeo-HAP (Figure 34 c, e). Pore diameter with the average size of 12.26 nm suggests the presence of mesopore particles. Having a large specific surface area, pore volume, and high porosity in the synthesized nanocomposite provides a large number of active will be able to interact more efficiently.

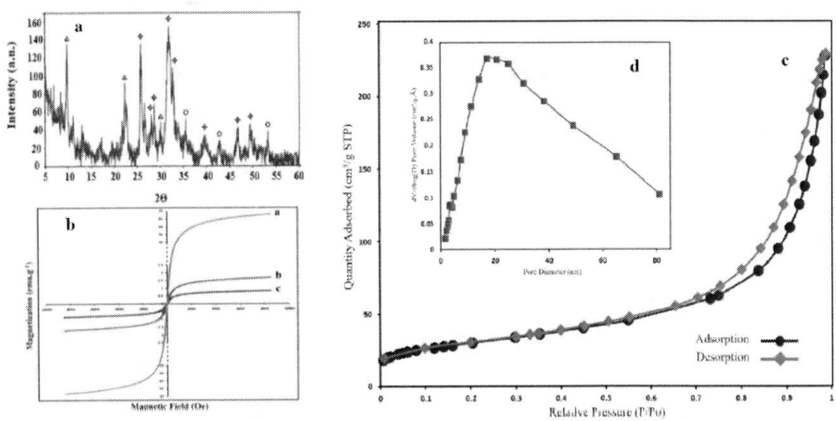

Figure 34. (a) XRD and (b) VSM and (c) N_2 adsorption-desorption isotherms for MZeo-HAP.

The effect of initial pH, contact time, adsorbent dosage, and initial dye concentration was investigated. The maximum dye removal was attained at pH value about 3 and the equilibrium contact time of 30 min. At dye concentration of 80 ppm, the highest adsorption capacity of RO5, RO16, and CR was 79.21, 72.59, and 103.10 mg/g, respectively. The pH is an important parameter in the adsorption process efficiency because it highly affects the surface charge of MZeo-HAP and also the degree of ionization and speciation of dyes during the reactions. Therefore, the effect of solution initial pH on the removal of dyes by the MZeo-HAP was studied. According to Figure 35a, clear that acidic conditions provide a suitable environment for dyes removal. By increasing the initial pH value from 2 to 9, the adsorption capacity of RO16, RO5, and CR decreased contiguously from 28.29 to 17.41 mg/g, 24.87

to 4.80 mg/g, and 28.97 to 1.43 mg/g, respectively. Thus, limited adsorption of dyes was observed in neutral and alkaline pH. It is known that pH_{pzc} refers to the pH at which the electrical charge density on a surface is zero. At pH > pHpzc, the surface becomes negatively charged, favoring the adsorption of cationic species. On the other hand, adsorption of anionic species will be favored at pH < pH_{pzc}. Figure 35b shows the zeta potential of MZeo-HAP, indicating that the synthesized adsorbent owned negatively surfaced charged, favoring the adsorption of cationic functional groups. It shows that by altering the initial pH values from 9 to 2, zeta potential value increased from -24.90 to -1.54 mV. HAP contains two types of binding sites, positively charged calcium (Ca^{2+}) and negatively charged phosphate (PO_4^{3-}) groups [119]. Similarly, there are exchangeable adsorption sites over the external surface of the zeolite, silanol (Si – OH) and aluminol (Al – OH) groups. Depending on the solution pH, positive or negative sites can be achieved according to the reactions below [120].

$$Al - OH + H^+ \leftrightarrow Al - OH_2^+$$

$$Si - OH + H^+ \leftrightarrow Si - OH_2^+$$

$$Al - OH + H^- \leftrightarrow Al - O^- + H_2O$$

$$Si - OH + H^- \leftrightarrow Si - O^- + H_2O$$

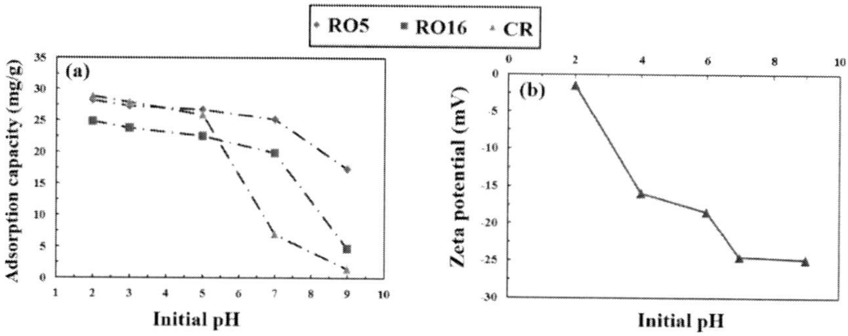

Figure 35. The effect of (a) pH, (b) pHzpc of MZeo-HAP.

Based on the previous studies, CR is an anionic azo dye, having a pKa value of 4.1; thus, CR would be positive in pH under 4.1 [121]. In fact, at lower pH values CR becomes an acidic cationic dye containing the amino functional groups (–NH_2) and =N_2 atoms protonated to ammonium ions such as $-NH_3^+$ and $=N^+$ [122]. In contrast, at alkaline medium CR has a negative charge due to the presence of sulfonic groups. At neutral pH CR has a red color, however at strong acidic pH the color changes to dark blue because of the $\pi - \pi$ interaction of azo group. In case of low pH (acidic media), the surface of MZeo-HAP has a negative charge and this is favorable for adsorption of CR dye because increase the electrostatic attraction between the positively charged dye and negatively charged adsorbent (Figure 36). Similar adsorption mechanism can be reported as like as CR, in which positive charges on the dye at lower pH reacts with negatively charged adsorbent to make a chemical binding via electrostatic interaction between adsorbate and adsorbent. Furthermore, at acidic media H^+ ions compete with positive functional groups of CR to be adsorbed by MZeo-HAP, resulting a diminish in CR removal efficiency. Reactive dyes, including RO16 and RO5 are known to ionize to a high degree in aqueous solutions to form colored anions due to the sulfonate groups in their structures. RO16 and RO5 have two and three sulfonate groups ($-SO_3^-$), respectively in which these groups are tending to dissociate to have negative surface charges in aquatic media. Low adsorption of reactive dyes at alkaline condition was attributed to the deprotonation of surface groups, leading to an electrostatic repulsion between negatively charged sites of the adsorbent and negative anionic dyes. The presence of magnetic particles in the MZeo-HAP was also important in terms of reactive dyes removal.

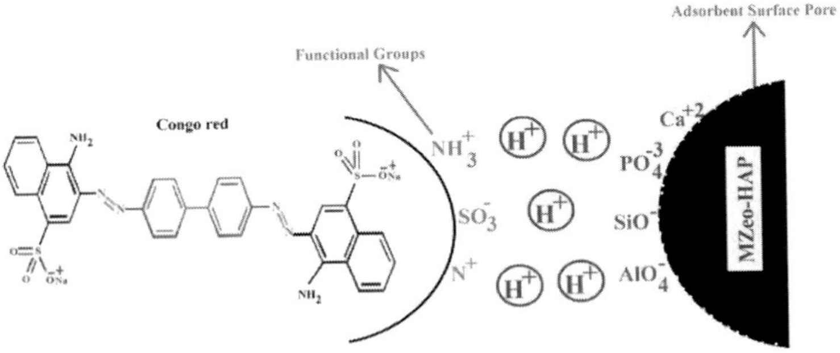

Figure 36. Schematic illustration for CR adsorption by MZeo-HAP via electrostatic attraction at low pH.

Thermodynamic parameters such as ΔG°, ΔS° and ΔH° were estimated for the adsorption study and suggested that the adsorption was endothermic and maximum dye adsorption occurred at 40ºC. Adsorption equilibrium data obeys Freundlich model with correlation coefficient ranging from 0.9632 to

0.9903, indicating the chemisorption nature of adsorption. Similarly, pseudo second order model is the most appropriate model to describe adsorption kinetic.

Table 7. Thermodynamics parameters calculated for the examined dyes

Dye	-ΔG° (Kj/mol)					ΔH° (Kj/mol)	ΔS° (J/mol)
	18	25	30	35	40		
RO5	1786	2108	2338	2068	2798	11.60	45.47
RO16	507	684	811	937	1064	6.85	25.30
CR	2869	3063	3202	3340	3479	5.20	27.73

Table 8. Obtained parameters from studied isotherm models

Isotherm model	Dyes		
	RO5	RO16	CR
	Langmuir		
q_m (mg/g)	239.39	90.09	40.60
K_L (L/mg)	0.006	0.12	0.23
R^2	0.9553	0.8167	0.9021
R_L	<1	<1	<1
	Freundlich		
K_F (L/mg)	1.62	15.29	8.22
n	1.06	2.08	1.91
R^2	0.9601	0.9446	0.9477

Table 9. Obtained parameters from PFO, PSO, and Elovich kinetics models

Kinetics model	Dyes		
	RO5	RO16	CR
	PFO		
K_1 (1/min)	0.429	0.409	0.249
$q_{e,cal}$ (mg/g)	24.54	26.91	28.22
$q_{e,exp}$ (mg/g)	24.93	27.29	28.77
R^2	0.6916	0.9161	0.9113
	PSO		
K_2 (g/mg.min)	0.042	0.036	0.012
$q_{e,cal}$ (mg/g)	25.70	28.17	29.19
$q_{e,exp}$ (mg/g)	24.93	27.29	28.77
R^2	0.9225	0.9705	0.9803
	Elovich		
α (mg/g.min)	47679	58340	102
β (g/mg)	0.537	0.496	0.226
R^2	0.9161	0.9167	0.9795

These finding confirm the MZeo-HAP as a promising candidate for dye adsorption in terms of water purification.

Antimicrobial Effect of HAP Composites

A multitude of ceramic materials with biomedical applications were synthesized. Among them, hydroxyapatite (HAP) is largely used in orthopedics and dentistry, owing to its exceptional biocompatibility and bioactivity [123-126]. Crystallographic and chemical investigations showed synthetic HAP to be similar to the natural one [127]. Since synthetic HAP is thermodynamically stable at physiological pH [128] and is osteoconductive, it was used for bone or dental reconstruction. Proteins, amino acids, and other organic substances are easily adsorbed on HAP, which in turn might favor the adsorption and replication of bacteria on HAP [129]. As the resistance of microorganisms toward antibiotics is increasing, there is a continuous preoccupation of scientists in finding new bioactive compounds, with high biocompatibility and antibacterial properties, such as nanometrically silver [130].

In this aim, Maria Tomoaia-Cotisel et al. have synthesized for the first time a multi-substituted hydroxyapatite, containing zinc, gold and silver in its inner HAP structure. In addition, this triple substituted HAP was mixed with silver nano particles (AgNPs) and a new HAP-Ag material was obtained, in which AgNPs are homogeneously adsorbed on the surface of nano HAP particles. Moreover, new nano composites have been developed by incorporating different amounts of HAP and/or HAP-Ag into an organic matrix, made by polymerization of monomers, such as (2,2-bis[4-(2-hydroxy-3-methacryloyloxypropoxy)-phenyl] propane (bis-GMA) (60%) and (triethylene glycol Di methacrylate) TEGDMA (40%) [131].

Figure 37a presents the XRD pattern of the nano HAP sample and of the nano HAP sample with 9 wt% of AgNPs. They show the crystallinity degree of these powders While in the HAP sample only the diffraction pattern characteristic to pure non substituted HAP is evidenced, in the HAP-Ag sample the diffraction pattern characteristic for Ag is also revealed. Therefore, the crystallinity of silver in the nanoparticles is also pronounced.

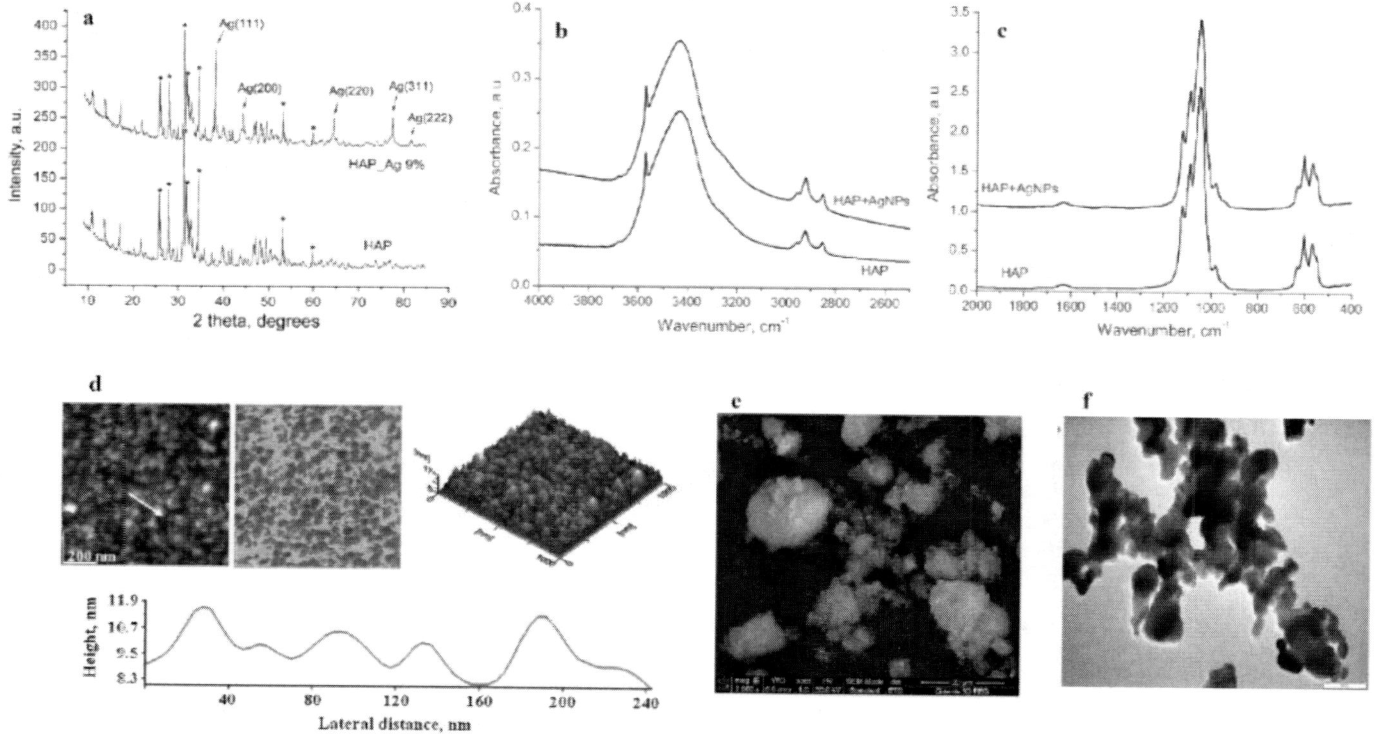

Figure 37. XRD, FT-IR, AFM, SEM and TEM of HAP particles and for the particles with AgNPs.

The FTIR spectra of HAP without and with adsorbed AgNPs are compared in Figure37 b, c. For clarity reasons, the spectral domain was split into two sections: from 4000 to 2500 cm^{-1}, and from 2000 to 400 cm^{-1} (with different scales in the two parts). The spectra are almost identical, showing that the adsorption of AgNPs did not modify the internal structure of the HAP. In the spectra of all the samples, the absorbance bands corresponding to the vibrations of PO_4 and HO groups could be identified. The broad band (maximum at 3435 cm^{-1}) is due to the OH^-... O stretching vibrations of absorbed water with hydrogen bonding in the samples. Another band also due to absorbed water has the maximum at 1630 cm^{-1}. The first band overlaps with the band of structural OH^- (from hydroxyapatite) at 3571 cm^{-1}. A band due also to structural OH^- (liberation vibration) is observed at 631 cm^{-1}. They suggest the presence of a number of OH^- group in hydroxyapatite. AFM images for the HAP particles without or with silver nanoparticles, on samples prepared by vertical adsorption on glass for 5 min are given in Figure 37d. SEM and TEM images for the multi-substituted HAP particles and for the particles with AgNPs adsorbed at their surface (HAP-Ag) are given in Figures 37e, f. The average size (diameter) of the HAP particles with AgNPs the average size was found to be 65 ± 17 nm.

Antimicrobial effectiveness tests were considered in the agar plates were made at 24 and 48 h. The inhibitory effect was manifested by the formation of circular, clear zones of inhibition around the disks, their size depending on the AgNPs content. No effect was observed for the control samples (pure polymer, or polymer with multi substituted HAP with Zn, Ag and Au as filler, in the ratio 1:1, but without AgNPs). The diameters of the inhibition zones (mm) were measured. Since the differences between the diameters of a zone did not exceed 1 mm, we considered the largest size observed in one of the three plates (Table 10).

Table 10. Silver content of the disk samples and the diameter of the zones of inhibition at 48 h

Tested strains	AgNPs content and diameter of the zones of inhibition (mm)						
	2%	2.5%	3%	3.6%	4.5%	5.4%	0%
Escherichia coli	10	12	12	12	14	15	0
Staphylococcus aureus	11	12	13	13	15	16	0
Staphylococcus spp.	0	8	10	10	12	14	0
Bacillus cereus	0	10	11	11	12	14	0
Candida albicans	10	11	12	12	15	15	0

In the case of E. coli, the size of the inhibition area was between 10 and 15 mm, as there is no clear demarcation between the edge of the area and the culture. Therefore, the contour of the zones of inhibition appears a little irregular (Figure 38a). For S. aureus the zones of inhibition were in the 11-16 mm range. It can be assessed that this species of staphylococcus presented the highest sensitivity against AgNPs, showing the largest zones of inhibition (Figure 38b). The Staphylococcus spp, species was less sensitive as compared with the reference strain. It was not inhibited at 2 wt% AgNPs content, but the inhibitory effect was observed in the other levels, the largest area (14 mm) appears at 4.5 wt% AgNPs (Figure 38c). For B. cereus (Figure 38d) there was also no inhibitory effect at 2 wt% AgNPs, but inhibition was present at higher concentration (14 mm at 5.4 wt% AgNPs). For C. albicans, readings were made at 48-72 h, since yeast strains develop slower as compared to bacteria (Figure 38e). There was an inhibitory effect for all concentrations (zones of inhibition from 10 to 15 mm). Excepting E. coli, where the zones of inhibition were not well delimited, for all the other tested microorganisms, these zones were circular and clear-cut, meaning the inhibition was total. After keeping the plates at +4°C, the zones of inhibition retained the same size. Removing the disks, the medium beneath them was transparent, without culture. The inoculation of bacteria in the glucose broth from the zones of inhibition (after removing the disks) was tested. After incubation at 37°C for 48 hours and a new verification after 7 days, it was stated that the inoculated medium remained sterile (no culture was developed). From these observations we conclude that the induced antimicrobial effect is bactericidal (Figure 38f). This study gives an understanding on the silver ions release from organic matrix composites (OMCs) containing multi-substituted HAP (0.2 wt% Zn, 0.25% Ag and 0.025 wt% Au), AgNPs and polymer (obtained from bis-GMA and TEGDMA). The data show significant antimicrobial effects of the slow Ag^+ ions release in vitro against the five pathogens.

The mechanisms about their action mostly agreed that they can bind to proteins on the bacterial wall and may enter in cytoplasm. Silver ions are considered highly reactive on bacteria. They can bind to bacterial DNA and RNAs and inhibit bacterial replication. Thus, they will cause structural changes in bacteria leading to cell distortion and death. They preferably do attack the respiratory chain, cell division finally leading to cell death. These results suggest that by using our composites, the silver ions release can be controlled and in vivo might improve the safety of using AgNPs being mixed with HAP and organic matrix in nano composites. Therefore, silver ions are released from AgNPs, and they determine the bactericidal activity when AgNPs are incorporated in carriers, like HAP and organic matrix. The exposure of HAP/AgNPs/organic matrix composites to the pathogens allowed silver ions to accumulate around the disks of composites and diffuse in agar, assuring an inhibition zone.

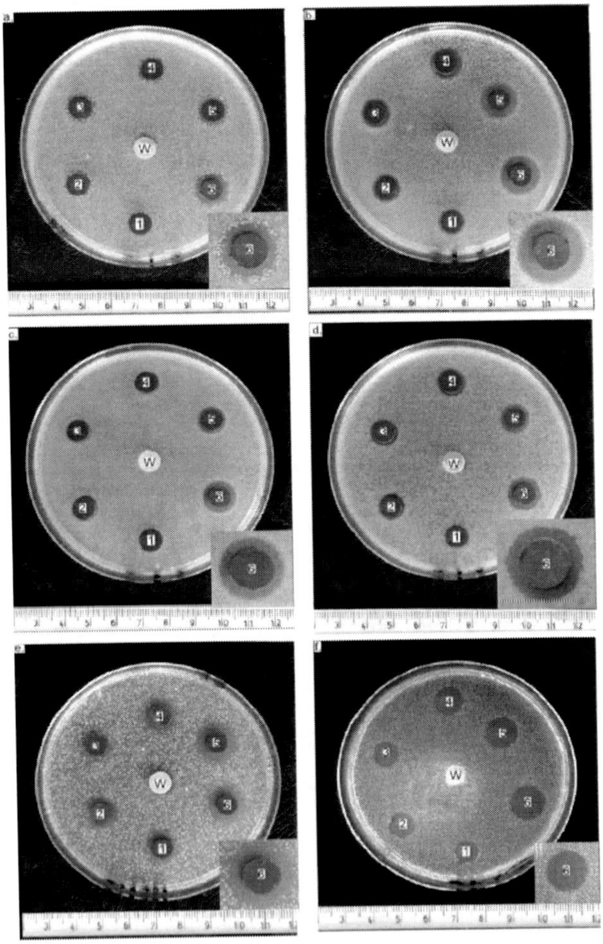

Figure 38. Inhibition zones for *Escherichia coli* ATCC 10536 (a), *Staphylococcus aureus* ATCC 6583 P (b), *Staphylococcus* spp. (gangrenous mastitis) (c), *Bacillus cereus* ATCC 14579,(d), *Candida albicans* ATCC 90028 (e), and after removing the disks for *Staphylococcus aureus* (f). Disks are numbered according to their silver nanoparticles content: disk 1:2%; disk 2: 2.5%; disk 3: 3%; disk 4: 3.6%; disk 5: 4.5%; disk 6: 5.4%; W stands for the control samples: two control disks were used (Table 10).

In other work, Ion N. Mihailescu and et al. composites were synthesized from Bovine hydroxyapatite (BHA) and BHA blended with clinoptilolite (CLIN) and alumina (Al_2O_3) coatings using pulsed laser deposition (PLD) with a KrF*excimer laser source (λ = 248 nm, $\tau_{FWHM} \leq$ 25 ns)[132]. Physical–chemical characteristics and the potential use of coatings for preventing bacteria adhesion and biofilm formation were investigated. SEM images are presented in Figure 39. Intimately aggregated round-shaped micrometric particulates of ~0.2−6.3 are visible. Al_2O_3 coatings (Figure 39 a) exhibit a rather rough and porous morphology, with quasi-equiaxial grains (~0.35−0.45 μm, Figure 39 f). They agglomerate in clusters, unevenly distributed on the surface. In the case of CLIN coatings, the aggregation of round-shaped particulates (~0.35−6.3 μm in size) is less extensive, so that one can easily discern their shape and morphology (Figure 39 b). At higher magnification, a smooth morphology was revealed for CLIN particulates (Figure 39 g). In the case of BHA based coatings, both isolated particulates and matrix they form by aggregation (Figure 39 c−e) are composed of closely packed nanosized grains (Figure 39 h−j). The nanograin size was seemingly reduced when Al_2O_3 and CLIN additives are introduced (Figure 39 h vs Figure 39 i, j). One may therefore suggest that a decrease in crystallinity occurs for the composite coatings.

The XRD and grazing incidence X-ray diffraction (GIXRD) patterns of targets and coatings are presented in Figure 40. Al_2O_3 coating preserved the crystalline features of the target. Both of them consisted of a well crystallized α-alumina phase (ICDD: 00-010-0173), with randomly oriented crystallites (Figure 40 a, b). The zeolite target contains, as the major component, a clinoptilolite-type compound (Figure 40c). The prominent diffraction maxima of the CLIN target are well matched by a $(Ca, K, Na)_5Al_6Si_{30}O_{72}\cdot 18\ H_2O$-like phase (ICDD: 00-047-1870). The minor line shifts and modification of the intensity ratios advocate for slight stoichiometric deviations from this mineral form. A few secondary phases were noticed, which are typical impurities in zeolite deposits. They have either volcanic (SiO_2 tridymite (ICDD: 01-071-0261) or calcium mica (ICDD: 00-046-0744)) or metamorphic origin (muscovite $KAl_3Si_3O_{11}\cdot H_2O$ (ICDD: 00-002-0058)). PLD CLIN films were amorphous even after the thermal treatment (Figure 40d), as indicated by the broad hump centered at $2\theta \approx 22°$, characteristic to a glassy SiO_2-rich compound. Both the BHA target and coating contain a highly crystallized monophasic HA-like hexagonal phase (ICDD: 00-009-0432) (Figure 40e, f). The Al_2O_3 and CLIN increased concentration in PLD targets was supported by the intensity augmentation of their corresponding diffraction maxima (Figure 40e).

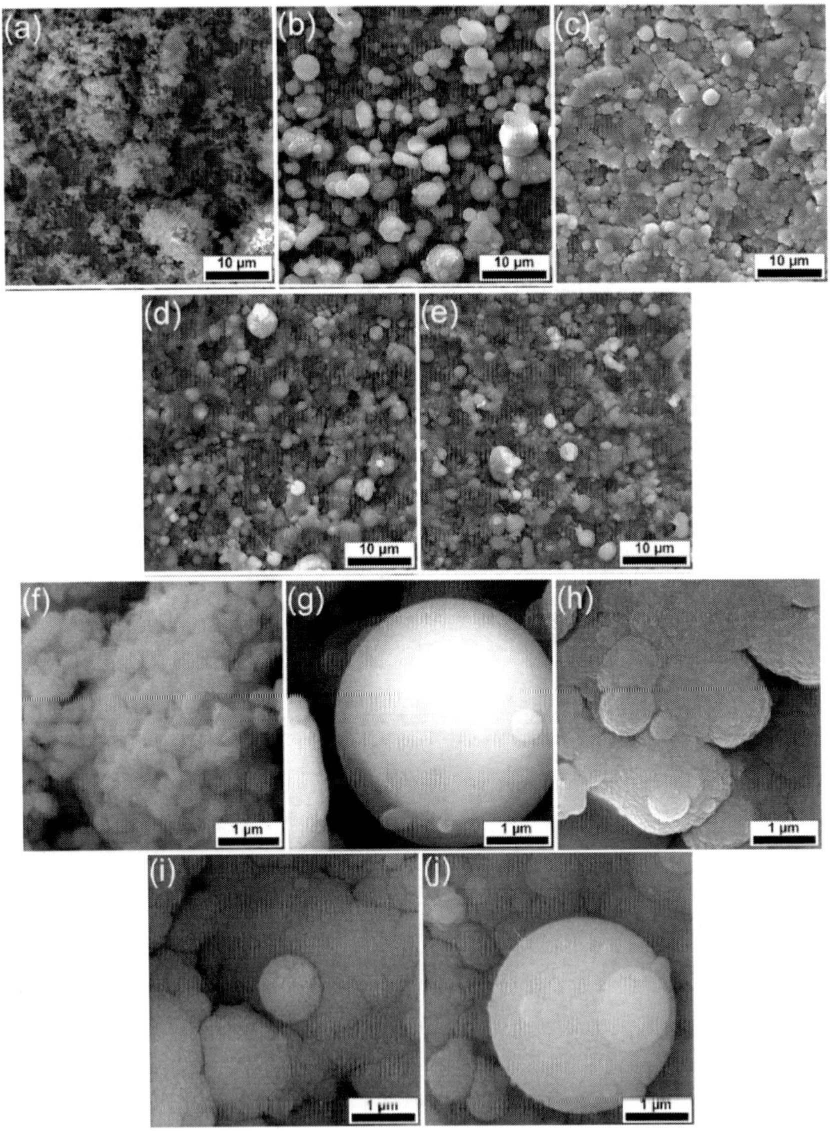

Figure 39. (a-e) Top-view general and (f–j) detailed SEM micrographs of PLD coatings deposited onto Ti substrates: (a, f) Al2O3; (b, g) CLIN;(c, h) BHA; (d, i) BHA-Al2O3-CLIN-5; and (e, j) BHA-Al2O3-CLIN-15. Note: with orange arrows are indicated some CLIN particulates.

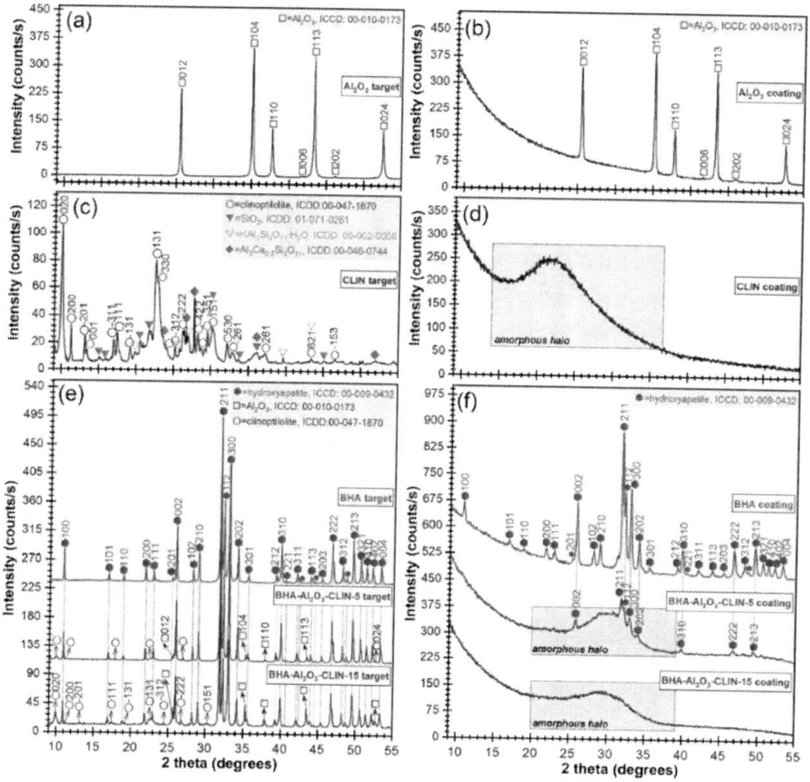

Figure 40. XRD patterns in (a, c, e) symmetric and (b, d, f) grazing incidence geometries, for (a, c, e) targets and (b, d, f) PLD coatings deposited onto Si substrates, respectively.

The antibacterial activity of PLD coatings was assessed from two points of view: (i) resistance of surface to initial bacterial adhesion (via the quantification of viable cells at T0 and after 6 h of contact) and (ii) persistence of the antibacterial effect due to the release of antimicrobial substances from coatings (by the quantification of viable cells at 24, 48, and 72 h).Tested Gram-positive and Gram-negative bacterial strains are among the most common etiological agents of microbial bio film-associated infections. Gram-positive S. aureus is frequently involved in orthopedic infections, [133-134] while E. faecalis is associated with dental implants and urinary catheter infections [135]. Gram-negative E. coli is one of the most prominent causes of orthopedic implant-associated infections, while P. aeruginosa is causing a wide range of medical device- and tissue biofilm-associated infections [136].

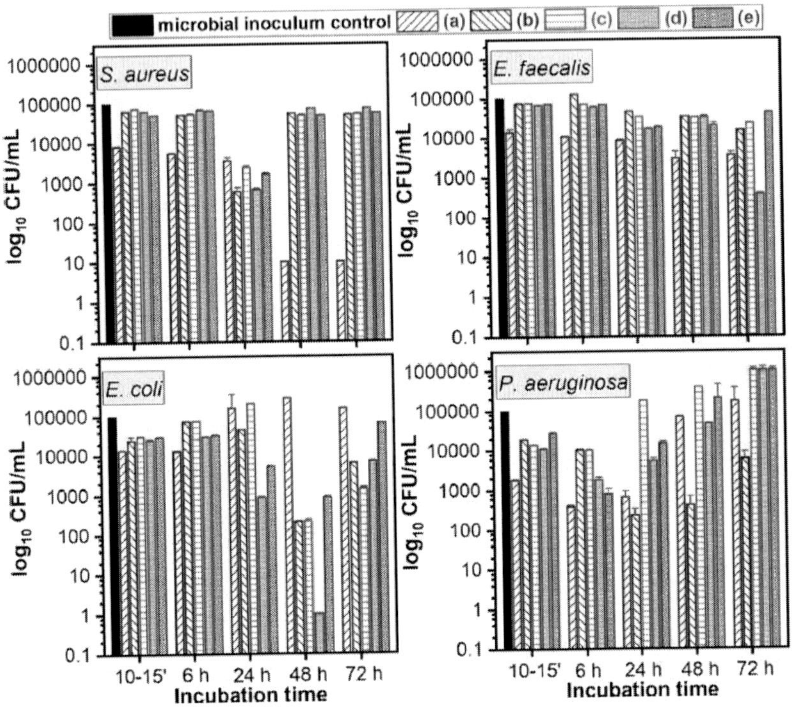

Figure 41. Logarithmic values of CFU/mL of the four tested bacterial strains, grown for different time intervals in the presence of the (a) Al_2O_3, (b) CLIN, (c) BHA, (d) BHA-Al_2O_3-CLIN-5, and (e) BHAAl_2O_3- CLIN-15 coatings deposited by PLD onto Ti substrates.

Logarithmic values of the VCC (CFU/mL) of bacterial cultures grown after different time intervals in the presence of coatings are displayed in Figure 41. Delineation of pure BHA and blended with CLIN and Al_2O_3 coatings, capable to resist against the initial bacterial adhesion and biofilm development, is reported. Films were fabricated using PLD and consisted of closely aggregated/fused round shaped micrometric particulates of ~0.2–6.3 μm. The transfer of counterparts in the composite coatings was confirmed. All samples exhibited excellent biocompatibility features. Al_2O_3 coatings proved to be most active against S. aureus and E. faecalis, BHA-Al_2O_3-CLIN-5 and BHA-Al_2O_3-CLIN-15 samples against E. coli, and all CLIN-containing coatings against P. aeruginosa biofilms. The composite coatings preserved their antibacterial activity over the entire duration of the experiment (up to 72 h), demonstrating the potential of PLD coatings to prevent both the initiation, as well as the maturation, of bacterial biofilms. They can thereby prevent nosocomial infections and inhibit thc burdcn of peri-implant infections. The excellent cytocompatibility correlated with the efficient antibacterial activity suggests that these novel coatings, derived from the proposed sustainable and/or inexpensive materials, could stand for a pioneering contribution to the progress of advanced medical devices.

Conclusion

The nanocomposite of HAP/Inorganic materials were prepared and characterized using various methods such as XRD, FT-IR, VSM, TGA/DTA, BET, SEM, TEM and EDAX. The results showed that with introducing of HAP to inorganic materials, the nanosize crystals of the synthesized HAP became more dispersed onto the external surface of inorganic materials external surface and/or incorporated within the channels and pores of it. It did not cause significant reducing in the crystallinity or collapse in the HAP or inorganic materials structure. The nanocomposite of HAP/inorganic materials showed a good efficiency in F^-, heavy metals and dyes removal from aqueous solution. To find the best condition for removal of pollutant, optimum conditions such as adsorbent dose, pH, contact time, initial concentration of aqueous solution and the effect of temperature on the adsorption ions should be studied. In addition, the consider of equilibrium adsorption data by different isotherm such as Temkin isotherm, Freundlich isotherm, Langmuir and the D-R isotherm model and also kinetic model can be helpful to study of removal of pollutants. Results of antibacterial activity show that HAP/inorganic materials nanocomposites are active against most common Gram-positive and Gram-negative bacteria making it useful for wastewater treatment.

Acknowledgment

Thanks to Dr. S. S. Taghavi and M. Noormohammad Beigi and my students (H. Ghaed Rahmat, N. Gavanmard, M. Yousefi and E. Jafari) which supporting and help me to complete this chapter.

References

[1] Teodosiu C, Wenkert R, Tofan L, Paduraru C (2014) Advances in preconcentration/removal of environmentally relevant heavy metal ions from water and wastewater by sorbents based on polyurethane foam. *Reviews in Chemical Engineering*. 30:403-20. https://doi.org/10.1515/revce-2013-0036.

[2] Aloulou W, Aloulou H, Khemakhem M, Duplay J, Daramola MO, Amar RB (2020) Synthesis and characterization of clay-based ultrafiltration membranes supported on natural zeolite for removal of heavy metals from wastewater. *Environmental Technology & Innovation*. 18:100794. https://doi.org/10.1016/j.eti.2020.100794.

[3] Gopalakannan V, Viswanathan N (2015) Development of nano-hydroxyapatite embedded gelatin biocomposite for effective chromium (VI) removal. *Industrial & Engineering Chemistry Research*. 54:12561-9. https://doi.org/10.1021/acs.iecr.5b01224.

[4] Nagaraj A, Munusamy MA, Ahmed M, Kumar SS, Rajan M (2018) Hydrothermal synthesis of a mineral-substituted hydroxyapatite nanocomposite material for fluoride removal from drinking water. *New Journal of Chemistry*. 42:12711-21. https://doi.org/10.1039/C8NJ02401D.

[5] Gao S, Sun R, Wei Z, Zhao H, Li H, Hu F (2009) Size-dependent defluoridation properties of synthetic hydroxyapatite. *Journal of Fluorine Chemistry*. 130:550-6. https://doi.org/10.1016/j.jfluchem.2009.03.007.

[6] Sani T, Gómez-Hortigüela L, Pérez-Pariente J, Chebude Y, Díaz I (2016) Defluoridation performance of nano-hydroxyapatite/stilbite composite compared with bone char. *Separation and Purification Technology*. 157:241-8. https://doi.org/10.1016/j.seppur.2015.11.014.

[7] Zendehdel M, Shoshtari-Yeganeh B, Cruciani G (2016) Removal of heavy metals and bacteria from aqueous solution by novel hydroxyapatite/zeolite nanocomposite, preparation, and characterization. *Journal of the Iranian Chemical Society*. 13:1915-30. https:// doi.org /10.1007/s13738-016-0908-9.

[8] Zhan Y, Lin J, Li J (2013) Preparation and characterization of surfactant-modified hydroxyapatite/zeolite composite and its adsorption behavior toward humic acid and copper (II). *Environmental science and pollution research* 20:2512-26. https:// doi.org /10.1007/s11356-012-1136-1.

[9] Zendehdel M, Shoshtari-Yeganeh B, Khanmohamadi H, Cruciani G (2017) Removal of fluoride from aqueous solution by adsorption on NaP: HAp nanocomposite using response surface methodology. *Process Safety and Environmental Protection* 109:172-91. https://doi.org/10.1016/j.psep.2017.03.028.

[10] Kim SS, Park MS, Jeon O, Choi CY, Kim BS (2006) Poly (lactide-co-glycolide)/hydroxyapatite composite scaffolds for bone tissue engineering. *Biomaterials* 27:1399-409. https://doi.org/10.1016/j.biomaterials.2005.08.016.

[11] Li M, Xiong P, Yan F, Li S, Ren C, Yin Z, Li A, Li H, Ji X, Zheng Y, Cheng Y (2018) An overview of graphene-based hydroxyapatite composites for orthopedic applications. *Bioactive Materials*. 3:1-8. https://doi.org/10.1016/j.bioactmat.2018.01.001.

[12] Li ZL, Zhang ZY, Zhao TW, Meng CY, Zhang QY, Wang MM (2020) In-situ fabrication of zeolite imidazole framework@hydroxyapatite composite for dispersive solid-phase extraction of benzodiazepines and their determination with high-performance liquid chromatography-VWD detection. *Microchimica Acta.* 187:540-550. https://doi.org/10.1007/s00604-020-04517-y.

[13] Roopalakshmi S, Ravishankar R, Belaldavar S, Prasad RV, Phani AR (2017) Investigation of structural and morphological characteristic of hydroxyapatite synthesized by sol-gel process. *Materials Today: Proceedings.* 4:12026-31. https://doi.org/10.1016/j.matpr.2017.09.126.

[14] Pepla E, Besharat LK, Palaia G, Tenore G, Migliau G (2014) Nano-hydroxyapatite and its applications in preventive, restorative and regenerative dentistry: a review of literature. *Annali di stomatologia* 5:108-114.

[15] Metco O, *Material Product Data Sheet Hydroxyapatite Powders*, 2018, pp. 1–3.

[16] Sadat-Shojai M, Khorasani MT, Dinpanah-Khoshdargi E, Jamshidi A (2013) Synthesis methods for nanosized hydroxyapatite with diverse structures. *Acta biomaterialia* 9:7591-621. https://doi.org/10.1016/j.actbio.2013.04.012.

[17] Szcześ A, Hołysz L, Chibowski E (2017) Synthesis of hydroxyapatite for biomedical applications. *Advances in colloid and interface science* 249:321-30. https://doi.org/ 10.1016/j.cis.2017.04.007.

[18] Pai S, Kini SM, Selvaraj R, Pugazhendhi A (2020) A review on the synthesis of hydroxyapatite, its composites and adsorptive removal of pollutants from wastewater. *Journal of Water Process Engineering* 38:101574. https://doi.org/ 10.1016/j.jwpe.2020.101574.

[19] Boudemagh D, Venturini P, Fleutot S, Cleymand F (2019) Elaboration of hydroxyapatite nanoparticles and chitosan/hydroxyapatite composites: a present status. *Polymer Bulleti*n 76:2621-53. https://doi.org/10.1007/s00289-018-2483-y.

[20] Bigi A, Fini M, Bracci B, Boanini E, Torricelli P, Giavaresi G, Aldini NN, Facchini A, Sbaiz F, Giardino R (2008) The response of bone to nanocrystalline hydroxyapatite-coated Ti13Nb11Zralloy in an animal model. *Biomaterials* 29:1730–1736. https ://doi.org/10.1016/j.bioma terials.2007.12.011.

[21] Khalid M, Mujahid M, Amin S, Rawat RS, Nusair A, Deen GR (2013) Effect of surfactant and heat treatment on morphology, surface area and crystallinity in hydroxyapatite nanocrystals. *Ceram Int* 39:39–50. https://doi.org/10.1016/ j.ceram int.2012.05.090.

[22] Murray MGS, Wang J, Ponton CB, Marquis PM (1995) An improvement in processing of hydroxyapatite ceramics. *J Mater Sci* 30:3061–3074. https://doi.org/ 10.1007/bf012 09218.

[23] Riman RE, Suchanek WL, Byrappa K, Chen CW, Shuk P, Oakes CS (2002) Solution synthesis of hydroxyapatite designer particulates. *Solid State Ion* 151:393–402. https://doi.org/10.1016/s0167 -2738(02)00545-3.

[24] Cho JS, Kang YC (2008) Nano-sized Hydroxyapatite powders prepared by flame spray pyrolysis. *J Alloys Compd* 464:282–287. https://doi.org/10.1016/j.jallc om.2007.09.092.

[25] Brundavanam RK, Poinern GE, Fawcett D (2013) Modelling the crystal structure of a 30 nm sized particle based hydroxyapatite powder synthesised under the

influence of ultrasound irradiation from X-ray powder diffraction data. *American Journal of Materials Science* 3:84-90. DOI: 10.5923/j.materials.20130304.04.

[26] Zhang D, Luo H, Zheng L, Wang K, Li H, Wang Y, Feng H (2012) Utilization of waste phosphogypsum to prepare hydroxyapatite nanoparticles and its application towards removal of fluoride from aqueous solution. *Journal of hazardous materials* 241:418-26. https://doi.org/10.1016/j.jhazmat.2012.09.066.

[27] Banarjee A, Bandyopadhyay A, Bose S (2007) Hydroxyapayite nanopowders: synthesis, densification and cell-material interactions. *Mater. Sci. Eng. C* 27:729-35. https://doi.org/10.1016/j.msec.2006.07.010.

[28] Sadat-Shojai M, Khorasani MT, Dinpanah-Khoshdargi E, Jamshidi A (2013) Synthesis methods for nanosized hydroxyapatite with diverse structures. *Acta biomaterialia.* 9:7591-621. https://doi.org/10.1016/j.actbio.2013.04.012.

[29] Koteswara R, Mallikarjun M (2014) Effective low cost adsorbents for removal of fluoride from water: A review. *International Journal of Science and Research* 3:1-5.

[30] Chen N, Zhang Z, Feng C, Sugiura N, Li M, Chen R (2010) Fluoride removal from water by granular ceramic adsorption. *Journal of colloid and interface science* 348:579-84. https://doi.org/10.1016/j.jcis.2010.04.048.

[31] Mahramanlioglu ME, Kizilcikli I, Bicer IO (2002) Adsorption of fluoride from aqueous solution by acid treated spent bleaching earth. *Journal of Fluorine Chemistry.* 115:41-7. https://doi.org/10.1016/S0022-1139(02)00003-9.

[32] Zhao X, Wang J, Wu F, Wang T, Cai Y, Shi Y, Jiang G (2010) Removal of fluoride from aqueous media by Fe3O4@ Al (OH) 3 magnetic nanoparticles. *Journal of Hazardous Materials.* 173:102-9. https://doi.org/10.1016/j.jhazmat.2009.08.054.

[33] Jafari A, Mahvi AH, Godini H, Rezaee R, Hosseini SS (2014) Process optimization for fluoride removal from water by Moringa oleifera seed extract. *Fluoride.* 47:152-60.

[34] Meenakshi S, Viswanathan N (2007) Identification of selective ion-exchange resin for fluoride sorption. *Journal of colloid and interface science* 308:438-50. https://doi.org/10.1016/j.jcis.2006.12.032.

[35] Ghorai S, Pant KK (2004) Investigations on the column performance of fluoride adsorption by activated alumina in a fixed-bed. *Chemical Engineering Journal* 98:165-73. https://doi.org/10.1016/j.cej.2003.07.003.

[36] Li, Y.H., Wang, S., Zhang, X., Wei, J., Xu, C., Luan, Z., Wu, D., 2003.Sorption of divalent metal ions from aqueous solution byoxidized carbon nanotubes and nanocages: review. *Mater. Res. Bull.* 38, 469–476.

[37] Medellin-Castillo NA, Leyva-Ramos R, Ocampo-Perez R, Garcia de la Cruz RF, Aragon-Pina A, Martinez-Rosales JM, Guerrero-Coronado RM, Fuentes-Rubio L (2007) Adsorption of fluoride from water solution on bone char. *Industrial & Engineering Chemistry Research.* 46:9205-12. https://doi.org/10.1021/ie070023n.

[38] Jun KC, Raman AA, Buthiyappan A (2020) Treatment of oil refinery effluent using bio-adsorbent developed from activated palm kernel shell and zeolite. *RSC advances* 10:24079-94. DOI: 10.1039/D0RA03307C.

[39] Hammari LE, Laghzizil A, Barboux P, Lahlil K, Saoiabi A (2004) Retention of fluoride ions from aqueous solution using porous hydroxyapatite: structure and

conduction properties. *Journal of hazardous materials.* 114:41-4. https://doi.org/10.1016/j.jhazmat.2004.06.032.

[40] Jiménez-Reyes M, Solache-Ríos M (2010) Sorption behavior of fluoride ions from aqueous solutions by hydroxyapatite. *Journal of hazardous materials.* 180:297-302. https://doi.org/10.1016/j.jhazmat.2010.04.030.

[41] Sundaram CS, Viswanathan N, Meenakshi S (2008) Defluoridation chemistry of synthetic hydroxyapatite at nano scale: equilibrium and kinetic studies. *Journal of Hazardous Materials.* 155:206-15. https://doi.org/10.1016/j.jhazmat.2007.11.048.

[42] Feng L, Xu W, Liu T, Liu J (2012) Heat regeneration of hydroxyapatite/attapulgite composite beads for defluoridation of drinking water. *Journal of hazardous materials* 221:228-35. https://doi.org/10.1016/j.jhazmat.2012.04.040.

[43] Pandi K, Viswanathan N (2014) Synthesis of alginate bioencapsulated nano-hydroxyapatite composite for selective fluoride sorption. *Carbohydrate polymers* 112:662-7. https://doi.org/10.1016/j.carbpol.2014.06.029.

[44] Motsi T, Rowson NA, Simmons MJ (2009) Adsorption of heavy metals from acid mine drainage by natural zeolite. *International journal of mineral processing* 92:42-8. https://doi.org/10.1016/j.minpro.2009.02.005.

[45] Sani T, Gómez-Hortigüela L, Pérez-Pariente J, Chebude Y, Díaz I (2016) Defluoridation performance of nano-hydroxyapatite/stilbite composite compared with bone char. *Separation and Purification Technology.* 157:241-8. https://doi.org/10.1016/j.seppur.2015.11.014.

[46] Gómez-Hortigüela L, Perez-Pariente J, Chebude Y, Diaz I (2014) Controlled growth of hydroxyapatite on the surface of natural stilbite from Ethiopia: application in mitigation of fluorosis. *RSC Advances* 4:7998-8003. DOI https://doi.org/10.1039/C3RA46866F.

[47] Gómez-Hortigüela L, Pinar AB, Pérez-Pariente J, Sani T, Chebude Y, Díaz I (2014) Ion-exchange in natural zeolite stilbite and significance in defluoridation ability. *Microporous and Mesoporous Materials* 193:93-102. https://doi.org/10.1016/j.micromeso.2014.03.014.

[48] Zendehdel M, Rezaeian K, Rezaei A, Jalalvandi S (2022) Synthesis and Characterization of a Low-Cost and Eco-friendly Hydroxyapatite/Clinoptilolite/NH_2 Adsorbent for Simultaneous Removal of Cr (VI) and F−. *Silicon* 14:1-7. https://doi.org/10.1007/s12633-021-01649-5.

[49] Mateus AY, Barrias CC, Ribeiro C, Ferraz MP, Monteiro FJ (2008) Comparative study of nanohydroxyapatite microspheres for medical applications. Journal of Biomedical Materials Research Part A: *An Official Journal of The Society for Biomaterials, The Japanese Society for Biomaterials, and The Australian Society for Biomaterials and the Korean Society for Biomaterials* 86:483-93. https://doi.org/ 10.1002/ jbm.a. 31634.

[50] Zhan Y, Lin J, Li J (2013) Preparation and characterization of surfactant-modified hydroxyapatite/zeolite composite and its adsorption behavior toward humic acid and copper (II). *Environmental science and pollution research* 20:2512-26. https://doi.org/10.1007/s11356-012-1136-1.

[51] Zhan Y, Lin J, Li J (2013) Preparation and characterization of surfactant-modified hydroxyapatite/zeolite composite and its adsorption behavior toward humic acid

and copper (II). *Environmental science and pollution research* 20:2512-26. https://doi. org/ 10.1007/s11356-012-1136-1

[52] Caputo D, Liguori B, Colella C (2008) Some advances in understanding the pozzolanic activity of zeolites: The effect of zeolite structure. *Cement and Concrete Composites* 30:455-62. https://doi.org/10.1016/j.cemconcomp.2007.08. 004.

[53] Aloulou W, Aloulou H, Khemakhem M, Duplay J, Daramola MO, Amar RB (2020) Synthesis and characterization of clay-based ultrafiltration membranes supported on natural zeolite for removal of heavy metals from wastewater. *Environmental Technology & Innovation.* 18:100794. https://doi.org/10.1016/j.eti. 2020. 100794.

[54] Elouear Z, Bouzid J, Boujelben N, Feki M, Jamoussi F, Montiel A (2008) Heavy metal removal from aqueous solutions by activated phosphate rock. *Journal of hazardous materials.* 156:412-20. https://doi.org/10.1016/j.jhazmat.2007.12.036.

[55] Nagaraj A, Munusamy MA, Ahmed M, Kumar SS, Rajan M (2018) Hydrothermal synthesis of a mineral-substituted hydroxyapatite nanocomposite material for fluoride removal from drinking water. *New Journal of Chemistry.* 42:12711-21. https:// doi. org/ 10. 1039/C8NJ0 2401D.

[56] Tor A, Danaoglu N, Arslan G, Cengeloglu Y (2009) Removal of fluoride from water by using granular red mud: batch and column studies. *Journal of hazardous materials.* 164:271-8. https:// doi. org/ 10. 1016/j. jhazm at. 2008. 08.011.

[57] Poinern GE, Ghosh MK, Ng YJ, Issa TB, Anand S, Singh P (2011) Defluoridation behavior of nanostructured hydroxyapatite synthesized through an ultrasonic and microwave combined technique. *Journal of Hazardous Materials.* 185:29-37. https://doi.org/10.1016/j.jhazmat.2010.08.087.

[58] Shi J, Li H, Lu H, Zhao X (2015) Use of carboxyl functional magnetite nanoparticles as potential sorbents for the removal of heavy metal ions from aqueous solution. *Journal of Chemical & Engineering Data.* 60:2035-41. https://doi.org/10.1021/jc5011196.

[59] Sundaram CS, Viswanathan N, Meenakshi S (2009) Defluoridation of water using magnesia/chitosan composite. *Journal of Hazardous Materials.* 163:618-24. https://doi.org/10.1016/j.jhazm at.2008.07.009.

[60] Yu X, Tong S, Ge M, Zuo J (2013) Removal of fluoride from drinking water by cellulose@ hydroxyapatite nanocomposites. *Carbohydrate polymers.* 92:269-75. https:// doi. org/ 10. 1016/j. carbp ol. 2012. 09. 045.

[61] Ejhieh AN, Khorsandi M (2010) Photodecolorization of Eriochrome Black T using NiS–P zeolite as a heterogeneous catalyst. *Journal of Hazardous Materials.* 176:629-37. https://doi.org/10.1016/j.jhazmat.2009.11.077.

[62] Huo Z, Xu X, Lü Z, Song J, He M, Li Z, Wang Q, Yan L (2012) Synthesis of zeolite NaP with controllable morphologies. *Microporous and Mesoporous Materials.* 158:137-40. https://doi.org/10.1016/j.micromeso.2012.03.026.

[63] Sani T, Adem M, Fetter G, Bosch P, Diaz I (2016) Defluoridation performance comparison of nano-hydrotalcite/hydroxyapatite composite with calcined hydrotalcite and hydroxyapatite. *Water, Air, & Soil Pollution.* 227:1-8. doi 10.1007/s11270-016-2786-2.

[64] Sato T, Fujita H, Endo T, Shimada M, Tsunashima A (1988) Synthesis of hydrotalcite-like compounds and their physico-chemical properties. *Reactivity of Solids.* 5:219-28. https://doi.org/10.1016/0168-7336(88)80089-5.

[65] Fan X, Parker DJ, Smith MD (2003) Adsorption kinetics of fluoride on low cost materials. *Water Research.* 37:4929-37. https://doi.org/10.1016/j.watres.2003.08.014.

[66] McCann HG (1953) Reactions of fluoride ion with hydroxyapatite. *J. biol. Chem.* 201:247-59.

[67] Margeta K, Zabukovec Logar N, Šiljeg M, Farkas A (2013) Natural zeolites in water treatment–how effective is their use. *Water treatment* 5: 81-112. https://doi.org/10.5772/50378.

[68] Rashed MN (2001) Lead removal from contaminated water using mineral adsorbents. *Environmentalist.* 21:187-95. https://doi.org/10.1023/A:1017931404249.

[69] King P, Rakesh N, Beenalahari S, Kumar YP, Prasad VS (2007) Removal of lead from aqueous solution using Syzygium cumini L.: equilibrium and kinetic studies. *Journal of Hazardous Materials.* 142:340-7. https://doi.org/10.1016/j.jhazmat.2006.08.027.

[70] Naseem R, Tahir SS (2001) Removal of Pb (II) from aqueous/acidic solutions by using bentonite as an adsorbent. *Water research.* 35:3982-6. https://doi.org/10.1016/S0043-1354(01)00130-0.

[71] Ali ME, Hoque ME, Safdar Hossain SK, Biswas MC (2020) Nano adsorbents for wastewater treatment: next generation biotechnological solution. *International Journal of Environmental Science and Technology.* 17:4095-132. https://doi.org/10. 1007/s13762- 020- 02755-4.

[72] Mahmoud ME, Yakout AA, Halbas AM, Osman MM (2016) Remediation of Cr (VI) via combined self-reduction and adsorption bychemically modified carbon sorbents. *Turkish Journal of Chemistry.* 40:906-20. https://doi.org/10.3906/kim-1603-16.

[73] Kurniawan TA, Chan GY, Lo WH, Babel S. Comparisons of low-cost adsorbents for treating wastewaters laden with heavy metals. *Science of the total environment.* 366:409-26.

[74] Sulaymon AH, Ebrahim SE, Mohammed-Ridha MJ (2013) Equilibrium, kinetic, and thermodynamic biosorption of Pb (II), Cr (III), and Cd (II) ions by dead anaerobic biomass from synthetic wastewater. *Environmental Science and Pollution Research.* 20:175-87. doi:10.1007/s11356-012-0854-8.

[75] Lee SM, Laldawngliana C, Tiwari D (2012) Iron oxide nano-particles immobilized and material in the treatment of Cu (II), Cd (II) and Pb (II) contaminated waste waters. *Chem Eng Journal* 195–196:103–111. https://doi.org/10.1016/j.cej.2012.04.075.

[76] Rivera-Utrilla J, Sánchez-Polo M, Gómez-Serrano V, Álvarez PM, Alvim-Ferraz MC, Dias JM (2011) Activated carbon modifications to enhance its water treatment applications. An overview. *Journal of hazardous materials.* 187(1-3):1-23. https://doi.org/10.1016/j.jhazmat.2011.01.033.

[77] Wang S, Yu D (2010) Adsorption of Cd (II), Pb (II), and Ag (I) in aqueous solution on hollow chitosan microspheres. *Journal of applied polymer science.* 118:733-9. https://doi.org/10.1002/app.32496.

[78] Chen YH, Li FA (2010) Kinetic study on removal of copper (II) using goethite and hematite nano-photocatalysts. *Journal of colloid and interface science.* 347:277-81. https://doi.org/10.1016/j.jcis.2010.03.050.

[79] Eloussaief M, Benzina M (2010) Efficiency of natural and acid-activated clays in the removal of Pb (II) from aqueous solutions. *Journal of hazardous materials.* 178:753-7. https://doi.org/10.1016/j.jhazmat.2010.02.004.

[80] Hamidpour M, Kalbasi M, Afyuni M, Shariatmadari H, Holm PE, Hansen HC (2010) Sorption hysteresis of Cd (II) and Pb (II) on natural zeolite and bentonite. *Journal of hazardous materials.* 181:686-91. https://doi.org/10.1016/j.jhazmat.2010.05.067.

[81] Dong L, Zhu Z, Qiu Y, Zhao J (2010) Removal of lead from aqueous solution by hydroxyapatite/magnetite composite adsorbent. *Chemical Engineering Journal.* 165:827-34. https://doi.org/10.1016/j.cej.2010.10.027.

[82] Chang YC, Chang SW, Chen DH (2006) Magnetic chitosan nanoparticles: Studies on chitosan binding and adsorption of Co (II) ions. *Reactive and Functional Polymers.* 66:335-41. https://doi.org/10.1016/j.reactfunctpolym.2005.08.006.

[83] Yuan X, Xing W, Zhuo SP, Han Z, Wang G, Gao X, Yan ZF (2009) Preparation and application of mesoporous Fe/carbon composites as a drug carrier. *Microporous and Mesoporous Materials.* 117:678-84. https://doi.org/10.1016/j.micromeso.2008.07.039.

[84] Yantasee W, Warner CL, Sangvanich T, Addleman RS, Carter TG, Wiacek RJ, Fryxell GE, Timchalk C, Warner MG (2007) Removal of heavy metals from aqueous systems with thiol functionalized superparamagnetic nanoparticles. *Environmental science & technology.* 41:5114-9. https://doi.org/10.1021/es0705238.

[85] Matsuda M, Kaminaga A, Hayakawa K, Takisawa N, Miyajima T (2009) Surfactant binding by humic acids in the presence of divalent metal salts. *Colloids and Surfaces A: Physicochemical and Engineering Aspects.* 347:45-9. https://doi.org/10.1016/j.colsurfa.2008.11.037.

[86] Tipping E (2002) *Cation binding by humic substances.* Cambridge University Press; 30. books.google.com.

[87] Yang H, Masse S, Rouelle M, Aubry E, Li Y, Roux C, Journaux Y, Li L, Coradin T (2015) Magnetically recoverable iron oxide–hydroxyapatite nanocomposites for lead removal. *International journal of environmental science and technology.* 12:1173-82. doi 10.1007/s13762-014-0514-2.

[88] Singh KS, Everett DH, Haul RW, Moscou L, Pierotti RA, Rouquerol J, Siemieniewska T (1985) Reporting physisorption data for gas/solid systems with special reference to the determination ofsurface area and porosity. *Pure Appl Chem* 57:603–619. https://doi.org/10.1351/pac198557040603.

[89] Achelhi K, Masse S, Laurent G, Roux C, Laghzizil A, Saoiabi A, Coradin T (2011) Ultrasound-assisted synthesis of mesoporouszirconia-hydroxyapatite nano-

composites and their dual surface affinity for Cr3?/Cr2O72- ions. *Langmuir* 27:15176–15184. https://doi.org/10.1021/la2029643.

[90] Lower SK, Maurice PA, Traina SJ, Carlson EH (1998) Aqueous Pb sorption by hydroxylapatite: application of atomic force microscopy to dissolution, nucleation, and growth studies. *Am Miner* 83:147–158.

[91] Tripathya SS, Bersillona JL, Gopal K (2006) Adsorption of Cd^{2+} on hydrous manganese dioxide from aqueous solutions. *Desalination,* 194: 11–21. https://doi.org/10.1016/j.desal.2005.10.023.

[92] Kanungo S B, Tripathy S S, Rajeev (2004) Adsorption of Co, Ni, Cu, and Zn on hydrous manganese dioxide from complex electrolyte solutions resembling sea water in major ion content. *Journal of Colloid and Interface Science.* 269: 1–10. https://doi.org/10.1016/S0021-9797(03)00464-8.

[93] Dong L, Zhu Z, Qiu Y, Zhao J. Removal of lead from aqueous solution by hydroxyapatite/manganese dioxide composite. *Frontiers of Environmental Science & Engineering.* 2016 Feb;10(1):28-36. https://doi.org/10.1007/s11783-014-0722-5.

[94] Terdkiatburana T, Wang S, Tadé MO (2008) Competition and complexation of heavy metal ions and humic acid on zeolitic MCM-22 and activated carbon. *Chemical Engineering Journal.* 139:437-44. https://doi.org/10.1016/j.cej.2007.08.005.

[95] Foroughi F, Hassanzadeh-Tabrizi SA, Amighian J, Saffar-Teluri A (2015) A designed magnetic $CoFe_2O_4$–hydroxyapatite core–shell nanocomposite for Zn (II) removal with high efficiency. *Ceramics International.* 41:6844-50. http://dx.doi.org/10.1016/j.ceramint.2015.01.133.

[96] Jian G, Fu Q, Zhou D (2012) Particles size effects of single domain CoFe2O4 on suspensions stability. *Journal of Magnetism and magnetic Materials.* 324:671-6. https://doi.org/10.1016/j.jmmm.2011.08.036.

[97] Chao HP, Chen SH (2012) Adsorption characteristics of both cationic and oxyanionic metal ions on hexadecyltrimethyl ammonium bromide-modified NaY zeolite. *Chemical Engineering Journal.* 193:283-9.

[98] Guan HD, Bestland E, Zhu CY, Zhu HL, Albertsdottir D, Hutson J, Simmons CT, Ginic-Markovic M, Tao X, Ellis AV (2010) Variation in performance of surfactant loading and resulting nitrate removal among four selected natural zeolites. *J Hazard Mater* 183:616–621. https://doi.org/10.1016/j.jhazmat.2010.07.069.

[99] Zhan YH, Zhu ZL, Lin JW, Qiu YL, Zhao JF (2010) Removal of humic acid from aqueous solution by cetylpyridinium bromide modified zeolite. *J Environ Sci* 22:1327–1334.

[100] Wang J, Zheng S, Shao Y, Liu J, Xu Z, Zhu D (2010) Amino-functionalized Fe_3O_4@ SiO_2 core–shell magnetic nanomaterial as a novel adsorbent for aqueous heavy metals removal. *Journal of colloid and interface science.* 349:293-9. https://doi.org/10.1016/j.jcis.2010.05.

[101] Saadat A, Hajiaghababaei L, Badịei A, Ganjali M, Mohammadi- Ziarani G (2019) Amino functionalized silica coated Fe_3O_4 magnetic nanoparticles as a novel adsorbent for removal of Pb^{2+} and Cd^{2+}. *Pollution.* 5:847-57. https://doi.org/ 10.22059/poll.2019.274986.573.

[102] Alam E, Feng Q, Yang H, Fan J, Mumtaz S (2021) Synthesis of magnetic core-shell amino adsorbent by using uniform design and response surface analysis (RSM) and its application for the removal of Cu^{2+}, Zn^{2+}, and Pb^{2+}. *Environmental Science and Pollution Research.* 28:36399-414. https://doi.org/10. 1007/ s11356-020-11840-7.

[103] Zhan Y, Lin J, Li J (2013) Preparation and characterization of surfactant-modified hydroxyapatite/zeolite composite and its adsorption behavior toward humic acid and copper (II). *Environmental science and pollution research.* 20:2512-26. doi 10.1007/s11356-012-1136-1,

[104] Kong F, Zhang Y, Wang H, Tang J, Li Y, Wang S (2020) Removal of Cr (VI) from wastewater by artificial zeolite spheres loaded with nano Fe–Al bimetallic oxide in constructed wetland. *Chemosphere.* 257:127224. https://doi.org/10. 1016/ j.chemosphere.2020.127224.

[105] Wang XH, Liu FF, Lu L, Yang S, Zhao Y, Sun LB, Wang SG (2013) Individual and competitive adsorption of Cr (VI) and phosphate onto synthetic Fe–Al hydroxides. *Colloids and Surfaces A: Physicochemical and Engineering Aspects.* 423:42-9. https://doi.org/10.1016/j.colsurfa.2013.01.026.

[106] Stötzel C, Müller FA, Reinert F, Niederdraenk F, Barralet JE, Gbureck U (2009) Ion adsorption behaviour of hydroxyapatite with different crystallinities. *Colloids and Surfaces B: Biointerfaces.* 74:91-5. https://doi.org/10.1016/j.colsurfb.2009. 06.031.

[107] Sandrine B, Ange N, Didier BA, Eric C, Patrick S (2007) Removal of aqueous lead ions by hydroxyapatites: equilibria and kinetic processes. *Journal of Hazardous Materials.* 139:443-6. https://doi.org/10.1016/j.jhazmat.2006.02.039.

[108] Ohe T, Watanabe T, Wakabayashi K (2004) Mutagens in surface waters: a review. *Mutation Research/Reviews in Mutation Research.* 567:109-49. https://doi.org/ 10.1016/j.mrrev.2004.08.003.

[109] Katheresan V, Kansedo J, Lau SY (2018) Efficiency of various recent wastewater dye removal methods: A review. *Journal of environmental chemical engineering.* 6:4676-97. https://doi.org/10.1016/j.jece.2018.06.060.

[110] Natarajan S, Bajaj HC, Tayade RJ (2018) Recent advances based on the synergetic effect of adsorption for removal of dyes from waste water using photocatalytic process. *Journal of Environmental Sciences.* Mar 65:201-22. doi:https://doi.org/ 10.1016/j.jes.2017.03.011.

[111] Mudhoo A, Beekaroo D (2011) Adsorption of reactive red 158 dye by chemically treated Cocos nucifera L. shell powder. In *Adsorption of Reactive Red 158 Dye by Chemically Treated Cocos Nucifera L. Shell Powder.* 9:1-65. doi:https://doi.org/ 10.1007/978-94-007-1986-6_1.

[112] Daneshvar N, Aleboyeh A, Khataee AR (2005) The evaluation of electrical energy per order (EEo) for photooxidative decolorization of four textile dye solutions by the kinetic model. *Chemosphere.* 59:761-7. https://doi.org/10.1016/j.chemosphere. 2004.11.012.

[113] Yaseen DA, Scholz M (2019) Textile dye wastewater characteristics and constituents of synthetic effluents: a critical review. *International journal of*

environmental science and technology. 16:1193-226. doi:10.1007/s13762-018-2130-z.
[114] Beheshti F, MA Tehrani R, Khadir A (2019) Sulfamethoxazole removal by photocatalytic degradation utilizing TiO_2 and WO_3 nanoparticles as catalysts: analysis of various operational parameters. *International Journal of Environmental Science and Technology*. 16:7987-96. doi:10.1007/s13762-019-02212-x.
[115] Piri F, Mollahosseini A, Hosseini MM (2019) Enhanced adsorption of dyes on microwave-assisted synthesized magnetic zeolite-hydroxyapatite nanocomposite. *Journal of Environmental Chemical Engineering*. 7:103338. https://doi.org/10.1016/j.jece.2019.103338.
[116] Iqbal N, Kadir MA, Iqbal S, Abd Razak SI, Rafique MS, Bakhsheshi-Rad HR, Idris MH, Khattak MA, Raghavendran HR, Abbas AA (2016) Nano-hydroxyapatite reinforced zeolite ZSM composites: A comprehensive study on the structural and in vitro biological properties. *Ceramics International*. 42:7175-82. doi:https://doi.org/10.1016/j.ceramint.2016.01.107.
[117] Gupta N, Kushwaha K, MC Chattopadhyaya A (2011) Adsorption of cobalt (II) from aqueous solution onto hydroxyapatite/zeolite composite. *Advanced Materials Letters*. 2:309-12. doi:10.5185/amlett.indias.201.
[118] Giacometti J, Josić D (2013) Protein and peptide separations. In *Liquid chromatography* Jan 1. 149-184). Elsevier. doi: https://doi.org/10.1016/B978-0-12-415806-1.00007-3.
[119] Piri F, Mollahosseini A, Hosseini MM (2019) Enhanced adsorption of dyes on microwave-assisted synthesized magnetic zeolite-hydroxyapatite nanocomposite. *Journal of Environmental Chemical Engineering*. 7:103338. doi: https://doi.org/10.1016/S0927-5215(05)80007-4.
[120] Ausavasukhi A, Kampoosaen C, Kengnok O (2016) Adsorption characteristics of Congo red on carbonized leonardite. *Journal of Cleaner Production*. 134:506-14. doi:https://doi.org/10.1016/j.jclepro.2015.10.034.
[121] Yokwana K, Kuvarega AT, Mhlanga SD, Nxumalo EN (2018) Mechanistic aspects for the removal of Congo red dye from aqueous media through adsorption over N-doped graphene oxide nanoadsorbents prepared from graphite flakes and powders. *Physics and Chemistry of the Earth*, Parts A/B/C. 107:58-70. doi:https://doi.org/10.1016/j.pce.2018.08.001.
[122] Bagambisa FB, Joos U (1990) Preliminary studies on the phenomenological behavior of osteoblasts cultured on hydroxyapatite ceramics. *Biomaterials*. 11:50-6. https://doi.org/10.1016/0142-9612(90)90052-R.
[123] Wang M (2003) Developing bioactive composite materials for tissue replacement. *Biomaterials*. 24:2133-51. https://doi.org/10.1016/S0142-9612(03)00037-1.
[124] Chen QZ, Wong CT, Lu WW, Cheung KM, Leong JC, Luk KD (2004) Strengthening mechanisms of bone bonding to crystalline hydroxyapatite in vivo. *Biomaterials*. 25:4243-54. https://doi.org/10.1016/j.biomaterials.2003.11.017.
[125] LeGeros RZ (2008) Calcium phosphate-based osteoinductive materials. *Chemical reviews*. 108:4742-53. https://doi.org/10.1021/cr800427g.
[126] Posner AS (1985) The structure of bone apatite surfaces. *Journal of biomedical materials research*. 19:241-50. https://doi.org/10.1002/jbm.820190307.

[127] Nath S, Basu B, Sinha A (2006) A comparative study of conventional sintering with microwave sintering of hydroxyapatite synthesized by chemical route. *Trends in Biomaterials & Artificial Organs.* 19:93-8. https://www.researchgate.net/publication/45626318.

[128] Nayak AK, Bhattacharya A, Sen KK (2010) Hydroxyapatite-antibiotic implantable minipellets for bacterial bone infections using precipitation technique: preparation, characterization and in-vitro antibiotic release studies. *J. Pharm. Res.* 3:53-9.

[129] Panacek A, Kvitek L, Prucek R, Kolar M, Veccerova R, Pizurova N, Sharma VK, Nevecna TJ, Zbořil R (2006) Silver colloid nanoparticles: synthesis, characterization, and their antibacterial activity. *The Journal of Physical Chemistry B.* 110:16248-53. https://doi.org/10.1021/jp063826h.

[130] Mocanu A, Furtos G, Rapuntean S, Horovitz O, Flore C, Garbo C, Danisteanu A, Rapuntean G, Prejmerean C, Tomoaia-Cotisel M (2014) Synthesis; characterization and antimicrobial effects of composites based on multi-substituted hydroxyapatite and silver nanoparticles. *Applied surface science.* 298:225-35. https://doi.org/10.1016/j.apsusc.2014.01.166.

[131] Popescu-Pelin G, Ristoscu C, Duta L, Stan GE, Pasuk I, Tite T, Stan MS, Bleotu C, Popa M, Chifiriuc MC, Oktar FN (2020) Antimicrobial and cytocompatible bovine hydroxyapatite-alumina-zeolite composite coatings synthesized by pulsed laser deposition from low-cost sustainable natural resources. *ACS Sustainable Chemistry & Engineering.* 8:4026-36, https://dx.doi.org/10.1021/acssuschemeng.9b05031.

[132] Arciola CR, Campoccia D, Montanaro L (2018) Implant infections: adhesion, biofilm formation and immune evasion. *Nature reviews microbiology.* 16:397-409. https://doi.org/10.1038/s41579-018-0019-y.

[133] Oliveira WF, Silva PM, Silva RC, Silva GM, Machado G, Coelho LC, Correia MT (2018) Staphylococcus aureus and Staphylococcus epidermidis infections on implants. *Journal of Hospital Infection.* 98:111-7. https://doi.org/10.1016/j.jhin.2017.11.008.

[134] Flanagan D (2017) Enterococcus faecalis and dental implants. *Journal of Oral Implantology.* 43:8-11. https://doi.org/10.1563/aaid-joi-D-16-00069.

[135] Cremet L, Broquet A, Brulin B, Jacqueline C, Dauvergne S, Brion R, Asehnoune K, Corvec S, Heymann D, Caroff N (2015) Pathogenic potential of Escherichia coli clinical strains from orthopedic implant infections towards human osteoblastic cells. *Pathogens and disease.* 73: 1-10. https://doi.org/10.1093/femspd/ftv065.

[136] Jamal M, Ahmad W, Andleeb S, Jalil F, Imran M, Nawaz MA, Hussain T, Ali M, Rafiq M, Kamil MA (2018) Bacterial biofilm and associated infections. *Journal of the Chinese medical association.* 81:7-11. https://doi.org/10.1016/j.jcma.2017.07.012.

Chapter 3

Hydroxyapatite: Synthesis and Characterization

Safia Alleg[*] and Meriem Harkat
Laboratory of Magnetism and Spectroscopy of Solids (LM2S), Department of Physics, Badji Mokhtar Annaba University, Annaba, Algeria

Abstract

Nanocrystalline hydroxyapatite (HAp) has been successfully used for biomedical applications during the last decades owing to its chemical similarity with natural apatite, its remarkable biocompatibility, bio-affinity, and bioactivity that can be related to its crystal structure. HAp is the most stable, least soluble of all calcium orthophosphates, and is one of the most promising calcium phosphate biomaterials which can be obtained from chemical reagent, natural sources, animal-like dental enamel, corals, etc. Thermally stable HAp powders have been prepared using wet chemical route and co-precipitation method from various precursors such as naturally abundant seashell waste, calcium hydroxide $Ca(OH)_6$ and mono ammonium phosphate $NH_4H_2PO_4$ as calcium and phosphate sources, respectively. The crystal structure, morphology, elemental analysis, molecular bonding, and electric properties of the synthesized HAp powders have been studied using X-ray diffraction (XRD), scanning electron microscopy (SEM), energy-dispersive X-ray spectroscopy (EDS), Raman spectroscopy, Fourier transform infrared spectroscopy (FT-IR), and complex impedance spectroscopy. The effect of simulated body fluids immersion on the molecular bending and electric properties is also reported.

Keywords: hydroxyapatite, structure, microstructure, morphology, electric properties

[*] Corresponding Author's Email: safia.alleg@univ-annaba.dz.

Introduction

Hydroxyapatite (HAp) is the least soluble, and most stable of all calcium orthophosphates at normal temperatures and pH ranging between 4 and 12. It is an important compound for catalysis, industry of fertilizers and pharmaceutical products, protein chromatography applications, water treatment processes, etc. The HAp has been fruitfully employed for biomedical applications owing to its excellent properties such as the biocompatibility, bio-affinity, bioactivity, and high stability that can be attributed mainly to the nature of its crystallographic structure. Known as $Ca_{10}(PO_4)_6(OH)_2$, the HAp structure consists of a hexagonal network of PO_4^{3-} tetrahedral where Ca^{2+} ions are in the interstitial sites and OH^- ions columns oriented along the c-axis (Tofail et al. 2005, 117; Kay et al. 1964, 1050). The deficience of the HAp or its deviation from the stoichiometry can be related to the existence of vacancies and ion substitutes such as potassium, sodium, carbonates, nitrates, hydrogen phosphates and chloride, which are generally introduced into the precipitating system with the reactants. Those defects lead to significant variations in their crystallographic structure and morphology as compared to stoichiometric HAp. Additionally, HAp is one of the most promising calcium phosphate biomaterials which can be obtained from natural sources (Kumar et al. 2017, 3457), chemical reagent (Sunila et al. 2016, 411), animal-like dental enamel (Raya et al. 2015, 1), and corals (Rujitanapanich et al. 2014, 112).

In addition to the mechanical and biological properties, HAp has also received particular attention due to its electrical properties that are linked to the alignment of the columnar hydroxyl groups along the crystallographic c-axis (Piccirillo et al. 2015, 309; Deb et al. 2019, 7479). Two conduction mechanisms have been proposed in HAp (Horiuchia et al. 2018, 19): (i) proton conduction among neighboring OH^- ions according to the following equation:

$$OH^- + OH^- \leftrightarrow O^{2-} + HOH \qquad (1)$$

And (ii) proton jump based on the OH^- ions networking with the double bonded oxygen on PO_4^{3-} groups such as:

$$OH^- + PO_4^{3-} \leftrightarrow O^{2-} + HPO_4^{2-} \leftrightarrow O^{2-} + PO_4^{3-} + HOH \qquad (2)$$

Accordingly, protons play an important role since they must be delocalized between O_2 anions in channels or among the PO_4^{3-} groups. Different conducting species such as oxide ions (O^{2-}), protons (H^+), and lattice hydroxyl ions (OH^-) have been proposed to describe the conduction mechanism in HAp according to the activation energy values. The obtained activation energy values were about 2–2.1 eV for OH^- ion jumps along the c-axis (Badran et al. 2017, 85), 1.5 eV for O^{2-} (Royce 1974, 131), and 0.5 eV for H^+ conduction (Laghzizil et al. 2001, 953).

Low-cost HAp powders from naturally abundant seashell waste were prepared using the co-precipitation and hydrothermal methods. Crystal structure, morphology, and elemental analysis were investigated by X-ray diffraction (XRD), Raman spectroscopy, Fourier Transform Infra-Red (FT-IR) spectroscopy, scanning electron microscopy (SEM), and energy-dispersive X-ray spectroscopy (EDX), respectively. Alternating current conductivity and impedance measurements were studied using complex impedance spectroscopy. The response of the as-prepared HAp to immersion in a simulated body fluids (SBF) solution at 37°C for 5, 10 and 15 days is studied using Raman spectroscopy and complex impedance spectroscopy.

Crystal Structure

The HAp crystallises generally in the hexagonal structure with crystallographic space group $P6_3/m$ and lattice parameters $a = 0.942$ nm and $c = 0.688$ nm. A monoclinic pseudohexagonal structure, space group $P2_1/b$, $a = 0.9421$ nm, $b = 2a$, $c = 0.6881$ nm, and $\gamma = 120°$ has been reported for samples with nearly stoichiometric chlorine contents (less than 5 to 15% deficient). Their distorsions from the hexagonal structure are similar to those in chlorapatite including enlargement of that triangular array of oxygen atoms in which the hydroxyl hydrogen ion is approximately centered (Elliot et al., 1973).

The stoichiometric Hap is defined by a Ca/P molar ratio of about 1.67 and a relative density of 3.156. The cell unit of HAp contains 14 calcium ions, where 6 calcium ions are located within the unit cell, and 8 calcium ions are shared by adjacent unit cells. Among the $10PO_4^{3-}$ groups of a unit cell, two remain inside and eight at the periphery. All the eight OH^- groups remain at the edge of unit cell, shared by 4 unit cells each. Calcium atoms are positioned in two sites named Ca1 and Ca2 linked to each other by the tetrahedral arrangement which consists of phosphorous atoms related to four atoms of oxygen (O1, O2, O3 (2)) distanced by 1.534 Å. Ca1 is surrounded by six oxygen atoms (2 O1, 2 O2 and 2 O3) connected to phosphate groups, with mean distances of 2.409 Å, and related to three other atoms of oxygen (O3 (3)) distanced by 2.821 Å. Ca2 is bounded by six oxygen atoms (O1, O2, O3 (4)) and one atom of oxygen related to OH group; the mean distances between Ca2, O atoms and OH group are 2.461 Å and 2.720 Å, respectively (Figure 1).

The PO_4 tetrahedron are generated form the tetrahedral and octahedral vacancies. Then, tricalcium phosphate is formed *via* the occupation of some vacancies by Ca ions, and when vacancies are filled up eight calcium phosphates are formed. Finally the crystal HAp is formed by the enterance of hydroxyls into the structure.

Calcium Atoms Position

They are distributed between the two crystallographic sites Ca1 and Ca2. The Ca1 site is occupied by four Ca1 atoms, two of which are at the $z = 0$ position and the others at $z = 1/2$. They form hence, the columns parallel to c, respectively, at $x = 1/3$, $y = 2/3$ and $x = 2/3$, $y = 1/3$. The Ca2 site is occupied by six Ca2 atoms arranged in equilateral triangles and alternated at sides 1/4 and 3/4.

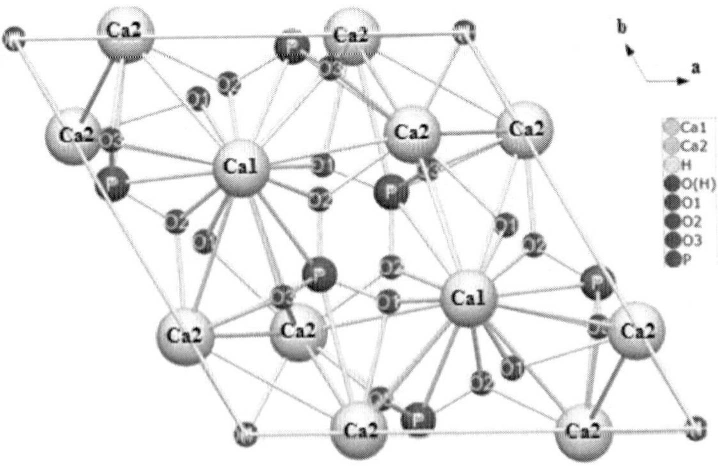

Figure 1. Representation of the HAp crystal structure showing the Ca1 and Ca2 environments.

Hydroxyl Ions Position

They are arranged in a column on the axis parallel to *c* in $z = 1/4$ and 3/4.

Phosphate Ion Position

The quasicompact arrangement of these phosphate groups forms the tetrahedral skeleton of this structure at the level $z = 1/4$ and 3/4. Tunnels play an important role in the physico-chemical properties of apatites because their presence allows apatites to behave either as ion exchangers or as compounds in which different ions can be substituted.

Synthesis Methods

HAp plays an important role in medical bone repair and replacement materials. However, the HAp shows low strength, low toughness, low brittleness, and low fatigue resistance in physiological environment. Consequently, it is difficult to meet the biochemical and physical conditions implanted in the human body. So, in order to improve the HAp properties, many studies have been conducted over the last decades. The preparation methods are continuously update and developed to achiev uniform composition, narrow particle size distribution, large specific surface area, superior performance and many other conditions (Guoqing 2019, 033057). Several methods have been used to prepare HAp such as the sol gel, wet chemistry technique, hydrothermal method, mechanochemical (solid state reaction), precipitation method, solvothermal method, spontaneous combustion method, micro-emulsion method, ultrasonic synthesis method, bionic method, and high-temperature reactions (Santhosh and Prabu 2013, 121; Javadinejad et al. 2017, 4401; Türk et al. 2019, 311; Laonapakul et al. 2021, 111333; Rujitanapanich et al. 2014, 112; Okada and Furuzono 2012, 1; Ramesha et al. 2015, 10434; Kramer and Podurgiel 2014, 145). Among the various methods, hydrothermal method, solvent thermal method and precipitation method are widely used due to their low cost, and best HAp comprehensive properties. Besides, the HAp nanoparticles can be prepared from animal, plant, and aquatic sources precursor, solvent, and surfactant materials to adjust their shape, size, stoichiometry, crystallinity, geometry, and particles agglomeration degree.

Among the various biogenic sources, seashells are composed principally of $CaCO_3$ (95–99%) in addition to some organic matter that displays excellent nanoscale regularity and strength. The seashells are present in nature in great amounts with various portions of proximate and mineral compositions from types to types, being dependent on the climatic conditions and habitation nature. The calcium phosphate can be obtained by the thermal dissociation of $CaCO_3$ into calcium oxide (CaO), and its further treatment with phosphoric acid (Deb et al. 2019, 7479; Laonapakul et al. 2021, 111333).

Hydrothermal Method

Hydrothermal preparation of HAp has become increasingly mature during the last decades (Kumar et al. 2018, 11257). It uses high temperature and high pressure aqueous solution as reaction medium, in high temperature and high pressure airtight pressure vessel (autoclave) by controlling the temperature of the solution within the autoclave to produce convection in supersaturated state formation. This method is simple with relatively low temperature, and moderate reaction conditions where the calcium precursors are $Ca(NO_3)_2$ and $Ca(OH)_2$, and unusual $CaCO_3$ or a mixture of phosphate and calcium precursors such as $Ca(H_2PO_4)_2$ or $CaHPO_4$. The inconvenient of this method is the longer synthesis time (many days) required for the obtention of pure HAp. Indeed, the obtained products have high crystallinity, small particle size, high purity, and controllable shape and size. It has been reported that the shape and size of hydroxyapatite can be regulated by changing the temperature, time and reaction content. According to the synthesis factors, high-purity HAp having different shapes, such as nano spherical, flake, small plate, hexagonal prism, rod-like, needle-like, and spine-like can be obtained (Kumar et al. 2017, 3457; Ines et al. 2009, 466; Zhang et al. 2009, 2725).

Co-Precipitation Method

In this method, an aqueous solution can be added by drop wise or directly to another aqueous solution and the mixture is then subjected to hot magnetic agitation in a temperature range of 25 to 100°C during 5 to 24 h. The pH of the solution is kept at about 8–11. Two types of precursors are commonly used: (i) $Ca(OH)_2$ and H_3PO_4 for a direct neutralization, but the produced

HAps are generally carbonated; and (ii) $Ca(NO_3)_2$ and $(NH_4)_2HPO_4$ or $(NH_4)H_2PO_4$ for a double decomposition by continuous precipitation. The nitrate and ammonium ions traces that can be found might be eliminated by heat treatment. Many parameters of the co-precipitation method such as the pH, synthesis temperature, raw materials, agitation rate, etc. can be controlled with high precision. It is possible to modify the parameters individually and to study their impact on the morphology, stoichiometry, catalytic activity, etc. of the final product (Harkat et al. 2021, 45).

Sol-Gel Method

The sol-gel process is a versatile wet chemical method used to synthesize bioceramic materials at relatively low temperatures, based on wet chemistry processing. This later involves the preparation of a sol, the gelation of the sol, and the removal of the liquid existing in fine interconnected channels within the gel. The sol-gel technology allows the preparation of HAp in the form of ultra-fine or spherical shaped powders (Türk et al. 2019, 311). Compared with the conventional ceramic routes, such as the co-precipitation, impregnation, the sol-gel exhibits many advantages, among them the low process temperature, high control of purity, the ability to control the composition on molecular scale, and the porosity to obtain high surface area materials, the homogeneity of the final product up to atomic scale, microstructure, and textural properties of the final material. The sol is obtained from the reaction of calcium citrate or calcium acetate with phosphoric acid and is aged into gel under certain conditions. The gel is dried at low temperature and calcined at high temperature under vacuum to obtain nanocrystalline HAp (Hench 1997, 604). The morphology, crystallization degree, purity, particle size, agglomeration, high surface activity, and performance of the obtained HAp powder can be affected by several parameters such as the initial concentration of the solution, aging time, calcination temperature, reaction atmosphere, hydrolyse time, etc. (Meng et al. 2002, 136)

Solid State Reaction Method

It is based on the decomposition of a mixed solid reactant at high temperature to produce new solids and gases (Piccirillo et al. 2015, 309). The solid-state reaction is a simple method in which the chemical precursors that contain calcium and phosphate are milled and calcined at an appropriate temperature to obtain the HAp. The mechanism of this method occurs by the solid diffusion of the ion from chemical precursors (calcium and phosphate) followed by a high-temperature process to initiate the reaction. By using the solid-state reaction, the calcium (CaO) and phosphate (P_2O_5) precursors are chemically combined to form the hexagonal structure of HAp. Due to the slight diffusion of ions during the solid-state reaction, the mechanochemical technique, which can improve the kinetics performance, can be used as an alternative to this process.

Mechanochemical Method

This technique is based on the solid state reaction using the milling process of calcium hydroxide ($Ca(OH)_2$), phosphorous pentoxide (P_2O_5), calcium fluoride (CaF_2), calcium pyrophosphate ($Ca_2P_2O_7$) and calcium carbonate. The milling process uses compression, shear, or friction *via* grinding and milling to induce a chemical transformation at room temperature (Javadinejad et al. 2017, 4401). The high ball-powder-ball and/or ball-powder-vials impact compression and the increase in local temperature inside the vials, during the milling process, lead to the chemical reaction and enhance the diffusion process. The mechanochemical process is usually conducted under an argon atmosphere to overcome the limitation in controlling the air and moisture-sensitive substance. The mechanochemical technique produces a well-defined structure compared to the solid-state method that produces a heterogeneous particle with an irregular shape. Several parameters might influence the end product in the mechanochemical method such as the rotation speed, milling time and ball-to-powder weight ratio.

Chemical Precipitation Method

It is one of the largest methods used for preparing HAp due to the large amount of HAp that can be synthesised at a reasonable cost (Harkat et al. 2021, 43). The chemical precipitation method undergoes several important processing parameters in order to synthesise the HAp. Initially, calcium and phosphate containing reagents are mixed, such as calcium hydroxide or calcium nitrate as the Ca^{2+} source and orthophosphoric acid or diammonium hydrogen phosphate as the PO_4^{3-} source, according to the molar ratio of the HAp. An unbalanced molar ratio in the chemical precursor could result in the formation of another compound. The mixture is then adjusted to a specific pH, and the temperature ranges from room temperature to the boiling point of water. Ensuing, the solution is stirred for ageing purposes, and finally, the precipitates are washed, filtered, and dried before being crushed into fine powder form. In the wet chemical precipitation process it is possible to vary the reaction temperature, acid addition rate (slow and rapid rates) and heat-treatment temperature to produce HAp. The reaction of $Ca(OH)_2$, and H_3PO_4 precursors can be proceeded by the maturation, decantation, and drying processes. The obtained precipitates exhibit a spherical-like morphology of the HAp.

Properties and Applications of HAp

HAp and tri-calcium phosphate (TCP) and their composites can be used in several biomedical applications such as bone substitution, hard tissue repairs (human bone and tooth), drug carrier, and bone graft owing to their good biocompatibility, bioactivity, porous structure and mechanical strength, osteoconductive and osteointegration properties (Haider et al. 2017, 7442; Fihri et al. 2017, 48). The HAp is an ideal substitute for a bone graft because it is a primary constituent of bone. Consequently, various material ceramics have been utilized to repair bone defects and to cover metal implant (usually stainless steel and titanium alloys) surfaces to change their morphological characteristics, and to enhance their biological activity and integration with the host bone in dentistry and orthopaedics (del Campo et al. 2017, 667).

Many studies have been devoted to the effect of pore size, shape and porosity on ossification, and to the influence of heterogeneous porosity on degradation, mechanical properties and angiogenesis after implantation to stimulate bone healing. Indeed, the pore size is very important for the pattern of scaffold degradation, while porosity, size, orientation, distribution and interconnectivity of the pores are linked to the permeability. Hence, a network of interconnected pores is necessary for bone substitutes to allow cell infiltration, nutrients and oxygen flow, migration, bone distribution, and elimination of discarded materials (Limmahakhun et al. 2017, 93). It has been reported that the pore structure of the bone scaffolds with pore sizes ranging from 200 to 800 μm are appropriate for bone substitute (Abbasi et al. 2020, 1) since larger porosity leads to further permeability and faster degradation. Whereas, smaller porosity may generate a hypoxic state and stimulate chondrogenesis instead of osteogenesis (Lin et al. 2019, 326). Furthermore, the morphology, pore size and pore homogeneity influence the degeneration of porous biomaterials. A greater compressive modulus is associated with smaller pore sizes, a gradient porosity and staggered orientated pores according to the scaffold's mechanical properties (Abbasi et al. 2020, 1). While, a greater pore size is ascribed to speedier degradation rate due to the larger acidic products spreading during degradation. The main benefit of a gradient porosity scaffolds is their capability to keep and recover their elastic properties after deformation, whereas the improvement of stable mechanical strength is linked to square pores.

Experimental Details

HAp Powder Synthesis

Nanocrystalline HAp powders have been prepared from calcium hydroxide $Ca(OH)_2.(H_2O)_6$, as a source of calcium, monoamonium dihydrogen phosphate $NH_4H_2PO_4$ as a source of phosphorous, and ammonia NH_4OH as a reaction promoter according to the following reaction:

$$Ca(OH)_2.(H_2O)_6 + NH_4H_2PO_4 \rightarrow Ca_{10}(PO_4)_6(OH)_2 \quad (3)$$

The as-dried HAp powder was crushed using an agate mortar and heat treated at different temperatures (600, 800 and 1000°C) for 2 h. The as-produced samples are named hereafter HT-600, HT-800, and HT-1000 for 600, 800 and 1000°C, respectively. Preparation details can be found in our previous paper (Harkat et al. 2021, 43).

The second type of HAp powders were synthesized from the collected Strombidae strombus seashells from the eastern region of Algeria (Sanhadja-Guerbes–Azzaba). The resulting $CaCO_3$ calcium carbonate powders were heat treated at 1000°C for 5 h to prepare lime calcium oxide (CaO) as follows:

$$CaCO_3 \rightarrow CaO + CO_2\uparrow \quad (4)$$

The formation of $Ca(OH)_2$ is due to moisture absorption from the atmosphere according to the following reaction:

$$CaO + H_2O \rightarrow Ca(OH)_2 \quad (5)$$

Further details on the preparation method of the HAp powders can be found in ref. (Harkat et al. 2022, 7693).

Characterization Methods

Structure and microstructure have been studied by X-ray diffraction (XRD) on a PanAlytical Emperian diffractometer equipped with a PIXcel1D-Medipix3 (256 channels) detector and using CuKα radiation (λ = 0.154059 nm). The measurements were executed in Bragg Brentano θ–θ geometry with a step size of 0.013°. The XRD patterns were analyzed by the Maud program (Lutterotti et al. 1992, 459) which is based on the Rietveld method (Rietveld 1969). The particle's morphology was observed by Zeiss Evo LS10 Jeol JSM 6500F microscope, and the elemental analysis was determined by energy dispersive X-ray spectrometry (EDX). Functional groups and molecular bonding properties were determined by Fourrier Transfrom Infrared, FT-IR, (Jasco 6300 spectrometer), and Raman (Labram HR 800 spectrometer) spectroscopies in the spectral range of 400–4000 cm^{-1}. An Agilent 4294A analyzer with 20 mV of signal amplitude was used to measure the electrical properties, under vacuum, and in the dark to avoid the contributions from photoconductivity and humidity effects. A thin film was deposited from highly pure silver on both sides of the pellet, by thermal evaporation, through a circular mask to allow a precise electrode's surface calculation. The electrical conductance was measured as a function of frequency in the temperature range of 300–680 K.

Immersion of HAp in SBF Solution

An acellular simulated body fluid, abbreviated often as SBF or Kokubo solution, with inorganic ion concentrations similar to those of human extracellular fluid has been developed by Kokubo et al. (Kokubo et al. 1990, 721; Kokubo T and Takadama 2006, 2907) in order to reproduce the formation of apatite on bioactive materials *in vitro*. The SBF can be used for evaluating the bioactivity of artificial materials *in vitro*.

The SBF solution was prepared according to Kokubo's formula by dissolving NaCl, $NaOHCO_3$, KCl, $K_2HPO_4.3H_2O$, $MgCl_2.6H_2O$, HCl, $CaCl_2.2H_2O$, and Na_2SO_4 into distilled water and 1kmol/m^3 HCl 6N to adjust the pH at 7.4. The SBF solution is colorless, transparent and close to the ion concentration within human blood plasma (Gamble 1967, 1). The ionic concentrations of the human blood plasma and SBF are given in Table 1.

Result and Discussion

Crystal Structure

Figure 2 displays the XRD patterns of the synthesized HAp starting from seashell wastes using the hydrothermal (Figure 2a) and co-precipitation (Figure 2b) methods. The presence of (200), (111), (002), (102), (120), (211), (112), (300), (202), (301), (212), (310), (311), (113), (203), (222), (312), (213), (321), (410), (402), and (004) diffraction peaks match well with the ICSD file of the hexagonal HAp structure for both the synthesized HAp powders. One notes the formation of a single HAp phase in the case of the co-precipitation method, but a mixture of HAp and β-Tricalcium phosphate (β-TCP) phase with chemical formula $Ca_3(PO_4)_2$ (Figure 3) in the case of the prepared HAp by the hydrothermal method. This is evidenced by the appearance of a new diffraction peak at $2\theta = 29.55°$ (Figure 2a) due to the fact that the hydrothermal method uses high temperatures. This result agrees well with those obtained by other researchers (Suparto et al. 2015, 27; Ramesha et al. 2015, 10434; Mansour et al. 2015, 991). Besides, the prepared HAp by hydrothermal method exhibits a good crystallinity as evidenced by the relatively high intensity of its diffraction peaks and their narrowing compared to those of the co-precipitation method.

Table 1. Ion concentration's of the SBF and human blood plasma (mMol/l)

	Na^+	K^+	Ca^{2+}	Mg^{2+}	Cl^-	HCO_3^-	HPO_4^{2-}
Human blood	142	5	2.5	1.5	103	27	1
SBF	142	5	2.5	1.5	147.8	4.2	1

The Rietveld refinement of the XRD patterns was performed by the MAUD program (Maud program) which is based on the Rietveld method

(Rietveld 1969). The diffraction peak profile can be affected by many parameters related to the sample (crystallite size, strain/stress, defects), and instrumental parameters (geometry, radiation source, etc.). The peak profile was well described by an asymmetric pseudo-Voigt function (Sánchez-Bajo and Cumbrera 1997, 427), while the instrumental parameters, comprising three Caglioti parameters for full-width at half-maximum (FWHM), two asymmetry factors and two parameters for Gaussian content of peak profiles were obtained from a standard material. First, the background of the diffraction pattern was modeled as a fifth-order polynomial function. Then, the lattice parameters (*a*, *b*, *c*), phase proportion, crystallite size, root mean square (*r.m.s*) microstrain, atomic positions (*x*, *y*, *z*), Debye–Waller factor (*B*), and site occupancy factor (*SOF*) were refined. The crystallite size and the root mean square microstrain (*r.m.s.*) were evaluated using the isotropic model.

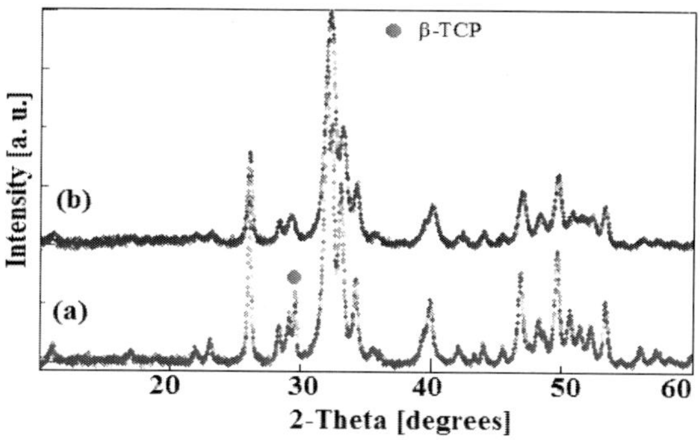

Figure 2. XRD patterns of the synthesized HAp powders from seashell wastes using two different preparation methods: (a) hydrothermal; and (b) co-precipitation.

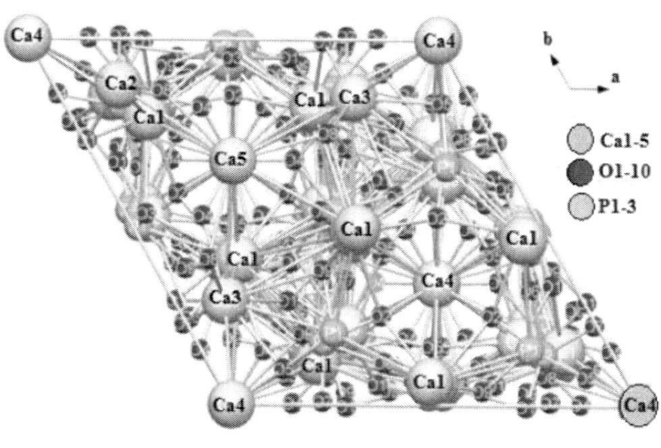

Figure 3. Crystal structure of β-TCP (β-$Ca_3(PO4)_2$) phase.

The satisfactory Rietveld refinement model is achieved through the minimization of the difference between the experimental and simulated diffraction patterns using the reliability parameters such as the weighted residual factor (Rwp), Bragg factor (R_B) and expected residual factor (Rexp). The best fitting model is confirmed by the goodness of fit (*GoF*) which represents an Rwp / Rexp ratio approaching the unity.

The Rietveld refinement of the XRD patterns was achieved with the hexagonal HAp structure, space group *P63/m*, and lattice constants $a = b =$ 9.432 Å, and c = 6.881 Å for the prepared HAp by the co-precipitation method (Figure 4), and a mixture of HAp and trigonal orthophosphate β-TCP (β-$Ca_3(PO4)_2$) phase with crystallographic space group *R3c:H* and lattice parameters $a = b$ = 10.4352 Å, and c = 37.4029 Å for the prepared HAp by the hydrothermal method (Figure 5). The structural and microstructural parameters are summarized in Table 2. The lattice parameters of the produced HAp are different for both the powders. Their relative deviations which are defined as $\Delta a = \frac{a-a_0}{a_0}$ and $\Delta c = \frac{c-c_0}{c}$, where a_0 and c_0 represent the lattice parameters of the ICSD file, a and c the calculated ones. Accordingly, the relative deviations reach as much as Δa = –0.09% and Δc = 0.37% for the synthesized HAp by the co-precipitation method, and Δa = –0.58% and Δc = –0.77% for the produced HAp using the hydrothermal method. The contraction of the crystal lattice is more important in the case of the hydrothermal method. Those differences are related to the conditions preparations. Besides, the higher crystallite size values of the HAp might be related to the higher temperature in the hydrothermal method. The formation of the secondary β-TCP structure might be related to the heterogeneity of the composition.

Table 2. The deduced lattice parameters (a, c), average crystallite size (<*L*>), r.m.s ($<\sigma^2>^{1/2}$), and weight fraction from the Rietveld refinement of the XRD patterns of the HAp

Method	Phases	$a \pm 10^{-4}$ (Å)	Δa (%)	$c \pm 10^{-4}$ (Å)	Δc (%)	<L> ±1 (nm)	r.m.s (%)	Fraction ±1 (%)
Co-precipitation	HAp	9.4233	–0.09	6.9069	0.37	31.0	0.42	100
Hydrothermal	HAp β-TCP	9.3765 10.3962	–0.58 –0.37	6.8275 37.4919	–0.77 0.24	47.3 51.0	0.30 0.11	94.4 5.6

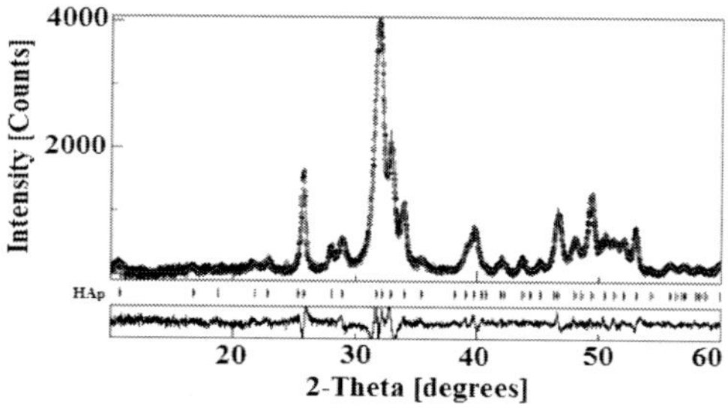

Figure 4. Rietveld refinement of the XRD pattern of the prepared HAp by the co-precipitation method.

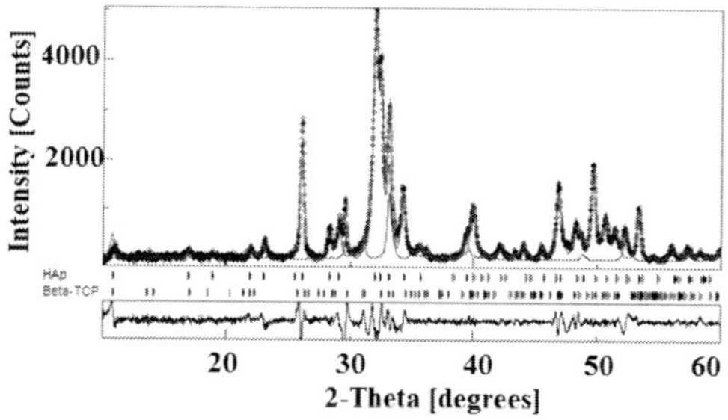

Figure 5. Rietveld refinement of the XRD pattern of the prepared HAp by the hydrothermal method.

The secondary β-TCP phase was observed after a heat treatment at 1000°C for 2 h of the produced HAp from commercial calcium hydroxide $Ca(OH)_2.(H_2O)_6$, monoamonium dihydrogen phosphate $NH_4H_2PO_4$ and ammonia NH_4OH (Harkat et al. 2021, 43). A biphasic composite microstructure of HAp and β-TCP was produced from mixtures of calcium phosphate (tribasic) and calcium hydroxide ($Ca(OH)_2$) powders by heat treatments in the temperature range of 600–1000°C for 8 h (Ramachandra et al. 1997, 511). It has been shown that beyond 600°C the HAp phase starts slowly transforming to β–TCP and transforms completely to pure β-TCP at 1000°C (Radin and Ducheyne 1993, 35). The results indicate that microcrystalline agglomerates of HAp and calcium deficient HAp transform to crystalline β-TCP, with a Ca/P ratio of 1.5, in the temperature range of 600–1000°C (Chaair et al. 1995, 895). Those observations were supported by infrared spectral characterization where the PO_4^{3-} peaks are located at 475, 570, 600, 962 and 1000–1130 cm^{-1}, and those corresponding to the librational and stretching frequencies of OH^- ions are situated at 630 and 3580 cm^{-1}, respectively.

Morphology

Figure 6 shows typical backscattered electron (BSE) morphologies of the as-prepared HAp powders from commercial precursors and heat treated at different temperatures. An irregular plate-like shape with a distribution of particles size where smaller and bigger particles coexist is observed. As can be also seen, the porosity is more important for the heat treated HAp at 1000°C. The EDX spectra (Figure 7) confirm the purity of the synthesized HAp, and the elemental EDX analysis (Table 3) reveals that the content of oxygen decreases and that of calcium increases after calcination for all temperatures, while thoe of phosphorous and carbon vary slightly. The Ca/P ratio varies between 2.55 and 3.15. The higher Ca/P ratio might be related to the composition of the HAp.

Table 3. Elemental EDX analysis of the prepared HAp powders

Sample	Oxygen (wt. %)	Calcium (wt. %)	Phosphorous (wt. %)	Carbon (wt. %)	Ca/P
As-prepared	46.64	27.70	10.84	14.82	2.55
HAp- T = 600°C	41.79	33.74	10.70	13.77	3.15
HAp- T = 800°C	33.23	39.14	13.24	14.39	2.95
HAp- T = 1000°C	40.04	37.07	12.08	10.82	3.06

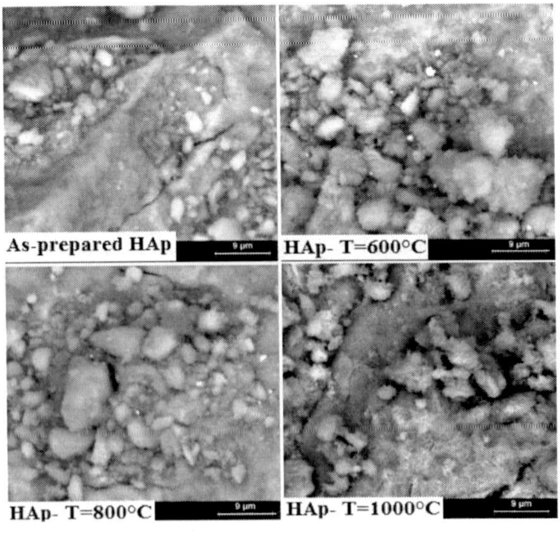

Figure 6. BSE Morphologies of the as-prepared and heat treated HAp at different temperatures.

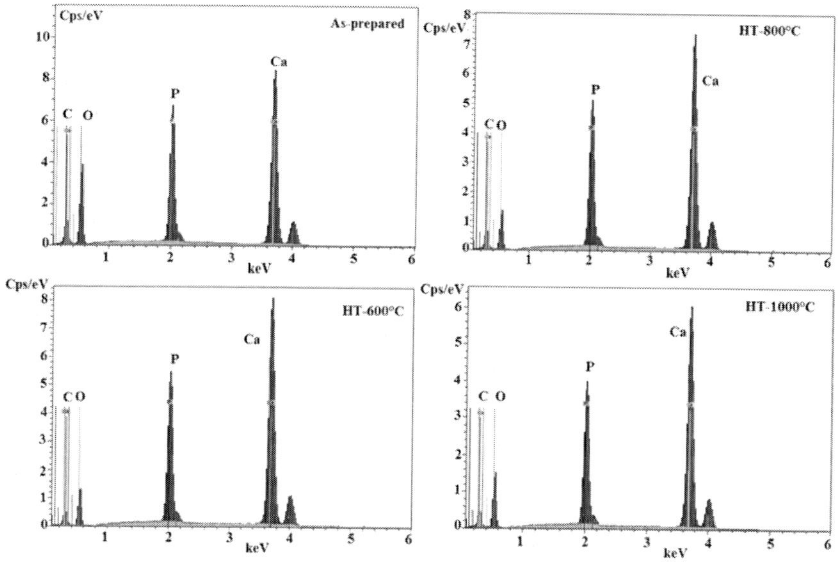

Figure 7. Elemental analysis spectra of the as-prepared and heat treated HAp at different temperatures.

The punctual elemental analysis (Figure 8) of the as-prepared HAp by the co-precipitation method reveals the presence of the main constituent calcium, phosphorous, and oxygen in addition to carbon. The EDX analysis reveals the heterogeneity of the HAp powders from one particle to another. Indeed, carbon content is about 4.5–20.18 wt.%, oxygen is about 31–43.78 wt.%, phosphorous is about 10.9–18.2 wt.%, calcium is about 27.8–44.5 wt.%, and the corresponding Ca/P ratio varies slightly between 2.36 and 2.54 (Table 4). The higher value of the Ca/P ratio of 1.8 can be attributed to the fact that the produced HAp is carbonated. The HAp Ca/P ratio value can vary between 1.2 and 2.0 depending on the site of the bone tissue or the stoichiometric apatite (Stanciu et al. 2007, 34).

Figure 9 displays the linear EDX analysis of the as-prepared HAp powders by the co-precipitation method. The concentration of each element varies from particle to particle. Besides, the higher content of carbon in the dark regions that corresponds to the distances $d_1 < 2$ μm and $d_2 = 10–20$ μm can be related to the sample holder.

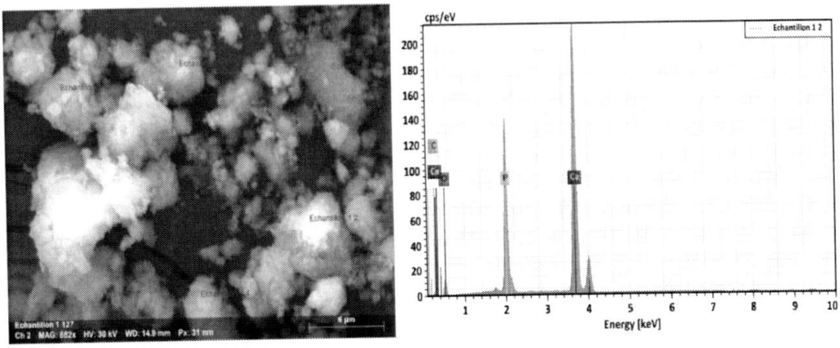

Figure 8. Typical SEM microgpah and EDX spectrum.

Table 4. Punctual elemental analysis of the synthesized HAp powders

Point	C (wt. %)	O (wt. %)	P (wt. %)	Ca (wt. %)	Ca/P
1	19.35	31.23	14.59	34.82	2.38
2	04.53	32.82	18.17	44.49	2.44
3	13.37	43.78	12.21	30.64	2.50
4	20.18	41.06	10.92	27.84	2.54
5	10.23	37.29	15.58	36.90	2.36
Mean	13.53	37.24	14.29	34.94	2.44
Sigma	6.52	5.32	2.85	6.40	
Sigma Mean	2.92	2.38	1.27	2.86	

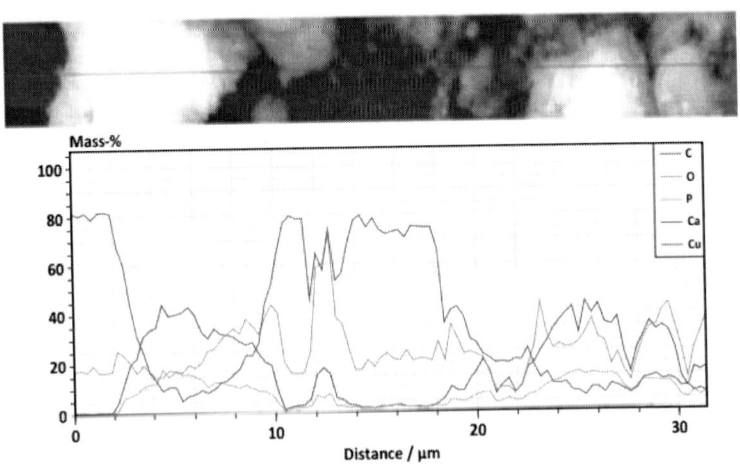

Figure 9. Linear EDX analysis of the as-prepared HAp powders from commercial precursors.

Spectroscopy Analysis

The Raman spectroscopy can be used to identify the calcium phosphates and confirm the formation of the HAp. Indeed, it is well known that the vibrational normal modes of the free tetrahedron lead to ν_1, ν_2, ν_3 and ν_4 frequencies related to the symmetric stretching of the PO bonds, doubly degenerate O-P-O bending modes (E symmetry), triply degenerate T2 mode involving asymmetric P-O stretching, and triply degenerate T2 modes of mainly O-P-O bending character, respectively.

The Raman spectroscopy results agree well with the XRD ones which reveal that the synthesized HAp is stable up to 800°C (Harkat et al. 2021, 43). Indeed, up to 800°C the Raman spectra (not shown here) reveal two peaks located at $\nu_1 = 962$ cm^{-1} (very strong) and $\nu_3 = 1047$ cm^{-1} that can be ascribed to symmetric and asymmetric stretching of P-O bond in phosphate, respectively. For the heat treated HAp at 1000°C, one observes the disappearance of the strong peak, in addition to the presence of new peaks at $\nu_2 = 418.22$ cm^{-1} and $\nu_4 = 594$ cm^{-1} that can be associated with the doubly and triply degenerated O-P-O bending character, respectively (Fowler et al. 1993, 1417). Those changes might be ascribed to the biphasic composite microstructure of HAp and β-TCP. Furthermore, the peaks situated at 230 and 418 cm^{-1} frequency can be attributed to the internal P-O group of the β-TCP phase. The Raman peak at 130 cm^{-1} is due to the external lattice mode. Likewise, the narrowing of the Raman peaks as the temperature increases can be related to ordering.

Figure 10 displays the Raman spectra of the prepared HAp from seashell wastes as a function of the immersion time in SBF. The Raman spectra show the same peaks with small variations in their positions. Indeed, all the spectra exhibit a dominant and very strong PO_4^{3-} peak located at 957–962 cm^{-1} assigned to a nondegenerated symmetric stretching mode, ν_1, of the P–O bond of the phosphate group. The position of the strong peak is the same for the as-synthesized and 5 SBF, while it shifts slightly to about ~957 cm^{-1} for 10 SBF and 15 SBF. Such shift might be related to small fluctuations in composition. The peak located at 1047 cm^{-1} is associated with the P–O bond's asymmetric stretching vibration mode ν_3 of PO_4^{3-} (Sauer et al. 1994, 414). The observed peak at 1324 cm^{-1} can be ascribed to the C–O bond of ν_3 CO_3 group. While, the detected peaks in the range of 3142–3576 cm^{-1} are correlated to the stretching vibration mode of the lattice hydroxyl O–H ions. The peaks located at 425-455 cm^{-1} can be ascribed to the doubly degenerate bending mode ν_2 of the PO_4 group (O–P–O bond). The situated peaks at 575-621 cm^{-1} can be linked to the triply degenerate bending mode ν_4 of the PO_4 group (O–P–O bond).

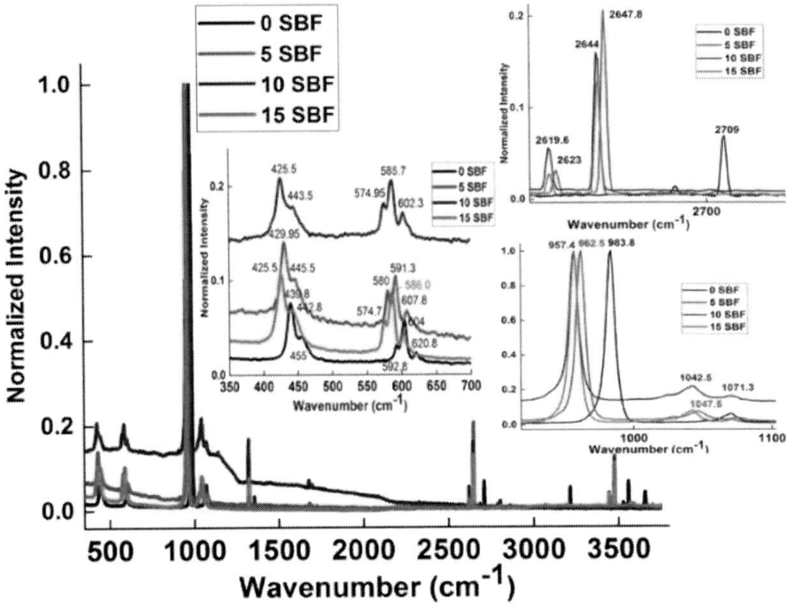

Figure 10. Raman spectra of the as-prepared HAp powders from seashell wastes, 5 SBF, 10 SBF, and 15 SBF.

It has been reported on the elimination of many peaks related to ν_4 PO_4 (606 cm^{-1}) and ν_3 phosphate groups at 1006, 1025 and 1100 cm^{-1}, and the appearance of two vibration modes (ν_3 and ν_4) of carbonate CO_3^{2-} and the OH^- groups in the FTIR spectra of the samples after immersion in SBF as a consequence of the formation of the HAp (Criseida Ruiz-Aguilar et al. 2022, 191).

Complex Impedance Analysis

AC Conductivity Behavior

Figures 11, 12 and 13 display the variation of AC conductivity of 5 SBF, 10 SBF and 15 SBF with frequency at various isothermal temperatures, respectively. The AC conductivity increases with increasing temperature which means that the process is thermally activated. Besides, three temperature regions can be detected as a function of frequency: (i) 300–320 K, (ii) 400–560 K and (iii) 580–680 K. The AC conductivity shows a linear increase with an increasing frequency above 10^3 Hz for 5 SBF and 10^4 Hz for 10 SBF, at 300 K, decreases slightly up to 10^3 Hz in the second region (400–560 K), and increases at higher frequency. Moreover, in the third region (580–680 K), the AC conductivity shows a plateau up to 10^4 Hz for 5 SBF and 10^3 Hz for 10 SBF and 15 SBF, decreases weakly (~10^5 Hz) and remains nearly constant. The hopping frequency corresponds to the point at which the slope of the conductivity curve changes (Santhosh et al. 2012, 276).

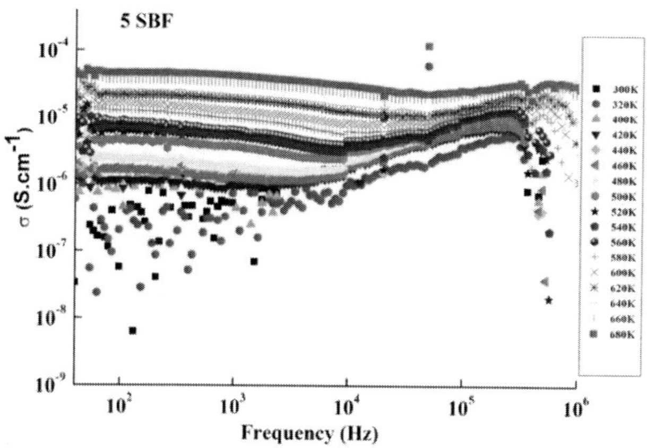

Figure 11. Evolution of the AC conductivity with frequency at various temperatures for 5 SBF.

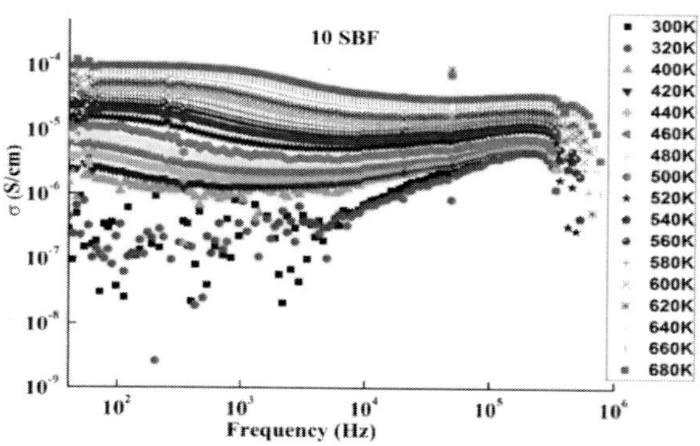

Figure 12. Evolution of the AC conductivity as a function of frequency at various temperatures for 10 SBF.

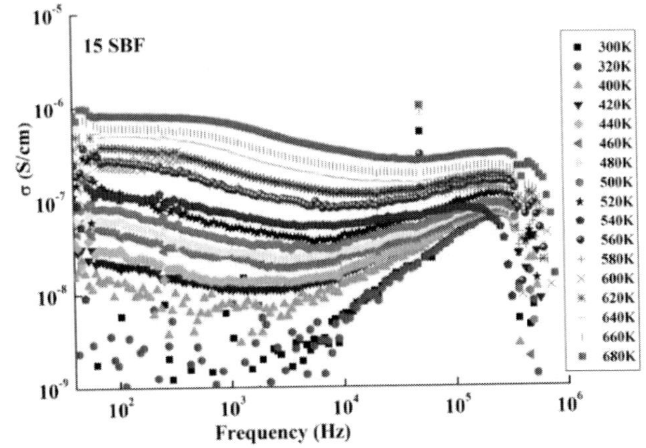

Figure 13. Evolution of the AC conductivity as a function of frequency at various temperatures for 15 SBF.

The variation in the AC conductivity with frequency at different temperatures can be related to the concentration of OH^- vacancies (Boutinguiza et al. 2012, 478). The conduction mechanism occurs owing to the formation of vacancies at hydroxyl positions in the crystal lattice (Orlovskii et al. 1996, 654). At higher frequencies, the increase in the AC conductivity can be associated with an increase in the free protons concentration, loss of bound water, hopping of charge carriers in finite clusters, and reduction of charge carrier displacement (Tămăşan et al. 2013, 2569). The migration of hydroxyl ions, through vacancies in the center of Ca ion triangle planes along with the *c*-axis of the hexagonal HAp, as well as the carrying of OH^- vacancies along with the *c*-axis give rise to the increase of the conduction mechanism in the HAp conductivity with increasing temperature (Kalil et al. 2002, 451). At higher temperatures, the hopping of H^+ ion to O^{2-} sites leads to the increase in AC conductivity. The conduction in HAp, at room temperature, might be related to its hygroscopic nature that leads to surface water absorption. The transfer of proton (H+) in absorbed water can be considered as one of the possible reasons for the room temperature conduction (Nagai and Nishino 1988, 1456).

The dehydration occurs at high temperature and low/intermediate frequencies giving rise to an increase of OH^- vacancies. Therefore, the diminution in AC conductivity with increasing frequency can be associated with the shrinking of charge carrying protons and the obliteration of the conduction paths. However, the conductivity reaches a constant value when the concentration of vacancies reaches equilibrium once the dehydration is completed. The occurrence of a plateau at high temperatures has been sturdily linked to the conductivity of the conductive channels in the low frequency region (Kalil et al. 2002, 451). Similarly, the dehydration of the HAp lattice structure can be attributed to the thermally activated protons jumping that become trapped onto OH^- lattice positions. However, the decrease in AC conductivity in the low frequency region can be correlated to the obstruction of the conduction jumping due to the applied field and HAp dehydration. The dehydration introduces vacancies at the hydroxyl site (ϒ) (Tofail et al. 2005, 117) which might hinder the movement of proton and weaken the electron hopping:

$$Ca_{10}(PO_4)_6(OH)_2 \rightarrow Ca_{10}(PO_4)_6(OH)_{2-2x}O_x[\square]_x + xH_2O(g)\uparrow \text{ (with } x<1) \quad (6)$$

The migration of OH^- and H^+ have been considered as conductive species at room and elevated temperatures, respectively, in adsorbed water (Tanaka et al. 2012, 815). Different HAp microstructures such as the deviation from purity and stoichiometry can lead to different dielectric responses against temperature and significant variations in the dielectric properties (Sajid Iqbal et al. 2021, 138507).

Activation Energy

The activation energy can be determined from the evolution of the AC conductivity with temperature using the following Arrhenius relation (Jonscher 1977, 673):

$$\sigma = \sigma_0 \exp\left(-{E_a}/{k_B T}\right) \quad (7)$$

Where σ_0 is a pre-exponent factor, T is the measurement temperature, E_a is the activation energy, and $k_B = 8.617\text{x}10^{-5}$ eV/K is the Boltzmann constant. The activation anergy, E_a, can be calculated from the slope of the linear fit of the log (σ) plot as a function of 1/T as shown in Figure 14.

The calculated E_a values vary between 0.187 and 0.241 eV for 5 SBF, 0.133 and 0.295 eV for 10 SBF, and 0.117 and 0.216 eV for 15 SBF. The relatively lower Ea values can be attributed to the contribution of the applied frequency to the conduction mechanism that improves the electronic jumps between localized states. Furthermore, the smaller Ea values on the one hand and the augmentation in AC conductivity with rising frequency on the other hand suggest that the conduction takes place through the hopping process (Nagai and Nishino 1988, 1456). Accordingly, the activation energy can be related to the mobility of charge carriers that are situated in empty sites. Therefore, the charge responsible for the ionic conduction mechanism in the produced HAp is H^+ proton (Tofail et al. 2005, 117; Bouhaouss et al. 2001, 743; Gittings et al. 2008; Prieto Valdes et al. 1995, 2174). The discrepancies between the obtained Ea values with those reported by Laghzizil et al. (1.21 eV) (Laghzizil et al. 2001, 953), Gitting et al. (Chavan et al. 2010, 224) (1.86–2.23 eV), and Yamishta et al. (0.7–1.4 eV) (Yamishta et al. 1995, 1191) might be attributed to the preparation conditions.

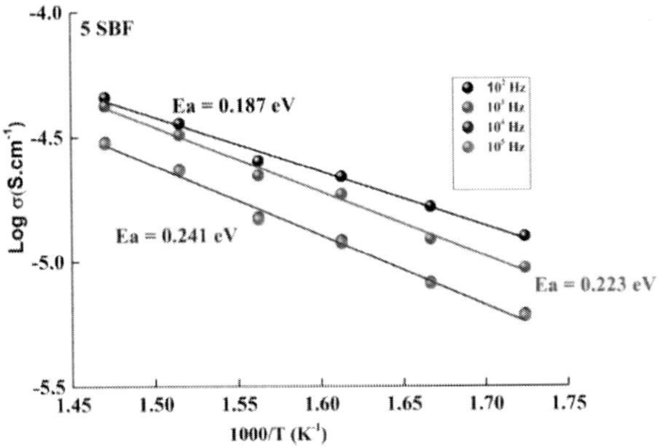

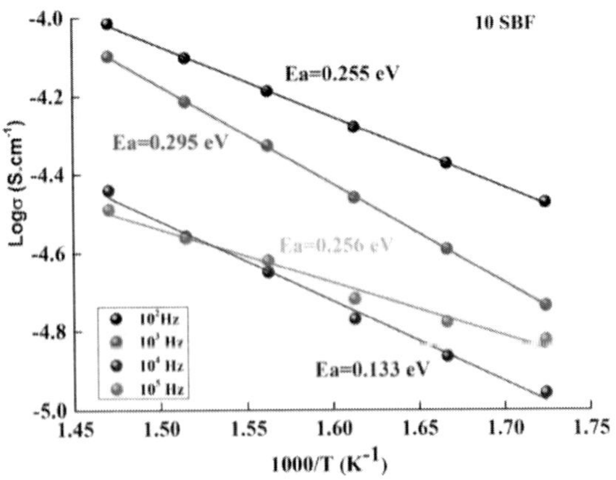

Figure 14. (Continued).

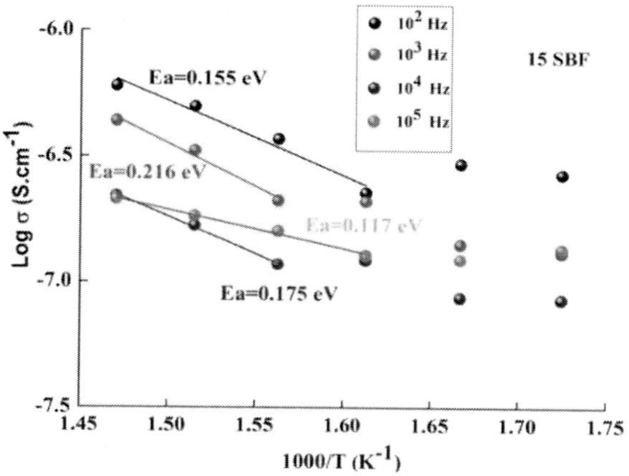

Figure 14. Arrhenius plots of the conductivity for 5 SBF, 10 SBF and 15 SBF samples.

Complex Electrical Modulus Analysis

The dielectric relaxation process as a function of temperature and frequency can be determined using the complex electric modulus (M* = M'+M") formalism according to Kohlrausch–Williams–Watts (KWW) equation (Dubey et al. 2014, 24601):

$$M^* = M_\infty(1 - \int_0^\infty \left(\frac{d\phi(t)}{dt}\right) \exp(-jwt)\, dt \qquad (8)$$

Where M_∞ is the asymptotic value of M'(ω), $\Phi(t) = \exp\left(-{}^{t}/_{(\tau_M)^\beta}\right)$ denotes the time variation of the electric field within the material where *β (0 < β< 1)* is the stretched exponent that characterizes the non-Debye behavior degree, and τ_M is the relaxation time. The real M' and imaginary M" parts of the electric modulus are displayed as a function of frequency in Figure 15. The increase of the real part M' with increasing frequency, for all temperatures, indicates that the conduction process is governed by the short range mobility of charge carriers. Moreover, the frequency dependence of the imaginary part M" (Figure 16) shows a single peak that shifts towards higher frequencies as the temperature increases predicting low relaxation time, and proves that the conduction process is thermally activated (Manna et al. 2016, 102853). The presence of the relaxation peaks in the M″ curves indicates that the conduction in the synthesized HAp is ionic.

Impedance Analysis

The impedance spectroscopy is related to the temperature and frequency dependence of the complex function $Z = Z' - jZ''$ where $j = \sqrt{(-1)}$, Z' and Z'' represent the real and imaginary parts of the complex quantity Z, respectively. The variation of Z′ and Z″ complex impedance with frequency are shown in Figures 17 and 18, respectively. The increase in temperature induces a diminution in the peaks' intensity and a shift in their position towards higher frequencies. This behavior can be linked to the thermal relaxation process. Moreover, the variation in the position and height of the peaks with increasing frequency suggests that the conduction mechanism is due to the hopping of charge carriers owing to thermal stimulation. Besides, the variation of the width confirms the presence of a temperature-dependent relaxation phenomenon. As frequency increases, Z″ value decreases at a given temperature.

The Nyquist plot is usually employed to determine the elements that describe the conduction mechanism in the HAp by differentiating the contribution of grain and grain boundary. The presence of two semicircles shows that the dielectric process obeys to a two relaxation phenomenon where the high frequency semicircle is associated with grain effect, and the low frequency semicircle is attributed to the grain boundary effect since the relaxation time of the grain boundary is much higher than that of the grain (Saha et al. 2017, 272). Figure 19 shows the Nyquist diagrams of 5 SBF, 10 SBF and 15 SBF in the temperature range of 580–680 K. The diagrams show the presence of an inductive loop in the low frequency region and a capacitive loop at the high/medium frequency region. In impedance spectroscopy, inductive loops are associated with complex multistep dynamics and are generally related to surface states in the analysis of electronic devices such as solar cells (Bisquert et al. 2006, 184). Furthermore, the loops pattern is temperature dependent. Hence, the diminution of the semicircle diameter can be related to the rise of the charge carriers' mobility (thermally activated conduction process). Besides, the non-perfect semicircle rises from many contributions and its depression as the temperature increases can be attributed to the heterogeneity of the HAp properties such as the nature of the interface, defects, crystallite size, porosity, irregularities, etc. as well as the spreading of the values of the same features at different sites over the electrode surface.

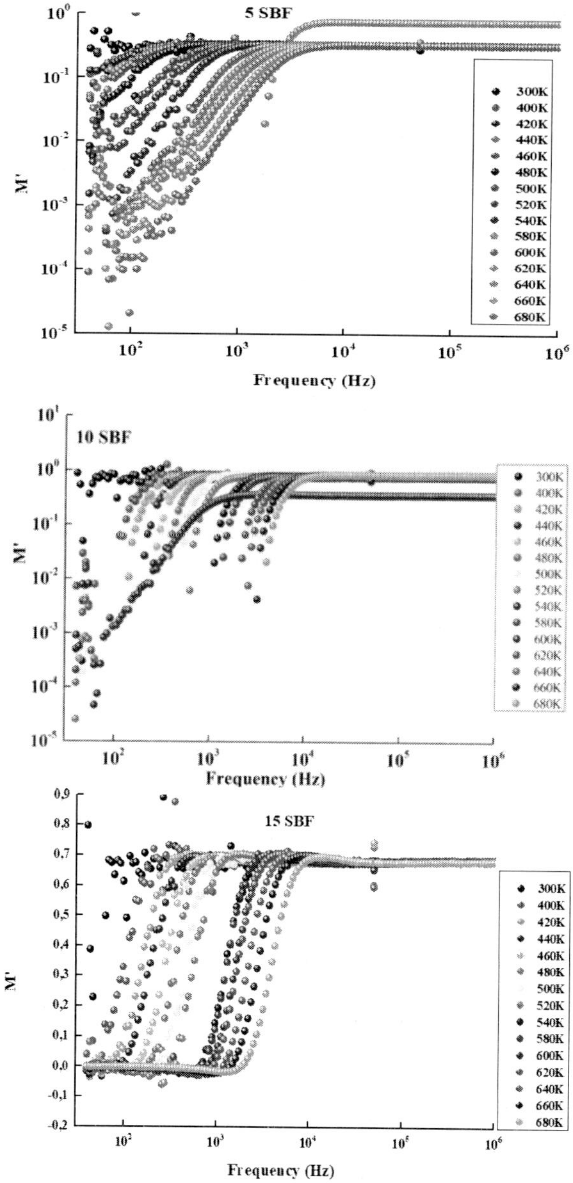

Figure 15. Variation of M' as a function of frequency at various temperatures for 5 SBF, 10 SBF, and 15 SBF.

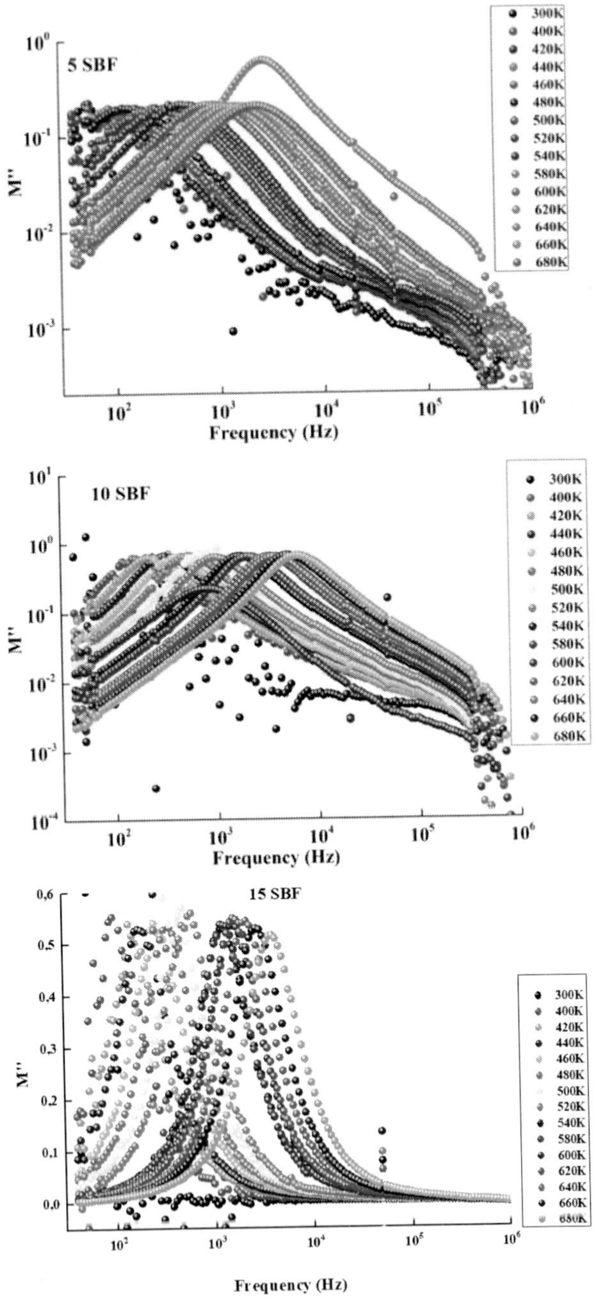

Figure 16. Variation of M″ with frequency at various temperatures for 5 SBF, 10 SBF, and 15 SBF.

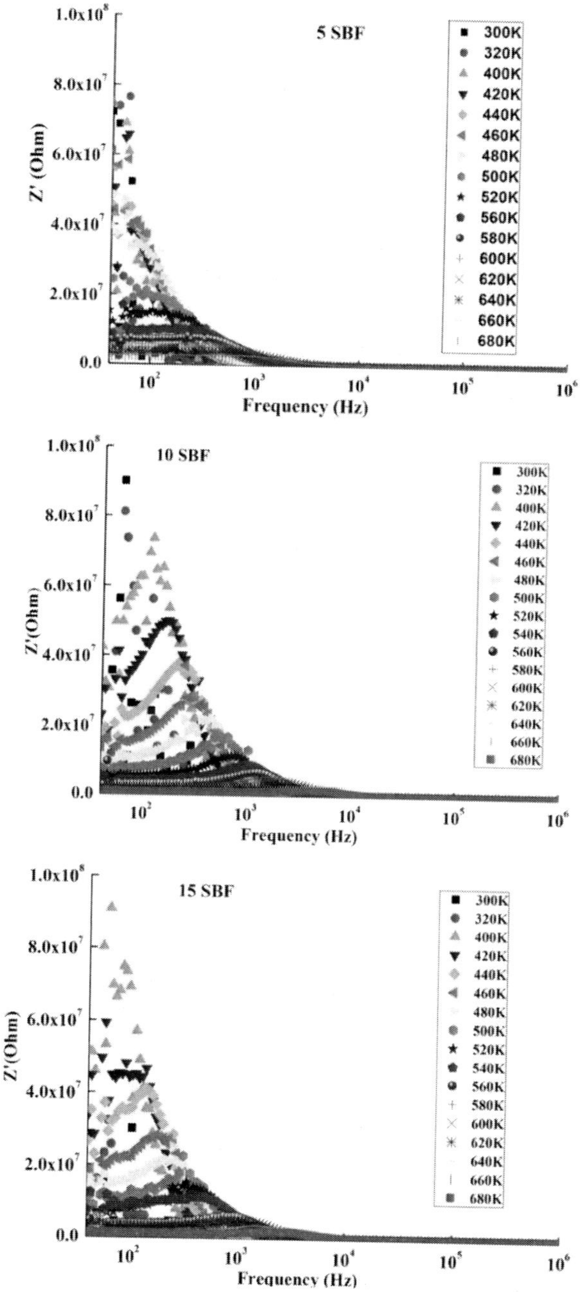

Figure 17. Variation of Z' with frequency and temperatures for 5 SBF, 10 SBF, and 15 SBF.

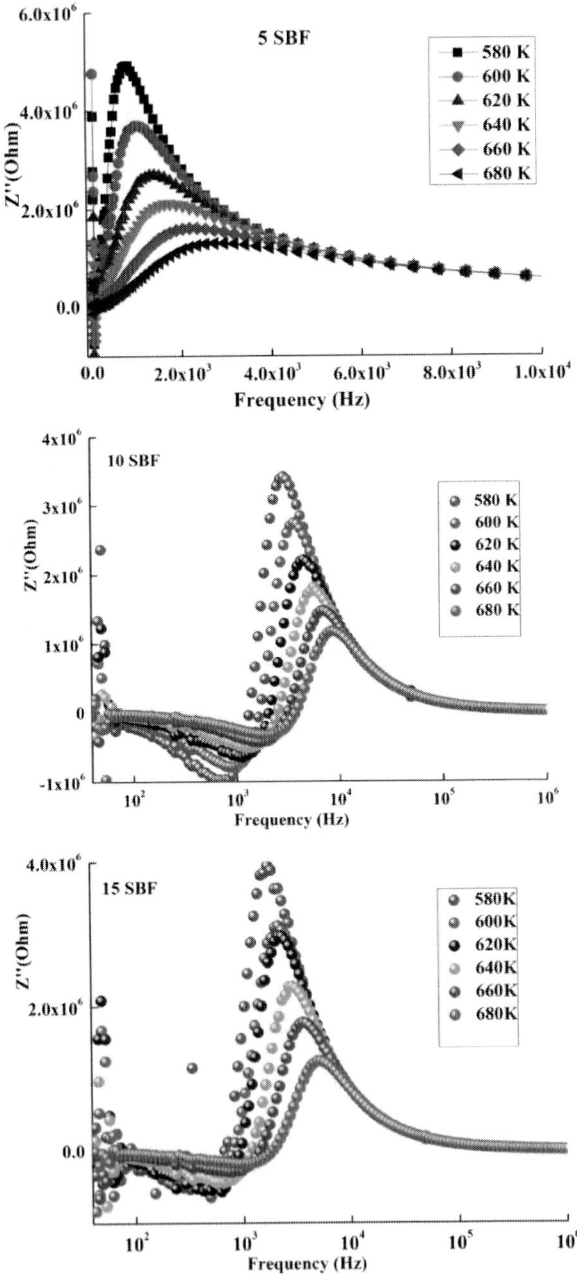

Figure 18. Variation of complex impedence (Z″) with frequency and temperatures for 5 SBF, 10 SBF, and 15 SBF.

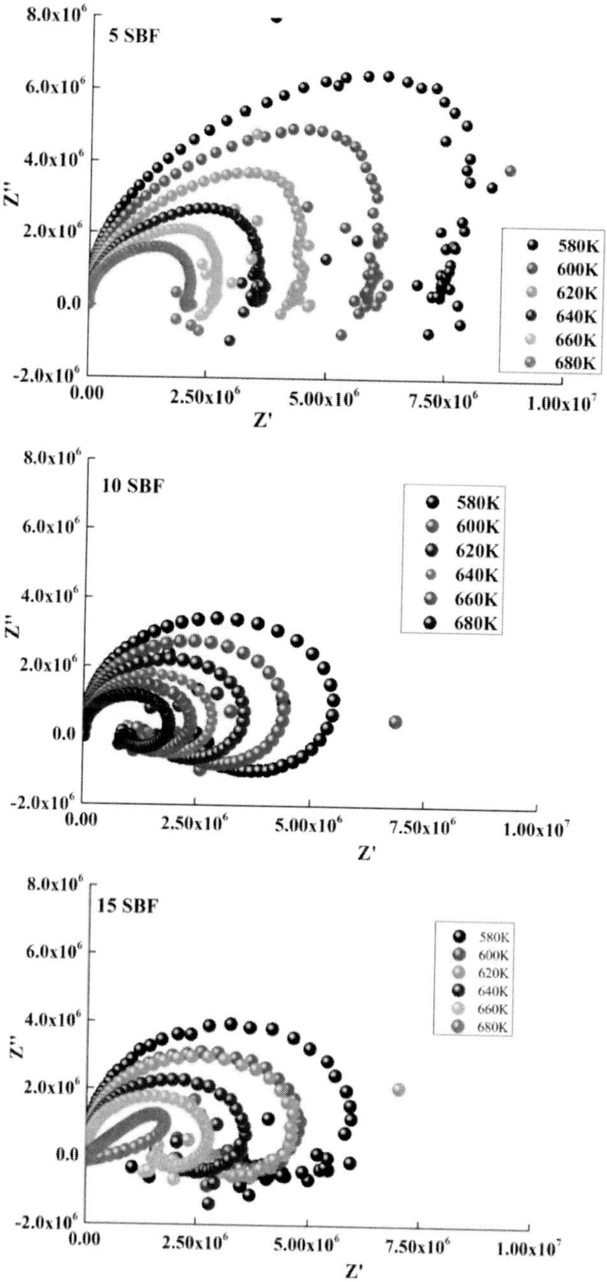

Figure 19. Complex impedance spectra of 5 SBF, 10 SBF and 15 SBF in the temperature range of 580–680 K.

Conclusion

Nanocrystalline HAp has been successfully prepared using hydrothermal and co-precipitation methods using various precursors such as naturally abundant seashell waste, calcium hydroxide $Ca(OH)_6$ and mono ammonium phosphate $NH_4H_2PO_4$ as calcium and phosphate sources, respectively. XRD, SEM, Raman spectroscopy, and complex impedance spectroscopy were used to investigate the structure, microstructure, morphology and electric properties of the synthesized HAp powders.

A single HAp phase is obtained with the co-precipitation method and a secondary β-TCP phase appears after a heat treatment at 1000°C for 2 h of the produced HAp from commercial calcium hydroxide. Besides, the β-TCP phase is observed in the as-prepared HAp from seashell wastes using the hydrothermal method. The effect of the preparation conditions is confirmed by the XRD. Indeed, the crystallite size of the produced HAp by the co-precipitation method is lower (31 nm) than that of the hydrothermal method (47 nm). Also, the relative deviations of the lattice parameters are more important for the hydrothermal method. The Raman spectroscopy results confirm the XRD ones. The HAp exhibits small and large size agglomerates with higher Ca/P ratio (between 2.55 and 3.15. Temperature and frequency dependent electric properties are studied at different temperatures using complex impedance analysis.The electrical properties are temperature and frequency dependent. The calculated E_a values vary between 0.187 and 0.241 eV for 5 SBF, 0.133 and 0.295 eV for 10 SBF, and 0.117 and 0.216 eV for 15 SBF. The electrical conduction is associated with the hopping process.

References

Abbasi, N., Hamlet, S., Love, R. M., and Nguyen, N. T. (2020). Porous scaffolds for bone regeneration. *Journal of Science: Advanced Materials and Devices*, 5, 1-9. https://doi.org/10.1016/j.jsamd.2020.01.007.

Abdel–Baset, T. A. (2021). Structure, Optical, Dielectric, and Magnetic Properties of $Zn_{0.99}M_{0.01}O$ (M = Ni, Fe, and Cd) nanoparticles. *Journal of Superconductivity and Novel Magnetism*, 34, 1259–1267. https://doi.org/10.1007/s10948-021-05843-9.

Badran, H., Yahia, I. S., Hamdy, M., and Awwad, N. S. (2017). Lithium-doped hydroxyapatite nano-composites: Synthesis, characterization, gamma attenuation coefficient and dielectric properties. *Radiation Physics and Chemistry*, 130, 85–91. https://doi.org/10.1016/j.radphyschem.2016.08.001.

Bisquert, J., Garcia-Belmonte, G., Pitarch, A., and Bolink, H. (2006). Negative capacitance caused by electron injection through interfacial states in organic light-emitting diodes. *Chemical and Physical Letters*, 422, 184−191. http://dx.doi.org/10.1016/j.cplett.2006.02.060.

Bouhaouss, A., Laghzizil, A., Bensaoud, A., Ferhat, M., Lorent, G., and Livage, J. (2001). Mechanism of ionic conduction in oxy and hydroxyapatite structures. *International Journal of Inorganic Materials*, 3, 743-747. https://doi.org/10.1016/S1466-6049(01)00054-X Specially.

Boutinguiza, M. J., Comesaña, R., Lusquiños, F., de Carlos, A., and León, B. (2012). Biological hydroxyapatite obtained from fish bones. *Material Science and Engineering C*, 32, 478–486. https://doi:10.1016/j.msec.2011.11.021.

Chaair, H., Heughebaert, J., and Heughebaert, M. (1995). *Journal of Materials Chemistry*, 5, 895.

Chavan, P. N., Bahir, M. M., Mene, R. U., Mahabole, M. P., and Khairnar, R. S. (2010). Study of nanobiomaterial hydroxyapatite in simulated body fluid: Formation and growth of apatite. *Materials Science and Engineering B*, 168, 224–230. https://doi.org/10.1016/j.mseb.2009.11.012.

Criseida Ruiz-Aguilar, Luz Eugenia Alcántara-Quintana, Ena Athenea Aguilar-Reyes, and U Olivares-Pinto. (2022). Fabrication, characterization, and *in vitro* evaluation of β-TCP/ZrO_2-phosphate-based bioactive glass scaffolds for bone repair. *Boletin de la Sociedad Espanola de Ceramica et Vidrio* [*Bulletin of the Spanish Society of Ceramics and Glass*], 61, 191-202. https://doi.org/10.1016/j.bsecv.2020.09.004.

Deb, P., and Deoghare, A. B. (2019). Effect of acid, alkali and alkali–acid treatment on physicochemical and bioactive properties of hydroxyapatite derived from Catla catla fish scales. *Arabian Journal for Science and Engineering*, 44, 7479–7490. https://doi.org/10.1007/s13369-019-03807-9.

del Campo, R., Savoini, B., Jordao, L., Muñoz, A., and Monge, M. A. (2017). Cytocompatibility, biofilm assembly and corrosion behavior of Mg-HAP composites processed by extrusion. *Material Science Engineering C*, 78, 667-673. https://doi.org/10.1016/j.msec.2017.04.143.

Dubey, A. K., Kakimoto, K., Obata, A., and Kasuga, T. (2014). Enhanced polarization of hydroxyapatite using the design concept of functionally graded materials with sodium potassium niobate. *RSC Advanced*, 4, 24601-24611. https://doi:10.1039/c4ra02329c.

Fihri, A., Len, C., Varma, R. S., and Solhy, A. (2017). Hydroxyapatite: A review of syntheses, structure and applications in heterogeneous catalysis. *Coordination Chemistry Reviews*, 347, 48-76. https://doi.org/10.1016/j.ccr.2017.06.009.

Fowler, B. O., Markovic´, M., and Brown, W. E. (1993). Octacalcium phosphate-carboxylates. 3. Infrared and Raman vibrational spectra. *Chemistry of Materials*, 5, 1417–1423.

Gamble, J. E. (1967). *Chemical anatomy, physiology and pathology of extracellular fluid*. Cambridge, MA: Harvard University Press, p. 1–17.

Gittings, J. P., Bowen, C. R., Turner, I. G., Dent, A. C. E., Baxter, F. R., and Chaudhuri, J. B. (2008). *Dielectric properties of hydroxyapatite based ceramics*. Cardiff University, Cardiff, UK. Published by Whittles Publishing Ltd. https://doi.org/10.1016/j.actbio.2008.08.012.

Gittings, J. P., Bowen, C. R., Dent, A. C. E., Turner, I. G., Baxter, F. R., and Chaudhuri, J. B. (2009). Electrical characterization of hydroxyapatite-based bioceramics. *Acta Biomaterialia*, 5, 743-754. https://doi.org/10.1016/j.actbio.2008.08.012.

Haider, A., Haider, S., Han, S. S., and Kang, I. K. (2017). Recent advances in the synthesis, functionalization and biomedical applications of hydroxyapatite: review. *RSC Advances*, 7, 7442-7458. https://doi.org/10.1039/C6RA26124H.

Harkat, M., Alleg, S., Chemam, R., Azzaza, S., Mokhtari, M., and Dhahri, E. (2021). Synthesis and characterization of thermally stable hydroxyapatite. *Annales de Chimie: Science des Matériaux*, 45, 43-50. https://doi:10.18280/acsm.450106.

Harkat, M., Alleg, S., Chemam, R. et al. (2022). Structure and Electric Characterizations of the Derived Nanocrystalline Hydroxyapatite from Strombidae Strombus Seashells. *Arabian Journal for Science and Engineering*, 47, 7693–7706. https://doi.org/10.1007/s13369-021-06556-w.

Horiuchia, N., Madokorob, K., Nozakia, K., Nakamuraa, M., Katayamab, K., and Nagaia Yamashita, A. K. (2018). Electrical conductivity of polycrystalline hydroxyapatite and its application to electret formation. *Solid. State Ionics*, 315, 19–25. https://doi.org/10.1016/j.ssi.2017.11.029.

Javadinejad, H. R., Saboktakin Rizi, M., Aghababaei Mobarakeh, E., and Ebrahimian, M. (2017). Thermal stability of nano-hydroxyapatite synthesized via mechanochemical treatment. *Arabian Journal for Science and Engineering*, 42, 4401–4408. https://doi.org/10.1007/s13369-017-2498-y.

Jonscher, A. (1977). The universal dielectric response. *Nature*, 267, 673–679. https://doi.org/10.1038/267673a0.

Kalil, M. Sh., Beheri, H. H., and Fattah, W. (2002). Structural and electrical properties of zirconia/hydroxyapatite porous composites. *Ceramics International*, 28, 451–458. https://doi:10.1016/S0272-8842(01)00118-3.

Kay, M. I., Young, R. A., and Posner, A. S. (1964). Crystal structure of hydroxyapatite. *Nature*, 204, 1050–1052. http://dx.doi.org/10.1038/2041050a0.

Kramer, E., and Podurgiel, J. (2014). Control of hydroxyapatite nanoparticle morphology using wet synthesis techniques: Reactant addition rate effects. *Materials Letters*, 131, 145-147. https://doi.org/10.1016/j.matlet.2014.05.105.

Kokubo, T., Kushitani, H., Sakka, S., Kitsugi, T., and Yamamuro, T. (1990). Solutions able to reproduce *in vivo* surface-structure changes in bioactive glass-ceramic A-W3. *Journal of Biomedical Materials Research*, 24(6), 721-734.

Kokubo, T., and Takadama, H. (2006). How useful is SBF in predicting *in vivo* bone bioactivity. *Biomaterials*, 27, 2907-2915.

Kumar, M. A., Chipara, A. C., Narayanan, N. T., Anumary, A., Sruthi, R., Thanikaivelan, P., Vajtai, R., Mani, S. A., and Ajayan, P. M. (2016). Three-dimensional porous sponges from collagen Biowastes. *ACS Applied Materilas Interfaces*, 8, 14836–14844. https://doi.org/10.1021/acsami.6b04582.

Kumar, G. S., Girija, E. K., Venkatesh, M., Karunakara, G., Kolesnikov, E., and Kuznetsov, D. (2017). One step method to synthesize flower-like hydroxyapatite architecture using mussel shell bio-waste as a calcium source. *Ceramics International*, 43, 3457–3461. https://doi:10.1016/j.ceramint.2016.11.163.

Kumar, G. S., Karunakaran, G., Girija, E. K., Kolesnikov, E., Van Minh, N., Gorshenkov, M. V., and Kuznetsov, D. (2018). Size and morphology-controlled synthesis of mesoporous hydroxyapatite nanocrystals by microwave-assisted hydrothermal method. *Ceramics International*, 44, 11257-11264. https://doi.org/10.1016/j.ceramint.2018.03.170.

Laghzizil, A., El herch, N., Bouhaoussa, A., Lorenteb, G., Coradinb, T., and Livageb, J. (2001). Electrical behavior of hydroxyapatites $M_{10}(PO_4)_6(OH)_2$ (M = Ca, Pb, Ba). *Materials Research Bulletin*, 36, 953–962. http://dx.doi.org/10.1016/S0025-5408(01)00576-1.

Louati, B., Guidara, K., and Gargouri, M. (2009). Dielectric and ac ionic conductivity investigations in the monetite. *Journal of Alloys and Compounds*, 472, 347. http://dx.doi.org/10.1016/j.jallcom.2008.04.050.

Li, M., Feteira, A., and Sinclair, D. C. (2005). Origin of the high permittivity in $(La_{0.4}Ba_{0.4}Ca_{0.2})(Mn_{0.4}Ti_{0.6})O_3$ ceramics. *Journal of Applied Physics*, 98(8), 084101. http://dx.doi.org/10.1063/1.2089159.

Limmahakhun, S., Oloyede, A., Sitthiseripratip, K., Xiao, Y., and Yan, C. (2017). 3D-printed cellular structures for bone biomimetic implants. *Additive Manufacturing*, 15, 93–101.

Lin, T. H., Wang, H. C., Cheng, W. H., Hsu, H. C., and Yeh, M. L. (2019). Osteochondral tissue regeneration using a tyramine-modified bilayered PLGA scaffold combined with articular chondrocytes in a porcine model. *International Journal of Molecular Science*, 20(2). https://doi.org/10.3390/ijms20020326.

Lutterotti, L., Scardi, P., and Maistrelli, P. (1992). LSI- computer program for simultaneous refinement of material structure and microstructure. *Journal of Applied Crystallography*, 25, 459-462. http://maud.radiographema.eu/.

Manna, A., Pramanik, S., Tripathy, A., Moradia, A., Radzi, Z., Pingguan-murphy, B., Hasnan, N., and Abu Osman, N. A. (2016). Development of biocompatible hydroxyapatite-Poly(ethylene glycol) core-shell nanoparticles as improved drug carrier: structural and electrical characterizations. *RSC Advanced*, 6, 102853-102868. https://doi.org/10.1039/C6RA21210G.

Nagai, M., and Nishino, T. (1988). Surface conduction of porous hydroxyapatite ceramics at elevated temperature. *Solid State Ionics*, 28-30, 1456-1461.

Nosenko, V. V. A., Yaremko, M., Dzhagan, V. M., Vorona, I. P., Romanyuk, Y. A., and Zatovsky, I. V. (2016). Nature of some features in Raman spectra of hydroxyapatite-containing materials. *Journal of Raman Spectroscopy*, 47, 726–730.

Orlovskii, V. P., Zakharov, N. A., and Ivanov, A. A. (1996). Structural transition and dielectric characteristics of high purity Hydroxyapatite. *Inorganic Materials*, 32 (6), 654-656.

Piccirillo, C., Pullar, R. C., Costa, E., Santos-Silva, A., Pintadoa, M. M., and Castro, P. M. L. (2015). Hydroxyapatite-based materials of marine origin: A bioactivity and sintering study. *Materials Science and Engineering*, 51, 309–315. https://doi.org/10.1016/j.msec.2015.03.020.

Prieto Valdes, J. J., Victorero Rodriguez, A., and Guevara Carrio, J. (1995). Dielectric properties and structure of hydroxyapatite ceramics sintered by different conditions.

Journal of Materials Research, 10, 2174-2177. https://doi.org/10.1557/JMR.1995.2174.

Prezas, P. R., Melo, B. M. G., Costa, L. C., Valente, M. A., Lança, M. C., Ventura, J. M., and Graça, M. P. F. (2017). TSDC and impedance spectroscopy measurements on hydroxyapatite, β-tricalcium phosphate and hydroxyapatite/β-tricalcium phosphate biphasic bioceramics. *Applied Surface Science*, 424, 28–38. https://doi:10.1016/j.apsusc.2017.02.225.

Radin, S. R., and Ducheyne, P. (1993). The effect of calcium phosphate ceramic composition and structure on *in vitro* behavior. II. Precipitation. *Journal of Biomedical Material Research*, 27, 35-45. https://doi.org/10.1002/jbm.820270106.

Ramesha, S., Natashaa, A. N., Tana, C. Y., Banga, L. T., Niakana, A., Purbolaksonoa, J., Chandranc, H., Chinga, C. Y., Rameshd, S., and Tenge, W. D. (2015). Characteristics and properties of hydroxyapatite derived by sol–gel and wet chemical precipitation methods. *Ceramics International*, 41(9), 10434-10441. https://doi.org/10.1016/j.ceramint.2015.04.105.

Raya, I., Mayasari, E., Yahya, A., Syahrul, M., and Latunra, A. I. (2015). Synthesis and characterizations of calcium hydroxyapatite derived from crabs shells (Portunus pelagicus) and its potency in safeguard against to dental demineralizations. *International Journal of Biomaterials*, 2015, 1–8. http://dx.doi.org/10.1155/2015/469176.

Rehwald, W., Kiess, H., and Binggeli, B. (1987). Frequency dependent conductivity in polymers and other disordered materials. *Z. Physik B – Condensed Matter*, 68, 143–148. https://doi.org/10.1007/BF01304219.

Rietveld, H. M. (1969). A profile refinement method for nuclear and magnetic structures. *Journal of Applied Crystallography*, 2, 65-71. https://doi.org/10.1107/S0021889869006558.

Royce, B. S. H. (1974). Field induced transport mechanisms in hydroxyapatite. *Annals of the New York Academy of Science*, 238, 131-138. https://doi.org/10.1111/j.1749-6632.1974.tb26783.x.

Rujitanapanich, S., Kumpapan, P., and Wanjanoi, P. (2014). Synthesis of hydroxyapatite from oyster shell via precipitation. *Energy Procedia*, 56, 112–117. https://doi.org/10.1016/j.egypro.2014.07.138.

Saha, S., Nandy, A., Pradhan, S. K., and Meikap, A. K. (2017). Electrical transport and dielectric modulus formalism of CuO doped ZrO 2 partially stabilized solid solution. *Materials Research Bulletin*, 88, 272–280. http://dx.doi.org/10.1016/j.materresbull.2017.01.003.

Sajid Iqbal, Muhammad Younas, Muhmoodul Hassan, Ho Jin Ryu, Mohsin Ali Raza Anjuma, M. (2021). Arshad Farhan, Muhammad Nadeem, Jong-Il Yun. Electronic, electrical, and dielectric analysis of Cr-doped hydroxyapatite. *Chemical Physics Letter*, 771, 138507. https://doi.org/10.1016/j.cplett.2021.138507.

Sánchez-Bajo, F., and Cumbrera, F. L. (1997). The Use of the Pseudo-Voigt Function in the Variance Method of X-ray Line-Broadening Analysis". *Journal of Applied Crystallography*, 30(4), 427–430. https://doi:10.1107/S0021889896015464.

Santhosh, S., and Prabu, S. B. (2012). Synthesis and characterization of nanocrystalline hydroxyapatite from sea shells. *International Journal of Biomedical Nanoscience and Nanotechnology*, 2, 276-283. https://doi:10.1504/ijbnn.2012.051221.

Sauer, G. R., Zunic, W. B., Durig, J. R., and Wuthier, R. E. (1994). Fourier transform Raman spectroscopy of synthetic and biological calcium phosphates. *Calcif Tissue International*, 54, 414–420. https://doi.org/10.1007/BF00305529.

Stanciu, P. L., Sandulescu, G. A., Savu, B., Stanciu, S. G., Paraskevopoulos, K. M., Chatzistavrou, X., and Kontonasaki, E. (2007). Investigation of the Hydroxiapatite growth on bioactive glass surface. *Journal of Biomedical Pharmaceutical Engineering*, 1, 34–39.

Sunila, B. R. and Jagannatham, M. (2016). Producing hydroxyapatite from fish bones by heat treatment. *Materials Letter*, 185, 411–414. https://doi:10.1016/j.matlet.2016.09.039.

Suparto, I. H., and Putri, D. K. (2015). Synthesis of Hydroxyapatite from Rice Fields Snail Shell (bellamyajavanica) through wet method and pore modification using chitosan. *Procedia Chemistry*, 17, 27-35. https://doi.org/10.1016/j.proche.2015.12.120.

Tămăşan, M., Ozyegin, L. S., Oktar, F. N., and Simon, V. (2013). Characterization of calcium phosphate powders originating from Phyllacanthus imperialis and Trochidae Infundibulum concavus marine shells. *Material Science and Engineering C*, 33, 2569–2577. https://doi.org/10.1016/j.msec.2013.02.019.

Tanaka, Y., Takata, S., Shimoe, K., Nakamura, M., Nagai, A., Toyama, T., and Yamashita, K. (2008). Conduction properties of non-stoichiometric hydroxyapatite whiskers for biomedical use. *Journal of the Ceramic Society of Japan*, 116, 815-821. https://10.2109/jcersj2.116.815.

Tavangar, M., Heidari, F., Hayati, R., Tabatabaei, F., Vashaee, D., and Tayebi, L. (2019). Manufacturing and characterization of mechanical, biological and dielectric properties of hydroxyapatite-barium titanate nanocomposite scaffolds. *Ceramics International*, 46, 9086-9095. https://doi:10.1016/j. ceramint.2019.12.157.

Tofail, S. A. M., Haverty, D., Stanton, K. T., and McMonagle, J. B. (2005). Structural order and dielectric behaviour of hydroxyapatite. *Ferroelectrics*, 319(1), 117– 123. http://dx.doi.org/10.1080/00150190590965523.

Türk, S., Altınsoy, İ., Efe, G. Ç., Ipek, M., Özacar, M., and Bindal, C. (2019). Effect of solution and calcination time on sol-gel synthesis of hydroxyapatite. *Journal of Bionic Engineering*, 16, 311–318. https://doi.org/10.1007/s42235-019-0026-3.

Chapter 4

Functionalized Hydroxyapatite in Bone Regeneration

Prasenjit Mukherjee[1], Subhasis Roy[1], Shubhamitra Chaudhuri[1], Rahul Deb Mukherjee[2], Biswanath Kundu[3] and Samit Kumar Nandi[4,*]

[1]Department of Veterinary Clinical Complex,
West Bengal University of Animal & Fishery Sciences, Kolkata, West Bengal, India
[2]Coochbehar Krishi Vigyan Kendra, Pundibari, Cooch Behar, West Bengal, India
[3]Bioceramics and Coating Division,
CSIR-Central Glass & Ceramic Research Institute, Kolkata, West Bengal, India
[4]Department of Veterinary Surgery & Radiology,
West Bengal University of Animal & Fishery Sciences, Kolkata, West Bengal, India

Abstract

From 1990 to 2019, there were globally 178 million new cases of fracture which accounts for a 33.4% increment. The epidemiological studies explored that in the last three decades the years lived with disabilities (YLD) have increased to 65.3%. The challenges of orthopedic surgery still deal with the dearth of satisfactory solutions for the healing of large osseous defects. Even though autografts are the gold standard, surgeons must rely on synthetic graft alternatives due to a lack of availability and donor site morbidity. Due to its close physio-biological and chemical similarity to natural bone, osteoconductive nature, and ease of adaptability with many other necessary elements for functionalization, calcium phosphate hydroxyapatite (HA, Ca10(PO4)6(OH)2) has gained considerable attention as a bone graft substitute since 1980. They

* Corresponding Author's Emails: samitnandi1967@gmail.com.; drsknandi@wbuafscl.ac.in.

In: What to Know about Hydroxyapatite
Editor: Isaac Lavoie
ISBN: 979-8-88697-523-9
© 2023 Nova Science Publishers, Inc.

consistently aid in bone growth on the implant surface. However, limited bioactivity is the main hindrance behind their wide use in orthopedic surgeries. In recent years, several strategies have been undertaken to improve the biological performances with the various ionic substitutions with the apatite. Functionalization is one of the key processes to alter the surface properties of materials aiming toward specific properties to be induced depending on the target tissue or to minimize any unwanted negative or hindrance factors. Functionalization of HA with specific biomolecules is the process by which the surface of HA is made more bioactive i.e., increment of bio integration and osteogenesis. The use of functionalized HA is not only restricted to bone tissue engineering, they are equally important for skin, cartilage, periodontal, vascular tissue, hepatic tissue, tendon, ligament as well as corneal tissue engineering. But, in this chapter, special emphasis has been given to the brief introduction of HA, basic properties, different synthesis techniques, summarized application in biomedical engineering in general, and especially the use and scope of clinical translational aspect of functionalized HA in bone tissue engineering.

Introduction

Bone is a mineralized structure of the body that provides mechanical support as well as helps in forming the body shape and locomotion (Webb and Tricker 2000). It also helps in mineral homeostasis and metabolic functions. Due to some extrinsic and intrinsic factors like trauma and pathological conditions, fractures may occur. Bone has a unique property to regenerate even in adulthood without even forming a scar and the pre-fracture properties of bone are restored (Giannoudis et al., 2007). Most small injuries or fractures heal spontaneously even without the involvement of surgical intervention (Packer and Colditz 1986). But where there is extensive bone loss in conditions like surgical resection of bone tumors, osteomyelitis, fractures with substantial tissue loss, and congenital deformities, segmental or large bone defect occurs (Verrier et al., 2016). But, in some conditions like diabetes, old age, bone neoplasms, and infection this bone defect healing becomes complicated where there is a necessity for auto graft application, which is considered the gold standard (Shafici et al., 2009; Bigham et al., 2011; Parizi et al., 2012). But, autografts have some disadvantages like morbidity and diminished cosmetic appearance of the donor site, pain, as well as less availability for large osseous defects, which have limited their application in clinical practice (Oryan et al., 2013). Allografts and xenografts also have some disadvantages like chances of immune rejection and disease transmission. Whatever may be the method of implantation, the basic aim of bone tissue engineering is to fill the defect area with functionally active new bone which produces fewer complications during the healing process (Oryan et al., 2012). Therefore, modern bone tissue engineering has been built with a triad of osteoconductive scaffolds incorporated with the osteogenic cells as well as factors having osteoinductive properties for the repair of large bone defects (Albrektsson and Johansson 2001; Andreassen and Oxlund 2001). To achieve this goal, many bone substitutes like ceramics, metals, polymers, and organic materials. have been studied to use in orthopedic surgeries (El-Ghannam 2005; Salinas et al., 2013; Zafar et al., 2019; Srinath et al., 2020).

Hydroxyapatite (HA) is a ceramic that has been extensively used as a bone graft substitute as natural bone is HA in its crystalline phase (Yoshikawa et al., 2009; Swetha et al., 2010; Zhou and Lee 2011; Venkatesan and Kim 2014; Cox et al., 2015). It is available in different forms as dense or porous and in blocks or granules. Recently, to enhance its bioactivity, surface modification is done in different ways which are called functionalization (Wilson and Hull 2008; Siniscalco et al., 2019). Hence, biofunctionalization is the process of surface modification HA with a specifically intended biological function, keeping the other properties of the material unaltered (Mas-Moruno 2018; Bhat et al., 2021). The bioactivity by functionalization is done to enhance osteogenic and cell adhesiveness, delivery of drugs, growth factors, cytokines, and stem cells as well as to exert antibacterial and immunomodulatory properties (Bauer et al., 2013; Agarwal and García 2015; Fernandez-Yague et al., 2015; Raphel et al., 2016). Functionalization should be aimed with the following criteria, viz. i) biocompatibility should be maintained, ii) non-cytotoxic, and iii) physical and biological properties should be unaltered. Several methods of functionalization have been investigated by researchers. Decanoic and hexanoic acid (Haider et al., 2017a), phosphoric acid (Pal et al., 2017; Tian et al., 2021b), oleic acid (Cardoso et al., 2016; Wang et al., 2021), sodium dodecyl sulfate (Collins Arun Prakash et al., 2022), carbon nanotubes and gelatin systems (Neffe et al., 2011; Yoon et al., 2014; Wang et al., 2015), and silanes (Wang et al., 2010; Rakmae et al., 2012; Nikbakht et al., 2021) are surface modifiers used by different scientists for functionalization of HA. Besides, grafting of polymers such as L-lactide (LA) (Hong et al., 2004a; Szustakiewicz et al., 2019), and poly(L-lactide) (PLLA) (Hong et al., 2005; Xu et al., 2007; Wang et al., 2010), as well as the incorporation of growth factors to HA, causes enhanced dispersion and cell proliferation on the scaffolds.

Properties and Limitations of Hydroxyapatite

Hydroxyapatite has gained considerable importance due to its wide application as a bone filler (Sotome et al., 2004; Kolanthai et al., 2016; Mudhafar et al., 2022), delivery of a drug, gene, or protein (Kim et al., 2004; Sotome et al., 2004; Boonsongrit et al., 2008; Uskoković and Uskoković 2011; Tram Do et al., 2012; Kolanthai et al., 2016; Mondal et al., 2018, 2018; Thirumalai 2018; Lara-Ochoa et al., 2021), coating material (Sun et al., 2001; Kim et al., 2004; Zhang 2013; Arcos and Vallet-Regí 2020), scaffolds (Pal and Pal 2006; Saiz et al., 2007; Fierz et al., 2008; Anita Lett et al., 2021; Bhat et al., 2021), soft tissue engineering (Ji et al., 2012; Mehedi Hasan et al., 2018). They are biocompatible, osteoconductive, and similar to the composition of the inorganic component of natural bone. The composition of HA is 39.68 wt% Ca, 18.45 wt% P and a Ca/P weight ratio is 2.15 (Boccaccini et al., 2021). Additionally, it contains magnesium fluoride (MgF^{+2}), chloride, sodium (Na^{+}), potassium (K^{+}), hydrogen phosphate (HPO_4^{2-}), and carbonate (CO_3^{2-}) (Cl^{-}) (Bilton 2012; Shavandi et al., 2015). HA has a hexagonal structure having P63/m space group number 176 where $\alpha = \beta = 90°$ and $\gamma = 120°$ (Shavandi et al., 2015). It can increase local Ca^{2+} concentration when applied as an implant, which in turn triggers the proliferation of osteoblast and mesenchymal stem cells (Zhao et al., 2017). As it is bioactive, non-immunogenic, biocompatible, and osteoconductive, hence it has wide application in bone tissue engineering (Murugan and Ramakrishna 2004; LeGeros 2008; Oliveira et al., 2017; Szcześ et al., 2017). Many variations of HA have been used clinically by several scientists viz. HA in pure form, HA with ion doping, and HA/polymer composites due to these properties.

Despite the advantages of using HA in bone tissue engineering, a few limitations exist:

1. Brittleness (Koons et al., 2020)
2. Cluster formation (Koons et al., 2020).
3. Low mechanical strength (Fiume et al., 2021)
4. Simple porous structure, i.e., grid-like arrangement of the macro channel (different from trabecular and cancellous bone) (Fiume et al., 2021)
5. Low biodegradability and limited fabrication technique for scaffold preparation (Mondal and Pal 2019).
6. Low drug-loading capacity and faster release rate during drug delivery (Lin and Chang 2015).

Different Synthesis Techniques for Making Functionalized HA

Researchers have reported different methods of synthesis or fabrication of functionalized HA for various end-use. These are summarised as follows:

Li et al. prepared a photocurable dental resin composite with improved antibacterial and mineralization properties by mixing different ratios of Zn and Sr doped HA into bisphenol A glycidyl methacrylate and triethylene glycol dimethacrylate mixture along with barium glass powder (Li et al., 2022). Tailorable Zn and Sr doped HA having nanorod like morphology, which was prepared by simple chemical coprecipitation method, served a superior functional filler in such applications of dental resin composite. Boanini et al., synthesized poly(ethylenimine) functionalized HA-supported platinum nano-particles with higher efficiency by the chemical reduction method of loading due to the presence of polyelectrolyte that interacted with Pt(II) metallic ions (Boanini et al., 2020). The composite showed cytotoxicity towards osteosarcoma cell lines, anti-oxidant properties, differentiation toward primary and matured osteoblastic cell lines with the indication of ALP and Collagen type 1 increase and reduction of ROS level. Kahjuria et al., used hydrothermal-cum-coprecipitation method to prepare very stable, uniform nanoparticles of nitrogen-doped carbon dots coupled with HA (Khajuria et al., 2018). Owing to this composite's ability to induce osteoblast differentiation and proliferation, this demonstrated advanced bone regeneration in Zebra fish jawbone. Deshmukh et al., investigated functionalized HA nanoparticles' ability for genetic transformation of yeast species (*Candida albicans*) and contributed to advancement in the field of biotechnology (Deshmukh et al., 2019). To deliver plasmid DNA (pDNA), they fabricated novel arginine-glucose-PEC functionalized HA nanoparticles as vehicles. The total system was non-toxic to yeast cells and did not require the preparation of efficient cells. Zantye et al., utilized a modified sol-gel method for making a self-fluorescent promising gene delivery system using cell-penetrating peptide-functionalized HA nanoparticles (R8HNPs) and checked for its therapeutic potential of delivering small interference RNA (siRNA) which has a potential in regenerative medicine (Zantye et al., 2021). To improve the present adsorbents' poor mechanical strength, biocompatibility, and adsorption capability, particularly during the hemoperfusion therapy for liver failure, Chai et al., proposed a new class of inorganic-organic co-crosslinked nanocomposite mesoporous adsorbent with remarkably better mechanical properties using vinyltriethoxysilane-functionalized HA nanoparticles and non-ionic styrene-divinylbenzene resins by using suspension polymerization method (Chai et al., 2021). Further, they verified and established the efficacy of the new system in *in vivo* pig model for directly removing bilirubin from whole blood. Verma et al., functionalized HA nanoparticles with glycine (amino acid) *in situ* by simple co-precipitation method for immobilizing doxorubicin covalently in a pH-responsive system with reduced side effects (Verma et al., 2018). They established dose and time-dependent cytotoxicity of doxorubicin-functionalized HA *in vitro* towards mouse fibrosarcoma cancer cells and demonstrated good cellular uptake of doxorubicin too. Youssef et al., proposed a new cost-effective bio-composite system for the treatment of osteomyelitis (Youssef et al., 2021). This was a bio-mixture of ciprofloxacin functionalized on a microwave-synthesized gold nanoparticle Zeolite-A and nano HA particle (mixed in different ratios). After thorough materials characterization, drug loading and release kinetics were investigated and fitted with the Higuchi square root model. Tian et al., fabricated blocks, laminates and flower balls of HA functionalized with two natural polysaccharide gums, i.e., tamarind seed polysaccharide gum and guar gum, in a single-step synthesis process using $Ca(NO_3)_2.4H_2O$ and $(NH_4)H_2PO_4$ as raw materials (Tian et al., 2021a). This was further conjugated with doxorubicin, studied for drug loading and elution kinetics, and showed excellent drug-loading performance. Linh et al., conjugated collagen and bone morphogenetic protein-2 on the surface of porous HA scaffolds and compared the adhesion and osteogenic development of human adipose-derived stem cells *in vitro* using these functionalized scaffolds (Linh et al., 2020). The authors observed enhanced osteogenic differentiation of cultured hADSCs. Zhang et al. adopted an electrophoretic deposition method to coat titanium substrate using antibacterial-functionalized Ag-loaded HA particles with Mg^{2+} as an inorganic binder (Zhang et al., 2021a). They anticipated this coating as potential in implant surgery because this will have an antibacterial function. While Oulguidoum et al., synthesized sulfonate-functionalized HA nanopowders by organo-functionalization using natural phosphate and Benzenedisulfonate molecules for heavy metals, such a Cd (II) removal in aqueous solution (Oulguidoum et al., 2019). The material was environment friendly and had excellent adsorption capacity, and shorter sorption reaction time. Verma et al., adopted a wet-chemical approach to prepare a colloidally stable nanoformulation of hydrophobic curcumin by encapsulating it into the hydrophilic network of gelatin and HA (Verma et al., 2021). The formulation showed a pH-dependent controlled release of curcumin and had better cytotoxicity towards A549 cancer cells in comparison with WI-26 VA4 normal cells. Cellular uptake of the new formulation was more in A549 cells than free curcumin. Hussain et al., reported development of one composite nanofibrous coating on implant by electrospinning technique for bone defects which not only worked as cell-instructive bioengineered degradable scaffold but also helped to restore pre-existing bone function (Hussain et al., 2022). There were 3 components of the composite, viz., Fe-doped HA nanorods as inorganic component, functionalized with collagen/polycaprolactone as organic component and FePN (iron phenolic network) as surface modifier, crosslinker and functional agent of nanorods. *In vitro* studies supported better bone cell attachment, proliferation, matrix mineralization and osteogenic gene expression while *in vivo* studies in rat model bone-implant zone revealed bioactive potential, avoiding ROS-mediated issues. Salamanna et al., developed, characterized and compared 2 bone graft materials, viz., alendronate functionalized HA nanocrystals (at increasing bisphosphonate content) and Sr substituted HA nanocrystals [with Sr/(Ca + Sr) ratio of 0.05 and 0.1] for posterolateral fusion rate in ovariectomized rats (Salamanna et al., 2019). Their study suggested an alternate biomaterial for spinal fusion especially for osteoporotic subjects. In another work, Vermal et al., prepared a promising nanocarrier targeted drug delivery system comprised of gelatin functionalized HA nanoparticles capable of binding both folic acid (a cancer cell targeting ligand) and anticancer drug doxorubicin hydrochloride, as gelatin is a protein which has both carboxyl and amine moieties (Verma et al., 2020). Folic acid was chemically conjugated to the nanoparticles by 1-Ethyl-3-(3-dimethylaminopropyl) carbodiimide (EDC)-N-hydroxysuccinimide (NHS) coupling reaction. *In vitro* study with said formulation was carried out and compared with different cell lines (e.g., HepG2, WEHI-164, WI-26 VA4, WRL-68 and KB) and also checked for targeted therapeutic efficacy of doxorubicin. Cirillo et al., functionalized Sr substituted HA with □-lactam integrin agonists for synergistic properties in promoting cell adhesion and activation of intracellular signalling restricting abnormal bone resorption (Cirillo et al., 2021). Authors also studied □-lactam release kinetics in aqueous medium from functionalized SrHA. They demonstrated better human primary mesenchymal stem cell adhesion, viability and differentiation towards osteoblastic lineage and enhanced inhibitory role of Sr on osteoclastic viability and differentiation. Zhang et al., worked in the field of heavy metal pollution and came up with amino trimethylphosphonate (ATMP) functionalized HA for efficient removal of Pb^{2+}, Cd^{2+} heavy metal ions by ion exchange and predominantly surface complexation (Zhang et al., 2021b). Authors proposed the mechanism of this complexation and inferred that phosphonic group of ATMP had strong affinity with HA.

Dixit et al., recently reported gum tragacanth (having both angiogenic and osteogenic characteristics) multi-layer functionalized nano HA through a silanized nHA intermediate (Dixit et al., 2022). They have shown osteogenic potential *in vitro* with MG-63 and hMSCs cell lines with increased cellular VEGF expression in MG-63. There was formation of tubular human umbilical vein endothelial cells which showed its angiogenic potential as well. Salahuddin et al., functionalized HA nanorods (with approx. 15 nm diameter and 65.24 nm length) with citrate by simple hydrothermal-precipitation process and loaded a common anti-malarial drug used in cancer therapy, artesunate (ART) for colon cancer therapies upto *in vitro* level (Salahuddin et al., 2022). This was further encapsulated into surfactant free PLGA nanoparticles, characterized in depth and studied for sustained release of ART at two colon cancer cell lines. Jiang et al., studied different HA material to understand the clinical application of HA as a drug carrier of popular anticancer drug doxorubicin (Jiang et al., 2022). For the purpose, they prepared single and binary ion doped HA (with Zn, Ba, F, Zn/F and Ba/F) by simple hydrothermal method, thoroughly evaluated the material, further functionalized with the drug doxorubicin and finally checked the drug loading capacity and release kinetics. They concluded Ba/F codoped HA had highest loading efficiency followed by Ba doped HA. In order to simultaneously diagnose and treat bone problems, Cipreste et al., developed a new approach., they selected HA nanoparticles, functionalized/covalently bonded with folic and medronic acid and further complexed with one radioisotope copper 64 for specific targeting to bone disorders, thus proposed a significantly potential theranostic alternative material (Cipreste et al., 2020). Sukhodub et al., used negatively charged sodium alginate functionalized HA nanoparticles to couple positively charged antibacterial agent chlorhexidine and studied the composite for local drug delivery in phosphate-buffered saline using HPLC method (Sukhodub et al., 2018). While Pandey et al., proposed one formulation for the treatment of incurable rheumatoid arthritis, an autoimmune inflammatory disorder (Pandey et al., 2021). They loaded methotrexate and teriflunomide into HA nanoparticles functionalized with hyaluronic acid ligand to accomplish CD44-mediated active targeted administration to collagen-induced arthritic rat knee joints. Though both methotrexate and teriflunomide are having severe hepatotoxicity, but the new formulation showed less cell viability than the drugs on RAW 264.7 macrophage cells *in vitro* and *in vivo*, while performing hepatotoxicity, liver histopathology and liver enzyme assay showed about 29% reduction in hepatotoxicity in comparison with conventional oral treatments. Almulaiky et al., investigated and used HA and

HA decorated ZrO_2 for immobilization of α-amylase enzyme at different pH and temperature, thus functionalized them in the extent of 69.7 and 84% respectively (Almulaiky et al., 2021). Immobilized α-amylase on HA-ZrO_2 showed comparatively higher catalytic activity than free α-amylase. Dogan et al., prepared polymer-ceramic nanocomposite and/or composite by solution removal method using PMMA and nano HA (Dogan et al., 2020). Further, biocompatibilities and effects of this functionalized material on some antioxidant enzymes were investigated. Nano bio-composites prepared in tetrahydrofuran medium were more thermally stable but less hemocompatible than which was prepared in acetone and both materials had insignificant negative effect on antioxidant enzymes. Enrich-Essvein et al., developed one alternative solution material for demineralized dentin for which they have developed functionalized nano HA with proanthocyanidin and studied for remineralization and collagen stabilization of such dentin (Enrich-Essvein et al., 2021). Different groups were selected for treatment of coronal dentin and samples were further studied for chemical, mineralogical, thermal and mechanical analysis. They came to the conclusion that nano HA-proanthocyanidin can be an alternate method to fortify demineralized dentin due to its capacity to enhance collagen stability and reinforcing. Lee et al., studied loading and release profile of anticancer drug curcumin nanoparticles (which is in otherwise having hydrophobic characteristics leading to low bioavailability in cancer cells) on nano HA via physical adsorption, manipulating the adsorption capacity of the formulation by functionalizing HA surface by different carboxylic acids, such as, lactic acid, tartaric acid and citric acid (Lee et al., 2019). Further, curcumin loaded HA were evaluated on breast cancer cell line MCF-7 with improved anti-cancer activities of curcumin. Fang et al., used a simple chemical precipitation technique for preparing HA functionalized with zoledronate hybrid nano-biomaterial (Fang et al., 2020). This was further utilized for the removal of Pb^{2+} and Cu^{2+}. The sorption mechanism was probed and it was found that ion exchange and dissolution-precipitation mechanism rather than surface complexation was predominant for Pb^{2+} on HA, while surface complexation was inferred for zoledronate-HA.

Biomedical and Clinical Applications of Functionalized Hydroxyapatite

The use of HA in bone tissue engineering started long back in 1969 (Levitt et al., 1969). The interactions between HA and different proteins are the subject of interest to researchers for their effective use as bioactive materials (Haider et al., 2017b). We will discuss the applications of HA in different subsections.

Drug Delivery Carrier

The inherent nature of the rough surface area of the HA favors its use as a carrier material. In addition to this property, HA also has two binding sites; 'C'(Ca^{+2}) and 'P'(PO_4^-), which again help to attach different proteins with the HA molecules(Kirkham et al., 2002). During the last few years, HA as a drug or growth factor carrier has been reviewed (Rimola et al., 2008; Haider et al., 2014a, 2014b) which reflects the surface modification of HA by glutamic acid and tagged growth factor(Hong et al., 2004b), surface modified HA by hydrazine bisphosphonate and oriented protein with better performance as compared with control (Yewle et al., 2012). HA aptamers are effective in the treatment of cancer (Li et al., 2012). HA as coating material is also

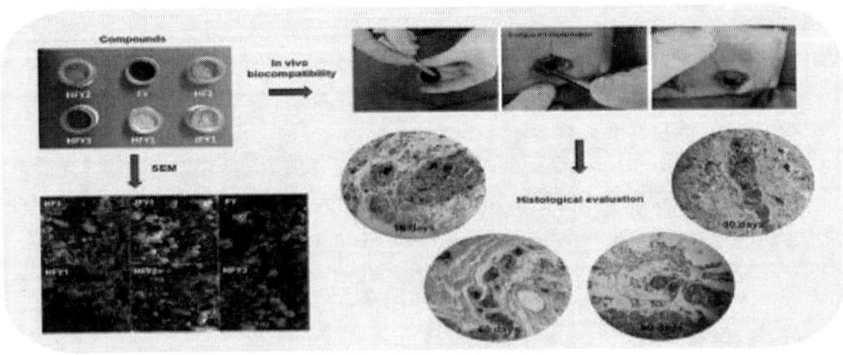

Figure 1. Schematic representation of synthetic hydroxyapatite compounds associated with magnetite nanoparticles for a drug delivery system in osteomyelitis treatment (Reproduced from Ferreira-Ermita et al., 2020 as open access article under a Creative Commons license).

investigated for delivery of drugs like dexamethasone and found suitable for sustained release of the drugs even up to 30 days (Son et al., 2013). The release of ibuprofen from mesoporous HA has been investigated with good dissolution and controlled release (Ibrahim et al., 2015, p. 2). Ciprofloxacin release from HA with magnetic nanoparticles in rabbit models showed remarkable positive effects on the control of osteomyelitis (Ferreira-Ermita et al., 2020). Other research has also been conducted to improve drug delivery by surface modification of HA with enhanced bioactivity (Loo et al., 2010; Koocheki et al., 2011; Lee et al., 2011; Yang et al., 2013a).

HA as Coating Material

As natural bone contains calcium phosphate in the form of HA crystals, hence synthetic HA is a good option for use in bone tissue engineering. Though it contains some impurities like Na^{+2} and Mg^{+2} ions, still its surface modifications have been proven to be excellent for bone tissue engineering. One unique property of HA is that it is not only osteoinductive but also osteoconductive between implant and bone (Orlovskii et al., 2002). The conventional techniques for bone defect management include autografting, allografting, and synthetic bone implants, but most of the time they resorb too early and remain ineffective (Hetrick and Schoenfisch 2006; Goodman et al., 2013). To overcome these problems, HA plays an important alternative role as it possesses signal carriers, morphogenetic signals, and responsive cells in the implant (Yu et al., 2003). The most common coating material is HA as it enhances bioactivity and biocompatibility with increased bio-integration (De Groot et al., 1987). Coating of HA by plasma spraying method on Ti implant has been proven to overcome the suboptimal fatigues of the implant (He et al., 2012). This is also seen that coating with HA has increased the shear strength of the implant near the normal bone (Geesink et al., 1988). Different techniques have been used to coat HA on implanting materials which include sputter coating, pulse laser deposition, dip coating, and hot pressing, and electrostatic sputter layer deposition (He et al., 2012; Goodman et al., 2013). Additionally, coating with HA on silicon nitride has also been investigated (Silva et al., 2008). Coating with HA stimulates the osteoid formation and hence increases the binding between bone and implant (LeGeros 2002). It has been found that coating with HA significantly reduces the release of metal ions in the implants and thereby reduces the chance of local toxicity and inflammatory reactions which in terms of healing and time to achieve a goal is better (Suchanek and Yoshimura 1998b).

HA Ceramics in Bone Tissue Engineering

Because HA closely resembles natural bone, it is frequently employed in composites for bone tissue creation. HA bioceramic possesses osteoinductive as well as osteoconductive properties and may be fabricated in various forms (powder, nanoparticles) with ammonium methacrylate, orthophosphate, polyoxyethylene lauryl ether (Dorozhkin 2013). Porous ceramics have been extensively used in bone tissue engineering (Uchida et al., 1992; Liu 1997, 1997; Hing et al., 1999; Paul and Sharma 1999). Various reinforced bioceramics in metals, nanoparticles, and fiber forms with HA have been studied for enhanced mechanical strength but have sometimes been found to have reduced biocompatibility. This problem was redressed by HA -Zr composite bioceramic and advocated for use both in dental and orthopedic applications (Ahn et al., 2001).

HA Composite

As it is evident from an earlier experiment that polymer can induce bone formation (Van De Putte and Urist 1965; Urist et al., 1967) and the implant-polymer interface plays a major role in bone regeneration (Kulkarni et al., 1971), extensive research on the use of different HA composite like PLGA (Zhang et al., 2009), PLA (Higashi ct al., 1986; Flahiff et al., 1996), poly-caprolactone (Dunn et al., 2001; Zhao et al., 2008) have been done with better (Yang et al., 2013b) osteoconductivity, osteointegration along with providing sufficient stability (Rimola et al., 2008; Haider et al., 2014a, 2014b). Accelerated bone formation was observed in insulin and BMP-2 immobilized HA-PLGA composite(Haider et al., 2014a). Pamidronic acid immobilized HA has been investigated with excellent results in osteoporosis where it was observed reduced osteoclastic activity due to pamidronic acid and enhanced osteoblastic role for HA (Ignjatović et al., 2001). Better composite strength and stability have been investigated for HA-PLLA and HA-Polyethylene(Wang et al., 1998), HA-Collagen (Reddi 1998; Bakoš et al., 1999; Dalby et al., 1999; Chevallay and Herbage 2000; Rhee et al., 2000; Orlovskii et al., 2002), HA-PMMA composite. The Composite of HA and chitosan has also been investigated with promising results in terms of improved mechanical strength (Chesnutt et al., 2009). Coral-derived HA functionalized with autologous periosteal cells showed better healing in the traumatic injury of thumbs in a human patient (Vacanti et al., 2001). In a similar study, coral-derived HA functionalized with insulin-like growth factor and BMP-2 showed an acceleration of bone growth in a rabbit bone defect model (Figure 2) (Nandi et al., 2015). Autologous bone marrow-HA composite is used to treat tumors of the tibia and femur with positive results (Morishita et al., 2006). Some of the commercially available HA products for bone tissue engineering have been presented in Table 1 (Dorozhkin).

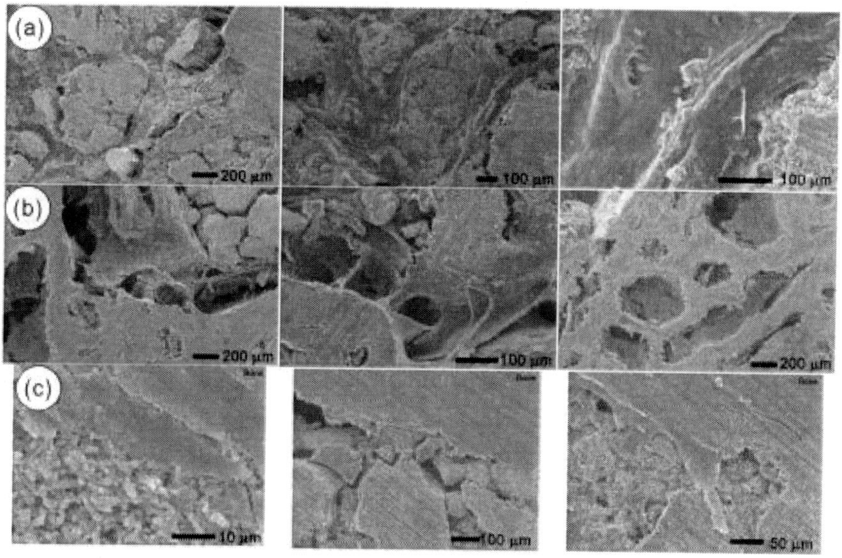

Figure 2. Scanning electron microscope photograph of the implanted bone sample interface showed less interfacial gap for (c) converted coral with BMP-2 followed by converted coral with IGF-1 and (a) group I (bare converted coral) after 90 days post-operatively.

Table 1. Available HA products for bone tissue engineering

Product name	Producing Company	Country
Actifuse	ApaTech	UK
ApaPore	ApaTech	UK
Apaceram	Pentax	Japan
Bonefil	Pentax	Japan
Bonetite	Pentax	Japan
Bonoceram	Sumitomo Osaka Cement	Japan
Bioroc	Depuy—Bioland	France
Cerapatite	Ceraver	France
BoneSource	Stryker Orthopaedics	NJ, USA
Calcitite	Zimmer	IN, USA
Osteograf	Ceramed	CO, USA

Future Direction

With the advancement of bioengineering, different graft substitutes have evolved and studied. Among these, HA is one of the most important candidates which successfully serve as a bone implant, coating material on the metallic implant, and vehicle for drug delivery. Despite many advantages, there are a few limitations to using HA in load-bearing orthopedic implants. Hence, the researchers are trying to develop the implant using HA with higher mechanical strength. For this purpose, polymers are used as a composite with HA to improve mechanical strength, elasticity, and toughness. Functionalization of HA nanoparticles with emulsifying and bridging agents as discussed earlier not only prevent aggregation but also help in the preparation of composite scaffolds, and the delivery of growth factors, genes, and drug in a controlled manner. Different synthesis methods of functionalization have been covered in this chapter, but an ideal technique that is most suitable for the preparation of biomimetic implants is still under investigation.

References

Agarwal R., García A. J. Biomaterial strategies for engineering implants for enhanced osseointegration and bone repair. *Adv Drug Deliv Rev.* 2015 Nov 1;94:53–62.

Ahn E. S., Gleason N. J., Nakahira A., Ying J. Y. Nanostructure Processing of Hydroxyapatite-based Bioceramics. *Nano Lett.* 2001 Mar 1;1(3):149–53.

Albrektsson T., Johansson C. Osteoinduction, osteoconduction and osseointegration. *Eur Spine J.* 2001 Oct 1;10(2):S96–101.

Almulaiky Y., Khalil N. M., El-Shishtawy R. M., Altalhi T., Algamal Y., Aldhahri M., Yousif Algamal, Musab Aldhahri, Sami A Al-Harbi, Esam S Allehyani, Muhammad Bilal, Mustafa M Mohammed. Hydroxyapatite-decorated ZrO2 for a-amylase immobilization: Toward the enhancement of enzyme stability and reusability. *Int J Biol Macromol.* 2021;167:299–308.

Andreassen T. T., Oxlund H. The effects of growth hormone on cortical and cancellous bone. *J Musculoskelet Neuronal Interact.* 2001 Sep;2(1):49–58.

Anita Lett J., Sagadevan S., Fatimah I., Hoque M. E., Lokanathan Y., Léonard E., Solhe F. Alshahateet, Romana Schirhagl, Won Chun Oh. Recent advances in natural polymer-based hydroxyapatite scaffolds: Properties and applications. *Eur Polym J.* 2021 Apr 5;148:110360.

Arcos D., Vallet-Regí M. Substituted hydroxyapatite coatings of bone implants. *J Mater Chem B.* 2020;8(9):1781–800.

Bakoš D., Soldán M., Hernández-Fuentes I. Hydroxyapatite–collagen–hyaluronic acid composite. *Biomaterials.* 1999 Jan 1;20(2):191–5.

Bauer S., Schmuki P., von der Mark K., Park J. Engineering biocompatible implant surfaces: Part I: Materials and surfaces. *Prog Mater Sci.* 2013 Apr 1;58(3):261–326.

Bhat S., Uthappa U. T., Altalhi T., Jung H. Y., Kurkuri M. D. Functionalized Porous Hydroxyapatite Scaffolds for Tissue Engineering Applications: A Focused Review. *ACS Biomater Sci Eng* [Internet]. 2021 Sep 9 [cited 2022 Aug 2]; Available from: https://doi.org/10.1021/acsbiomaterials.1c00438.

Bigham A. S., Shadkhast M., Bigham Sadegh A., Shafiei Z., Lakzian A., Khalegi M. R. Evaluation of osteoinduction properties of the demineralized bovine foetal growth plate powder as a new xenogenic biomaterial in rat. *Res Vet Sci.* 2011 Oct 1;91(2):306–10.

Bilton M. W. Nanoparticulate hydroxyapatite and calcium-based CO2 sorbents [Internet] [phd]. *University of Leeds;* 2012 [cited 2022 Aug 6]. Available from: https://etheses.whiterose.ac.uk/3847/?utm_term=wreo&utm_source=twitterfeed&utm_medium=twitter.

Boanini E., Torricelli P., Cassani M. C., Rubini K., Fini M., Pagani S., Adriana Bigi. Platinum nanoparticles supported on functionalized hydroxyapatite: Anti-oxidant properties and bone cells response. *Ceram Int.* 2020;46(11):19574–82.

Boccaccini A. R., Ma P. X., Liverani L. *Tissue Engineering Using Ceramics and Polymers.* Woodhead Publishing; 2021.

Boonsongrit Y., Abe H., Sato K., Naito M., Yoshimura M., Ichikawa H., Yoshinobu Fukumori. Controlled release of bovine serum albumin from hydroxyapatite microspheres for protein delivery system. *Mater Sci Eng B.* 2008 Feb 25;148(1):162–5.

Cardoso G. B. C., Maniglio D., Volpato F. Z., Tondon A., Migliaresi C., Kaunas R. R., Cecilia A. C. Zavaglia. Oleic acid surfactant in polycaprolactone/hydroxyapatite-composites for bone tissue engineering. *J Biomed Mater Res B Appl Biomater.* 2016;104(6):1076–82.

Chai Y., Liu Z., Du Y., Wang L., Lu J., Zhang Q., Wenyan Han, Tingting Wang, Yameng Yu, Lisha Sun, Lailiang Ou Hydroxyapatite reinforced inorganic-organic hybrid nanocomposite as high-performance adsorbents for bilirubin removal *in vitro* and in pig models. *Bioact Mater.* 2021;6(12):4772–85.

Chesnutt B. M., Viano A. M., Yuan Y., Yang Y., Guda T., Appleford M. R., Joo L. Ong, Warren O. Haggard, Joel D. Bumgardner. Design and characterization of a novel chitosan/nanocrystalline calcium phosphate composite scaffold for bone regeneration. *J Biomed Mater Res A.* 2009;88A(2):491–502.

Chevallay B, Herbage D. Collagen-based biomaterials as 3D scaffold for cell cultures: applications for tissue engineering and gene therapy. *Med Biol Eng Comput.* 2000 Mar 1;38(2):211–8.

Cipreste M., da Nova Mussel W., da Silva J., de Freitas Marques M., Batista R., Gastelois P., Waldemar Augusto de Almeida, Macedo, Edésia Martins Barros de Sousa. A new theranostic system for bone disorders: Functionalized folate-MDP hydroxyapatite nanoparticles with radiolabeled copper-64. *Mater Chem Phys.* 2020;254:123265.

Cirillo M., Martelli G., Boanini E., Rubini K., Di Filippo M., Torricelli P., Stefania Pagani, Milena Fini, Adriana Bigi, Daria Giacomini. Strontium substituted hydroxyapatite with b-lactam integrin agonists to enhance mesenchymal cells adhesion and to promote bone regeneration. *Colloids Surf B Biointerfaces.* 2021;200:111580.

Collins Arun Prakash V., Venda I., Thamizharasi V., Sathya E. A comparative study on microemulsion synthesis of hydroxyapatite powders by ionic and Non-Ionic surfactants. *Mater Today Proc.* 2022 Jan 1;51:1701–5.

Cox S. C., Thornby J. A., Gibbons G. J., Williams M. A., Mallick K. K. 3D printing of porous hydroxyapatite scaffolds intended for use in bone tissue engineering applications. *Mater Sci Eng C.* 2015 Feb 1;47:237–47.

Dalby M. J., Di Silvio L., Harper E. J., Bonfield W. *In vitro* evaluation of a new polymethylmethacrylate cement reinforced with hydroxyapatite. *J Mater Sci Mater Med.* 1999 Dec 1;10(12):793–6.

De Groot K., Geesink R., Klein C. P. A. T., Serekian P. Plasma sprayed coatings of hydroxylapatite. *J Biomed Mater Res.* 1987;21(12):1375–81.

Deshmukh K., Ramanan S. R., Kowshik M. A novel method for genetic transformation of C. albicans using modified-hydroxyapatite nanoparticles as a plasmid DNA vehicle. *Nanoscale Adv.* 2019;1(8):3015–22.

Dixit K., Kulanthaivel S., Agarwal T., Pal K., Giri S., Maiti T. K., Indranil Banerjee. Gum tragacanth modified nano-hydroxyapatite: An angiogenic-osteogenic biomaterial for bone tissue engineering. *Ceram Int.* 2022;48(10):14672–83.

Dogan S., Ozcan T., Dogan M., Turhan Y. The effects on antioxidant enzymes of PMMA/hydroxyapatite nanocomposites/composites. *Enzyme Microb Technol.* 2020;142:109676.

Dorozhkin S. V. Calcium Orthophosphate-Based Bioceramics. *Materials.* 2013 Sep;6(9):3840–942.

Dorozhkin S. V. *Medical Application of Calcium Orthophosphate Bioceramics.*

Dunn A. S., Campbell P. G., Marra K. G. The influence of polymer blend composition on the degradation of polymer/hydroxyapatite biomaterials. *J Mater Sci Mater Med.* 2001 Aug 1;12(8):673–7.

El-Ghannam A. Bone reconstruction: from bioceramics to tissue engineering. *Expert Rev Med Devices.* 2005 Jan 1;2(1):87–101.

Enrich-Essvein T., Rodriguez-Navarro A. B., Alvarez-Lloret P., Cifuentes-Jimenez C., Bolanos-Carmona M. V., Gonzalez-Lopez S. Proanthocyanidin-functionalized hydroxyapatite nanoparticles as dentin biomodifier. *Dent Mater.* 2021;37(9):1437–45.

Fang X., Zhu S., Ma J., Wang F., Xu H., Xia M. The facile synthesis of zoledronate functionalized hydroxyapatite amorphous hybrid nanobiomaterial and its excellent removal performance on Pb^{2+} and Cu^{2+}. *J Hazard Mater.* 2020;392:122291.

Fernandez-Yague M. A., Abbah S. A., McNamara L., Zeugolis D. I., Pandit A., Biggs M. J. Biomimetic approaches in bone tissue engineering: Integrating biological and physicomechanical strategies. *Adv Drug Deliv Rev.* 2015 Apr 1;84:1–29.

Ferreira-Ermita D. A. C., Valente F. L., Carlo-Reis E. C., Araújo F. R., Ribeiro I. M., Cintra C. C. V., Andréa P. B. Borges. Characterization and *in vivo* biocompatibility analysis of synthetic hydroxyapatite compounds associated with magnetite nanoparticles for a drug delivery system in osteomyelitis treatment. *Results Mater.* 2020 Mar 1;5:100063.

Fierz F. C., Beckmann F., Huser M., Irsen S. H., Leukers B., Witte F., O. Degistirici, A. Andronache, M. Thie, B. Müller. The morphology of anisotropic 3D-printed hydroxyapatite scaffolds. *Biomaterials.* 2008 Oct 1;29(28):3799–806.

Fiume E., Magnaterra G., Rahdar A., Verné E., Baino F. Hydroxyapatite for Biomedical Applications: A Short Overview. *Ceramics.* 2021 Dec;4(4):542–63.

Flahiff C. M., Blackwell A. S., Hollis J. M., Feldman D. S. Analysis of a biodegradable composite for bone healing. *J Biomed Mater Res.* 1996;32(3):419–24.

Geesink R., de Groot K., Klein C. Bonding of bone to apatite-coated implants. *J Bone Joint Surg Br.* 1988 Jan;70-B(1):17–22.

Giannoudis PV, Einhorn TA, Marsh D. Fracture healing: The diamond concept. *Injury.* 2007 Sep 1;38:S3–6.

Goodman S. B., Yao Z., Keeney M., Yang F. The future of biologic coatings for orthopaedic implants. *Biomaterials.* 2013 Apr 1;34(13):3174–83.

Haider A., Gupta K. C., Kang I. K. Morphological Effects of HA on the Cell Compatibility of Electrospun HA/PLGA Composite Nanofiber Scaffolds. *BioMed Res Int.* 2014a Feb 26;2014:e308306.

Haider A., Gupta K. C., Kang I. K. PLGA/nHA hybrid nanofiber scaffold as a nanocargo carrier of insulin for accelerating bone tissue regeneration. *Nanoscale Res Lett.* 2014b Jun 25;9(1):314.

Haider A., Haider S., Soo Han S., Kang I. K. Recent advances in the synthesis, functionalization and biomedical applications of hydroxyapatite: a review. *RSC Adv.* 2017a;7(13):7442–58.

Haider A., Haider S., Soo Han S., Kang I. K. Recent advances in the synthesis, functionalization and biomedical applications of hydroxyapatite: a review. *RSC Adv.* 2017b;7(13):7442–58.

He J., Huang T., Gan L., Zhou Z., Jiang B., Wu Y., Fang Wu, Zhongwei Gu Collagen-infiltrated porous hydroxyapatite coating and its osteogenic properties: *In vitro* and *in vivo* study. *J Biomed Mater Res A.* 2012;100A(7):1706–15.

Hetrick E. M., Schoenfisch M. H. Reducing implant-related infections: active release strategies. *Chem Soc Rev.* 2006 Aug 23;35(9):780–9.

Higashi S., Yamamuro T., Nakamura T., Ikada Y., Hyon S. H., Jamshidi K. Polymer-hydroxyapatite composites for biodegradable bone fillers. *Biomaterials.* 1986 May 1;7(3):183–7.

Hing K. A., Best S. M., Tanner K. E., Bonfield W., Revell P. A. Quantification of bone ingrowth within bone-derived porous hydroxyapatite implants of varying density. *J Mater Sci Mater Med.* 1999 Oct 1;10(10):663–70.

Hong Z., Qiu X., Sun J., Deng M., Chen X., Jing X. Grafting polymerization of l-lactide on the surface of hydroxyapatite nano-crystals. *Polymer.* 2004a Sep 3;45(19):6699–706.

Hong Z., Qiu X., Sun J., Deng M., Chen X., Jing X. Grafting polymerization of l-lactide on the surface of hydroxyapatite nano-crystals. *Polymer.* 2004b Sep 3;45(19):6699–706.

Hong Z., Zhang P., He C., Qiu X., Liu A., Chen L., Xiabin Jing Nano-composite of poly(l-lactide) and surface grafted hydroxyapatite: Mechanical properties and biocompatibility. *Biomaterials.* 2005 Nov 1;26(32):6296–304.

Hussain Z., Ullah I., Liu X., Shen W., Ding P., Zhang Y., Tong Gao, Mojtaba Mansoorianfar, Tian Gao, Renjun Peia. Tannin-reinforced iron substituted hydroxyapatite nanorods functionalized collagen-based composite nanofibrous coating as a cell-instructive bone-implant interface scaffold. *Chem Eng J.* 2022;438:135611.

Ibrahim A. R., Li X., Zhou Y., Huang Y., Chen W., Wang H., Jun Li. Synthesis of Spongy-Like Mesoporous Hydroxyapatite from Raw Waste Eggshells for Enhanced Dissolution of Ibuprofen Loaded via Supercritical CO2. *Int J Mol Sci.* 2015 Apr;16(4):7960–75.

Ignjatović N., Delijić K., Vukcević M., Uskoković D. The Designing of Properties of Hydroxyapatite/Poly-l-lactide Composite Biomaterials by Hot Pressing. *Int J Mater Res.* 2001 Feb 1;92(2):145–9.

Ji D. Y., Kuo T. F., Wu H. D., Yang J. C., Lee S. Y. A novel injectable chitosan/polyglutamate polyelectrolyte complex hydrogel with hydroxyapatite for soft-tissue augmentation. *Carbohydr Polym.* 2012 Aug 1;89(4):1123–30.

Jiang X., Zhao Y., Wang C., Sun R., Tang Y. Effects of physico-chemical properties of ions-doped hydroxyapatite on adsorption and release performance of doxorubicin as a model anticancer drug. *Mater Chem Phys.* 2022;276:125440.

Khajuria D. K., Kumar V. B., Gigi D., Gedanken A., Karasik D. Accelerated bone regeneration by nitrogen-doped carbon dots functionalized with hydroxyapatite nanoparticles. *ACS Appl Mater Interfaces.* 2018;10(23):19373–85.

Kim H. W., Knowles J. C., Kim H. E. Hydroxyapatite/poly(ε-caprolactone) composite coatings on hydroxyapatite porous bone scaffold for drug delivery. *Biomaterials.* 2004 Mar 1;25(7):1279–87.

Kirkham J., Brookes S. J., Shore R. C., Wood S. R., Smith D. A., Zhang J., Haifeng Chen, Colin Robinson. Physico-chemical properties of crystal surfaces in matrix–mineral interactions during mammalian biomineralisation. *Curr Opin Colloid Interface Sci.* 2002 Mar 1;7(1):124–32.

Kolanthai E., Ganesan K., Epple M., Kalkura S. N. Synthesis of nanosized hydroxyapatite/agarose powders for bone filler and drug delivery application. *Mater Today Commun.* 2016 Sep 1;8:31–40.

Koocheki S., Madaeni S. S., Niroomandi P. Application of hydroxyapatite nanoparticles in development of an enhanced formulation for delivering sustained release of triamcinolone acetonide. *Int J Nanomedicine.* 2011;6:825–33.

Koons G. L., Diba M., Mikos A. G. Materials design for bone-tissue engineering. *Nat Rev Mater.* 2020 Aug;5(8):584–603.

Kulkarni R. K., Moore E. G., Hegyeli A. F., Leonard F. Biodegradable poly(lactic acid) polymers. *J Biomed Mater Res.* 1971;5(3):169–81.

Lara-Ochoa S., Ortega-Lara W., Guerrero-Beltrán C. E. Hydroxyapatite Nanoparticles in Drug Delivery: Physicochemistry and Applications. *Pharmaceutics.* 2021 Oct;13(10):1642.

Lee K. W., Bae C. M., Jung J. Y., Sim G. B., Rautray T. R., Lee H. J., Tae-Yub Kwon, Kyo-Han Kim. Surface characteristics and biological studies of hydroxyapatite coating by a new method. *J Biomed Mater Res B Appl Biomater.* 2011;98B(2):395–407.

Lee W. H., Loo C. Y., Rohanizadeh R. Functionalizing the surface of hydroxyapatite drug carrier with carboxylic acid groups to modulate the loading and release of curcumin nanoparticles. *Mater Sci Eng C.* 2019;99:929–39.

LeGeros R. Z. Calcium phosphate-based osteoinductive materials. *Chem Rev.* 2008 Nov;108(11):4742–53.

LeGeros R. Z. Properties of osteoconductive biomaterials: calcium phosphates. *Clin Orthop.* 2002 Feb;(395):81–98.

Levitt S. R., Crayton P. H., Monroe E. A., Condrate R. A. Forming method for apatite prostheses. *J Biomed Mater Res.* 1969 Dec 1;3(4):683–4.

Li Y., Zhang D., Wan Z., Yang X., Cai Q. Dental resin composites with improved antibacterial and mineralization properties via incorporating zinc/strontium-doped hydroxyapatite as functional fillers. *Biomed Mater.* 2022;17(4):045002.

Li Z., Liu Z., Yin M., Yang X., Yuan Q., Ren J., and Xiaogang Qu. Aptamer-Capped Multifunctional Mesoporous Strontium Hydroxyapatite Nanovehicle for Cancer-Cell-Responsive Drug Delivery and Imaging. *Biomacromolecules.* 2012 Dec 10;13(12): 4257–63.

Lin K., Chang J. 1 - Structure and properties of hydroxyapatite for biomedical applications. In: Mucalo M., editor. *Hydroxyapatite (Hap) for Biomedical Applications* [Internet]. Woodhead Publishing; 2015 [cited 2022 Aug 9]. p. 3–19. (Woodhead Publishing Series in Biomaterials). Available from: https://www.sciencedirect.com/science/article/pii/B9781782420330000018.

Linh N. T. B., Abueva C. D. G., Jang D., Lee B. Collagen and bone morphogenetic protein-2 functionalized hydroxyapatite scaffolds induce osteogenic differentiation in human adipose-derived stem cells. *J Biomed Mater Res B Appl Biomater.* 2020;108(4):1363–71.

Liu D. M. Fabrication of hydroxyapatite ceramic with controlled porosity. *J Mater Sci Mater Med.* 1997 Apr 1;8(4):227–32.

Loo S. C. J., Moore T., Banik B., Alexis F. Biomedical Applications of Hydroxyapatite Nanoparticles. *Curr Pharm Biotechnol.* 2010 Jun 1;11(4):333–42.

Mas-Moruno C. 3 - Surface functionalization of biomaterials for bone tissue regeneration and repair. In: Barbosa M. A., Martins M. C. L., editors. *Peptides and Proteins as Biomaterials for Tissue Regeneration and Repair* [Internet]. Woodhead Publishing; 2018 [cited 2022 Aug 2]. p. 73–100. Available from: https://www.sciencedirect.com/science/article/pii/B9780081008034000036.

Mehedi Hasan M., Nuruzzaman Khan M., Haque P., Rahman M. M. Novel alginate-di-aldehyde cross-linked gelatin/nano-hydroxyapatite bioscaffolds for soft tissue regeneration. *Int J Biol Macromol.* 2018 Oct 1;117:1110–7.

Mondal S., Dorozhkin S. V., Pal U. Recent progress on fabrication and drug delivery applications of nanostructured hydroxyapatite. *WIREs Nanomedicine Nanobiotechnology.* 2018;10(4):e1504.

Mondal S., Pal U. 3D hydroxyapatite scaffold for bone regeneration and local drug delivery applications. *J Drug Deliv Sci Technol.* 2019 Oct 1;53:101131.

Morishita T., Honoki K., Ohgushi H., Kotobuki N., Matsushima A., Takakura Y. Tissue Engineering Approach to the Treatment of Bone Tumors: Three Cases of Cultured

Bone Grafts Derived From Patients' Mesenchymal Stem Cells. *Artif Organs.* 2006;30(2):115–8.

Mudhafar M., Zainol I., Alsailawi H. A., Aiza Jaafar C. N. Synthesis and characterization of fish scales of hydroxyapatite/collagen–silver nanoparticles composites for the applications of bone filler. *J Korean Ceram Soc.* 2022 Mar 1;59(2):229–39.

Murugan R., Ramakrishna S. Bioresorbable composite bone paste using polysaccharide based nano hydroxyapatite. *Biomaterials.* 2004 Aug 1;25(17):3829–35.

Nandi S. K., Kundu, B., Mukherjee, J., Datta, S. and Balla, V. K. Converted marine coral hydroxyapatite implants with growth factors: *In vivo* bone regeneration. *Materials Science and Engineering C* 2015; 49: 816-823

Neffe A. T., Loebus A., Zaupa A., Stoetzel C., Müller F. A., Lendlein A. Gelatin functionalization with tyrosine derived moieties to increase the interaction with hydroxyapatite fillers. *Acta Biomater.* 2011 Apr 1;7(4):1693–701.

Nikbakht A., Dehghanian C., Parichehr R. Silane coatings modified with hydroxyapatite nanoparticles to enhance the biocompatibility and corrosion resistance of a magnesium alloy. *RSC Adv.* 2021;11(42):26127–44.

Oliveira H. L., Da Rosa W. L. O., Cuevas-Suárez C. E., Carreño N. L. V., da Silva A. F., Guim T. N., Odir A. Dellagostin, Evandro Piva. Histological Evaluation of Bone Repair with Hydroxyapatite: A Systematic Review. *Calcif Tissue Int.* 2017 Oct 1;101(4):341–54.

Orlovskii V. P., Komlev V. S., Barinov S. M. Hydroxyapatite and Hydroxyapatite-Based Ceramics. *Inorg Mater.* 2002 Oct 1;38(10):973–84.

Oryan A., Alidadi S., Moshiri A. Current concerns regarding healing of bone defects. *Hard Tissue.* 2013;2(2):1–2.

Oryan A., Moshiri A., Sharifi P. Advances in injured tendon engineering with emphasis on the role of collagen implants. *Hard Tissue.* 2012;1(12):12.

Oulguidoum A., Bouyarmane H., Laghzizil A., Nunzi J. M., Saoiabi A. Development of sulfonate-functionalized hydroxyapatite nanoparticles for cadmium removal from aqueous solutions. *Colloid Interface Sci Commun.* 2019;30:100178.

Packer J. W., Colditz J. C. Bone Injuries: Treatment and Rehabilitation. *Hand Clin.* 1986 Feb 1;2(1):81–91.

Pal A., Maity S., Chabri S., Bera S., Chowdhury A. R., Das M., Arijit Sinha, Sumit Chabri. Mechanochemical synthesis of nanocrystalline hydroxyapatite from $\less$i$\greater$Mercenaria$\less$/i$\greater$ clam shells and phosphoric acid. *Biomed Phys Ampmathsemicolon Eng Express.* 2017 Jan;3(1):015010.

Pal K., Pal S. Development of Porous Hydroxyapatite Scaffolds. *Mater Manuf Process.* 2006 May 1;21(3):325–8.

Pandey S., Rai N., Mahtab A., Mittal D., Ahmad F. J., Sandal N., Yub Raj Neupane, Anita Kamra Verma, Sushama Talegaonkar. Hyaluronate-functionalized hydroxyapatite nanoparticles laden with methotrexate and teriflunomide for the treatment of rheumatoid arthritis. *Int J Biol Macromol.* 2021;171:502–13.

Parizi A. M., Oryan A., Shafiei-Sarvestani Z., Bigham A. S. Human platelet rich plasma plus Persian Gulf coral effects on experimental bone healing in rabbit model: radiological, histological, macroscopical and biomechanical evaluation. *J Mater Sci Mater Med.* 2012 Feb 1;23(2):473–83.

Paul W., Sharma C. P. Development of porous spherical hydroxyapatite granules: application towards protein delivery. *J Mater Sci Mater Med.* 1999 Jul 1;10(7):383–8.

Rakmae S., Ruksakulpiwat Y., Sutapun W., Suppakarn N. Effect of silane coupling agent treated bovine bone based carbonated hydroxyapatite on *in vitro* degradation behavior and bioactivity of PLA composites. *Mater Sci Eng C.* 2012 Aug 1;32(6):1428–36.

Raphel J., Holodniy M., Goodman S. B., Heilshorn S. C. Multifunctional coatings to simultaneously promote osseointegration and prevent infection of orthopaedic implants. *Biomaterials.* 2016 Apr 1;84:301–14.

Reddi A. H. Role of morphogenetic proteins in skeletal tissue engineering and regeneration. *Nat Biotechnol.* 1998 Mar;16(3):247–52.

Rhee S. H., Lee J. D., Tanaka J. Nucleation of Hydroxyapatite Crystal through Chemical Interaction with Collagen. *J Am Ceram Soc.* 2000;83(11):2890–2.

Rimola A., Corno M., Zicovich-Wilson C. M., Ugliengo P. Ab Initio Modeling of Protein/Biomaterial Interactions: Glycine Adsorption at Hydroxyapatite Surfaces. *J Am Chem Soc.* 2008 Dec 3;130(48):16181–3.

Saiz E., Gremillard L., Menendez G., Miranda P., Gryn K., Tomsia A. P. Preparation of porous hydroxyapatite scaffolds. *Mater Sci Eng C.* 2007 Apr 1;27(3):546–50.

Salahuddin N., Ibrahim E., El-Kemary M. Surfactant free-poly (lactic-co-glycolic acid) coated artesunate loaded citrate-functionalized hydroxyapatite nanorods as a nanocapsule for improving artesunate delivery and antitumor efficiency. *Ceram Int.* 2022;48(10):13452–63.

Salamanna F., Giavaresi G., Parrilli A., Torricelli P., Boanini E., Bigi A., Milena Fini. Antiresorptive properties of strontium substituted and alendronate functionalized hydroxyapatite nanocrystals in an ovariectomized rat spinal arthrodesis model. *Mater Sci Eng C.* 2019;95:355–62.

Salinas A. J., Esbrit P., Vallet-Regí M. A tissue engineering approach based on the use of bioceramics for bone repair. *Biomater Sci.* 2013;1(1):40–51.

Shafiei Z., Bigham A. S., Dehghani S. N., Torabi Nezhad S. Fresh cortical autograft versus fresh cortical allograft effects on experimental bone healing in rabbits: radiological, Histopathological and Biomechanical evaluation. *Cell Tissue Bank.* 2009 Feb 1;10(1):19–26.

Shavandi A., Bekhit A. E. D. A., Sun Z. F., Ali A. A Review of Synthesis Methods, Properties and Use of Hydroxyapatite as a Substitute of Bone. *J Biomim Biomater Biomed Eng.* 2015;25:98–117.

Silva C. C. G., Rigo E. C. da S., Marchi J., Bressiani A. H. de A., Bressiani J. C. Hydroxyapatite coating on silicon nitride surfaces using the biomimetic method. *Mater Res.* 2008 Mar;11:47–50.

Siniscalco D., Dutreilh-Colas M., Hjezi Z., Cornette J., El Felss N., Champion E., Chantal Damia. Functionalization of Hydroxyapatite Ceramics: Raman Mapping Investigation of Silanization. *Ceramics.* 2019 Jun;2(2):372–84.

Son J. S., Choi Y. A., Park E. K., Kwon T. Y., Kim K. H., Lee K. B. Drug delivery from hydroxyapatite-coated titanium surfaces using biodegradable particle carriers. *J Biomed Mater Res B Appl Biomater.* 2013;101B(2):247–57.

Sotome S., Uemura T., Kikuchi M., Chen J., Itoh S., Tanaka J., Tetsuya Tateishi, Kenichi Shinomiya. Synthesis and *in vivo* evaluation of a novel hydroxyapatite/collagen–alginate as a bone filler and a drug delivery carrier of bone morphogenetic protein. *Mater Sci Eng C*. 2004 Apr 1;24(3):341–7.

Srinath P., Abdul Azeem P., Venugopal Reddy K. Review on calcium silicate-based bioceramics in bone tissue engineering. *Int J Appl Ceram Technol.* 2020;17(5):2450–64.

Suchanek W., Yoshimura M. Processing and properties of hydroxyapatite-based biomaterials for use as hard tissue replacement implants. *J Mater Res.* 1998a Jan;13(1):94–117.

Suchanek W., Yoshimura M. Processing and properties of hydroxyapatite-based biomaterials for use as hard tissue replacement implants. *J Mater Res.* 1998b Jan;13(1):94–117.

Sukhodub L. F., Sukhodub L. B., Litsis O., Prylutskyy Y. Synthesis and characterization of hydroxyapatite-alginate nanostructured composites for the controlled drug release. *Mater Chem Phys.* 2018;217:228–34.

Sun L., Berndt C. C., Gross K. A., Kucuk A. Material fundamentals and clinical performance of plasma-sprayed hydroxyapatite coatings: A review. *J Biomed Mater Res.* 2001;58(5):570–92.

Swetha M., Sahithi K., Moorthi A., Srinivasan N., Ramasamy K., Selvamurugan N. Biocomposites containing natural polymers and hydroxyapatite for bone tissue engineering. *Int J Biol Macromol.* 2010 Jul 1;47(1):1–4.

Szcześ A., Hołysz L., Chibowski E. Synthesis of hydroxyapatite for biomedical applications. *Adv Colloid Interface Sci.* 2017 Nov 1;249:321–30.

Szustakiewicz K., Gazińska M., Kryszak B., Grzymajło M., Pigłowski J., Wiglusz R. J., Masami Okamoto. The influence of hydroxyapatite content on properties of poly(L-lactide)/hydroxyapatite porous scaffolds obtained using thermal induced phase separation technique. *Eur Polym J.* 2019 Apr 1;113:313–20.

Thirumalai J. Hydroxyapatite: Advances in Composite Nanomaterials, Biomedical Applications and Its Technological Facets. *BoD – Books on Demand*; 2018.

Tian C., Xu P., Jiang J., Han C. Preparation and drug-delivery study of functionalized hydroxyapatite based on natural polysaccharide gums with excellent drug-loading properties. *J Dispers Sci Technol.* 2021a;42(5):751–9.

Tian Z., Lu C., Zhou Y., Zhang Y., Wei W. Phosphoric acid-induced activation of sepiolite for enhanced As(III) adsorption: role of *in situ* deposition of nano-hydroxyapatite. *J Dispers Sci Technol.* 2021b Jul 21;0(0):1–13.

Tram Do T. N., Lee W. H., Loo C. Y., Zavgorodniy A. V., Rohanizadeh R. Hydroxyapatite nanoparticles as vectors for gene delivery. *Ther Deliv.* 2012 May;3(5):623–32.

Uchida A., Shinto Y., Araki N., Ono K. Slow release of anticancer drugs from porous calcium hydroxyapatite ceramic. *J Orthop Res.* 1992;10(3):440–5.

Urist M. R., Silverman B. F., Büring K., Dubuc F. L., Rosenberg J. M. The bone induction principle. *Clin Orthop.* 1967 Aug;53:243–83.

Uskoković V., Uskoković D. P. Nanosized hydroxyapatite and other calcium phosphates: Chemistry of formation and application as drug and gene delivery agents. *J Biomed Mater Res B Appl Biomater.* 2011;96B(1):152–91.

Vacanti C. A., Bonassar L. J., Vacanti M. P., Shufflebarger J. Replacement of an Avulsed Phalanx with Tissue-Engineered Bone. *N Engl J Med.* 2001 May 17;344(20):1511–4.

Van De Putte K. A., Urist M. R. 23 Osteogenesis in the Interior of Intramuscular Implants of Decalcified Bone Matrix. *Clin Orthop Relat Res.* 1965 Dec;43:257–70.

Venkatesan J., Kim S. K. Nano-Hydroxyapatite Composite Biomaterials for Bone Tissue Engineering - A Review. *J Biomed Nanotechnol.* 2014 Oct 1;10(10):3124–40.

Verma G., Gajipara A., Shelar S. B., Priyadarsini K. I., Hassan P. A. Development of water-dispersible gelatin stabilized hydroxyapatite nanoformulation for curcumin delivery. *J Drug Deliv Sci Technol.* 2021;66:102769.

Verma G., Shetake N. G., Barick K. C., Pandey B. N., Hassan P. A., Priyadarsini K. I. Covalent immobilization of doxorubicin in glycine functionalized hydroxyapatite nanoparticles for pH-responsive release. *New J Chem.* 2018;42(8):6283–92.

Verma G., Shetake N. G., Pandrekar S., Pandey B. N., Hassan P. A., Priyadarsini K. I. Development of surface functionalized hydroxyapatite nanoparticles for enhanced specificity towards tumor cells. *Eur J Pharm Sci.* 2020;144:105206.

Verrier S., Alini M., Alsberg E., Buchman S. R., Kelly D., Laschke M. W., M. D. Menger, W. L. Murphy, J. P. Stegemann, M. Schütz, T. Miclau, M. J. Stoddart, C. Evans. Tissue engineering and regenerative approaches to improving the healing of large bone defects. *Eur Cell Mater.* 2016 Jul 19;32:87–110.

Wang H., Chu C., Cai R., Jiang S., Zhai L., Lu J., Xingjiang Li and Shaotong Jiang Synthesis and bioactivity of gelatin/multiwalled carbon nanotubes/hydroxyapatite nanofibrous scaffolds towards bone tissue engineering. *RSC Adv.* 2015 Jun 15;5(66):53550–8.

Wang M., Joseph R., Bonfield W. Hydroxyapatite-polyethylene composites for bone substitution: effects of ceramic particle size and morphology. *Biomaterials.* 1998 Dec 1;19(24):2357–66.

Wang X., Song G., Lou T. Fabrication and characterization of nano-composite scaffold of PLLA/silane modified hydroxyapatite. *Med Eng Phys.* 2010 May 1;32(4):391–7.

Wang Y. chuan, Wang J. ning, Xiao G. yong, Huang S. yun, Xu W. li, Yan W., Yu-peng Lu. Investigation of various fatty acid surfactants on the microstructure of flexible hydroxyapatite nanofibers. *Cryst Eng Comm.* 2021 Oct 18;23(40):7049–55.

Webb J. C. J., Tricker J. A review of fracture healing. *Curr Orthop.* 2000;6(14):457–63.

Wilson O. C., Hull J. R. Surface modification of nanophase hydroxyapatite with chitosan. *Mater Sci Eng C.* 2008 Apr 1;28(3):434–7.

Xu X., Chen X., Liu A., Hong Z., Jing X. Electrospun poly(l-lactide)-grafted hydroxyapatite/poly(l-lactide) nanocomposite fibers. *Eur Polym J.* 2007 Aug 1;43(8):3187–96.

Yang Y. H., Liu C. H., Liang Y. H., Lin F. H., Wu K. C. W. Hollow mesoporous hydroxyapatite nanoparticles (hmHANPs) with enhanced drug loading and pH-responsive release properties for intracellular drug delivery. *J Mater Chem B.* 2013a Apr 24;1(19):2447–50.

Yang Y. H., Liu C. H., Liang Y. H., Lin F. H., Wu K. C. W. Hollow mesoporous hydroxyapatite nanoparticles (hmHANPs) with enhanced drug loading and pH-responsive release properties for intracellular drug delivery. *J Mater Chem B.* 2013b Apr 24;1(19):2447–50.

Yewle J. N., Wei Y., Puleo D. A., Daunert S., Bachas L. G. Oriented Immobilization of Proteins on Hydroxyapatite Surface Using Bifunctional Bisphosphonates as Linkers. *Biomacromolecules*. 2012 Jun 11;13(6):1742–9.

Yoon I. K., Hwang J. Y., Seo J. won, Jang W. C., Kim H. W., Shin U. S. Carbon nanotube-gelatin-hydroxyapatite nanohybrids with multilayer core–shell structure for mimicking natural bone. *Carbon*. 2014 Oct 1;77:379–89.

Yoshikawa H., Tamai N., Murase T., Myoui A. Interconnected porous hydroxyapatite ceramics for bone tissue engineering. *J R Soc Interface*. 2009 Jun 6;6(suppl_3):S341–8.

Youssef H. F., El-Kady A. M., Ahmed M. M., Abo-almaged H. H. Preparation and characterization of novel bio-composites based on functionalized zeolite and nano-hydroxyapatite for a potential application in bone treatment. *J Porous Mater*. 2021;28(6):1979–98.

Yu J. C., Ho W., Lin J., Yip H., Wong P. K. Photocatalytic Activity, Antibacterial Effect, and Photoinduced Hydrophilicity of TiO2 Films Coated on a Stainless Steel Substrate. *Environ Sci Technol*. 2003 May 1;37(10):2296–301.

Zafar M. J., Zhu D., Zhang Z. 3D Printing of Bioceramics for Bone Tissue Engineering. *Materials*. 2019 Jan;12(20):3361.

Zantye P., Shende S., Ramanan S. R., Talukdar I., Kowshik M. Design of a biocompatible hydroxyapatite-based nanovehicle for efficient delivery of small interference ribonucleic acid into mouse embryonic stem cells. *Mol Pharm*. 2021;18(3):796–806.

Zhang C., Uchikoshi T., Liu L., Iwanami-Kadowaki K., Uezono M., Moriyama K., Masanori Kikuchi. Antibacterial-functionalized Ag loaded-hydroxyapatite (HAp) coatings fabricated by electrophoretic deposition (EPD) process. *Mater Lett*. 2021a;297:129955.

Zhang P., Hong Z., Yu T., Chen X., Jing X. *In vivo* mineralization and osteogenesis of nanocomposite scaffold of poly(lactide-co-glycolide) and hydroxyapatite surface-grafted with poly(l-lactide). *Biomaterials*. 2009 Jan 1;30(1):58–70.

Zhang S., editor. *Hydroxyapatite Coatings for Biomedical Applications [*Internet]. Taylor & Francis; 2013 [cited 2022 Aug 6]. Available from: https://library.oapen.org/handle/20.500.12657/41266.

Zhang Y., Xia M., Wang F., Ma J. Experimental and theoretical study on the adsorption mechanism of Amino trimethylphosphate (ATMP) functionalized hydroxyapatite on Pb (II) and Cd (II). *Colloids Surf Physicochem Eng Asp*. 2021b;626:127029.

Zhao J, Guo LY, Yang XB, Weng J. Preparation of bioactive porous HA/PCL composite scaffolds. *Appl Surf Sci*. 2008 Dec 30;255(5, Part 2):2942–6.

Zhao R., Xie P., Zhang K., Tang Z., Chen X., Zhu X., Yujiang Fan, Xiao Yang, Xingdong Zhang. Selective effect of hydroxyapatite nanoparticles on osteoporotic and healthy bone formation correlates with intracellular calcium homeostasis regulation. *Acta Biomater*. 2017 Sep 1;59:338–50.

Zhou H., Lee J. Nanoscale hydroxyapatite particles for bone tissue engineering. *Acta Biomater*. 2011 Jul 1;7(7):2769–81.

Chapter 5

Modified Hydroxyapatite Applications in Catalysis: A Review

Kathlen C. E. S. Yokoo, MD,
Felipe Baião Ferreira, MD,
Valéria Stephanny Oliveira Amorim, MD
and Gustavo Paim Valença*, PhD

Laboratory for the Study of Adsorptive and Catalytic Processes (LEPAC),
School of Chemical Engineering, University of Campinas (UNICAMP),
Campinas, SP, Brazil

Abstract

Hydroxyapatite (HAP, $Ca_{10}(PO_4)_6(OH)_2$) and its applications in catalysis have attracted considerable attention in recent years. Mild acid and basic sites uniformly distributed on the surface of HAP makes it an attractive solid catalyst for condensation reactions with high conversion, selectivity, and stability. Thus, HAP is the subject of increasing interest in producing chemicals from oxygenated molecules such as ethanol. This chapter addresses the various modified hydroxyapatites applications in catalysis. The first part of the review will include HAP as a catalyst totally or partially substituted with different ions. The unique structure of HAP allows it to accept many anionic and cationic substituents. The second part will present HAP as the active phase support. Different metals supported on HAP showed improvement in the many catalytic reactions. More recent findings and the current challenges owing to the development of modified hydroxyapatite are summarized.

* Corresponding Author's Email: gustavo@feq.unicamp.br.

In: What to Know about Hydroxyapatite
Editor: Isaac Lavoie
ISBN: 979-8-88697-523-9
© 2023 Nova Science Publishers, Inc.

Keywords: hydroxyapatite, modified apatites, catalysis

Introduction

Hydroxyapatite, a mineral widely used in different applications in chemistry and biochemistry, has shown interesting catalytic properties and is emerging as an alternative catalyst in reactions involving oxygenated molecules. The range of reactions studied includes reactions of oxidation, reduction, formation of C–C, C–O, C-S and C–N bonds, cycloadditions, and multi-component reactions (Gruselle et al., 2015). The versatility of the apatite group in altering its chemical composition allows the adjustment of its properties and the development of specific materials applicable to a wide range of catalytic processes (Gruselle et al., 2015; Ho et al., 2018). This review describes apatites based on cationic and anionic substituted hydroxyapatites and their applications as catalysts in different reactions.

Cationic Modified Hydroxyapatites

Different groups or cations can substitute the phosphate group and the Ca^{+2} in apatites without a change in their crystal structure (Fihri et al., 2017). Apatites have a general formula $Me_{10}(XO_4)_6Y_2$, where Me represents divalent cations such as Ca^{2+}, Ba^{2+}, Mg^{2+}, Sr^{2+}, and Zn^{2+}. However, there are known apatites with monovalent, trivalent, and tetravalent cations such as Na^+, K^+, Al^{3+}, Re^{3+}, Ru^{3+}, U^{4+} and Th^{4+} (Ibrahim et al., 2020). Studies have shown that substituting Ca^{+2} in HAP with different cations results in solids with other catalytic properties (El-Aal et al., 2022). This section will review the different cationic modified hydroxyapatites, Na, Mg, Fe, Co, Ni, Cu, Sr, Ru, Ce, La, and Pb used as catalysts and their applications in many reactions.

Sodium

Hydroxyapatites and carbonate hydroxyapatites substituted with sodium, namely, NaHAP and Na-CHAP, respectively, were prepared using the co-precipitation method resulting in calcium deficient apatites. The surface area of the NaHAP was close to 80 m^2 g^{-1} and Na-CHAP solid reached

106 m^2 g^{-1}. The XRD patterns showed that all solids were crystalline HAP, without interference from any other crystalline phase. A 0.44 wt%Na/HAP was tested as a catalyst in the methanol thiolation reaction, at 400°C, and a conversion of 60% and a selectivity of close to 30% to methanethiol were reported. When the same reaction was carried out on HAP, a conversion lower than 10% and lower selectivity was observed. Additionally, conversion of 100% and selectivity of 60% to methanethiol were achieved with Na-CHAP under the same reaction conditions (Lamonier et al., 2011).

Magnesium

Calcium hydroxyapatite modified with magnesium, $Ca_{10-x}Mg_x(PO_4)_6(OH)_2$, with x ranging from 1 to 3, was synthesized, resulting in an atomic ratio (Mg+Ca)/P ranging from 1.59 (Ca_7Mg_3HAP) to 1.66 (CaHAP). The XRD patterns showed a single crystalline phase, however, as the Mg^{2+} concentration was increased, the crystallinity of the solid decreased. The surface area increased with increasing Mg^{2+} ions ranging from 82 to 116 m^2 g^{-1}. The solids were tested on the one-pot Biginelli reaction to produce 3,4-dihydropyrimidin-2-(1H)-one (DHPM). Firstly, the Biginelli reaction was carried out in absence of solvent for different magnesium concentration ions in the HAP structure, the results show that the increase of magnesium ions in the HAP structure leads to an increase in the reaction rate and the yield for DHPM product, (85% yield with Ca_7Mg_3HAP against 55% using $Ca_{10}HAP$, within 5h of reaction). The enhanced Lewis acidity promoted by the addition of magnesium ions is suggested to be the main cause of the increase in the product yield for the Biginelli reaction. Similar results were also observed in the presence of solvents (ethanol and toluene). After the catalytic tests, it was observed that the solution did not have Mg^{2+} ions present in it, and the recovered catalysts did not show any significant change in the magnesium content (quantified by ICP). The magnesium HAP catalysts proved to be easily recyclable, being used up to five times without a significant change in the yield of the Biginelli reaction (Moussa et al., 2020).

Iron

Iron substituted hydroxyapatite prepared with different amounts of iron, Fe(x)/HAP, with x (expressed in wt.%) being the weight percentage of iron in the catalyst, were synthesized by the cation exchange method. In all samples, the iron ion was present as Fe^{3+}. The Fe(x)/HAP solids were tested as a catalyst for the butan-2-ol conversion, with and without oxygen, and for the propane oxidative dehydrogenation. The conversion of butan-2-ol increased as the iron content increased when the reaction was carried out without oxygen. However, a minimum in the conversion was observed as the iron content increased when the reaction was carried out with oxygen. The highest yield of butane was 55% at 230ºC using Fe(5.40)/HAP when the reaction was carried out without oxygen. The increase in the yield was correlated to the increase in the number of acid sites with the introduction of Fe in the HAP structure. The yield of methyl ethyl ketone is lower than 2.5% for all amounts of iron. In the presence of oxygen. The yield of methyl ethyl ketone decreased as the amount of iron increased, becoming constant for x values above 1.7%, reaching 17% at 230ºC. Propane oxidative dehydrogenation results showed that the maximum yield achieved for propene was 6.2% at 550 °C and with a low amount of iron (0.52 wt.%) (Khachani et al., 2010).

The HAP substituted with Fe and later impregnated with Mo were synthesized to investigate the oxidative desulfurization of dibenzothiophene (DBT). The average pore diameter decreased from 17.6 to 9.4 nm as well as total pore volume enlarged from 0.10 to 0.33 $cm^3 g^{-1}$ with the growth of the Fe^{3+} substitution degree. A complete DBT conversion was achieved under relatively mild conditions: $H_2O_2/S = 4:1$ (molar), 0.5% wt catalyst, 70 °C, and 30 min. The amount of Mo and the presence of Fe^{3+} ions in the HAP lattice resulted in a significant increase in the conversion of DBT (Goldberg et al., 2021).

Cobalt

Co-substituted hydroxyapatite (Co/HAP) was synthesized and used as a catalyst in peroxymonosulfate (PMS) activation for eliminating Rhodamine B (RhB) in an aqueous solution. The XRD patterns showed the presence of the HAP crystalline phase and Co_3O_4 (predominantly on the surface). The degradation of RhB was 77.5%, 19.8%, and 16.9% for $CoFe_2O_4$, MnO_2, and Fe_3O_4, while it was 92.1% on Co/HAP. The surface hydroxyl group on the HAP played a synergetic effect on the Co/HAP activation of PMS. Therefore, HAP could be considered a catalytic promoter, which plays an important co-catalytic role in promoting the degradation of RhB in the PMS activation process (Pang et al., 2020).

Nickel

Nickel-calcium phosphate/hydroxyapatite, with a part of the nickel, in the form of Ni^{2+}, substituting or taking the position of Ca^{2+} in the structure of the apatite, and the phosphate was synthesized (Jun et al., 2004). The synthesized solids presented a range of mole ratios for Ca/PO_4 varying from 8.0/6 to 10/6, and for Ni/PO_4 from 0.5/6 to 3.0/6. The XRD characterizations indicated the presence of calcium phosphate, hydroxyapatite, NiO, and metallic Ni. It was observed that as the Ca/PO_4 ratio increased it resulted in the amount of HAP phase increasing and that the increase of Ni/PO_4 ratio led to an increase in the amount of NiO in the fresh catalyst. The XRD and XPS characterizations confirmed that part of the nickel was incorporated into the phosphate and HAP structure by replacing Ca^{2+} (Jun et al., 2004). The prepared catalysts were tested on the partial oxidation of methane, demonstrating promising results, considering that it achieved yields values close to the equilibrium. The catalyst with Ca/PO_4 mole ratio of 9.5/10 and Ni/PO_4 mole ratio of 2.5/6 achieved, at 1073 K, a conversion of 93%, a selectivity of 96% to CO, and a yield of 90% to H_2. The TEM results of the used samples showed the presence of many fine nickel particles, in needle and round shapes, that were not present in the fresh sample, suggesting that the high conversion and selectivity could be from the presence of the fine nickel metal particles (Jun et al., 2004).

The methane oxidation reaction under stoichiometric conditions over Pd and Ni on hydroxyapatite was studied. The Pd-Ni/HAP catalysts demonstrated higher CH_4 conversions, when compared to single metal catalysts, showing 100% conversion of CH_4 at 400°C. The presence of more active sites on the surface of the HAP and hence its greater availability for the reaction, as the two metals underwent oxidation/reduction reactions, therefore gave rise to the amount of oxygen to activate methane oxidation. The reaction proceeded *via* a redox cycle between a metal oxide and reduced metal. Oxidized metal showed higher conversions than the reduced metal and exhibited more stable CH_4 conversions. The oxygen species from the metal oxide were directly involved in the catalytic reaction. All the samples were resistant to the coking process under the conditions studied (Kamieniak et al., 2018).

Copper

Solid solutions of copper-substituted calcium hydroxyapatites, $Ca_{10-x}Cu_x(PO_4)_6(OH)_2$, synthesized by the co-precipitation method were tested for the discoloration of the naphthol blue black solutions. The substitution of Ca^{2+} by Cu^{2+} resulted in a reduction of the lattice parameters of HAP, a decrease in the crystallinity degree, crystal size, and thermal stability, and an increase in the surface area. The prepared solid replaced calcium with copper up to 15%, and it was observed that the rate of discoloration is directly related to the content of copper, achieving 20% color removal using a 5% copper substituted HAP and 40% with a 15% Cu-substituted HAP, both after 30 minutes of reaction. The removal of the color managed to reach 70% with the most substituted solid and is suggested that the competition effect between the copper active sites on the HAP surface and the organic dye are the cause for the higher removal results with the increase of substituted copper (Othmani et al., 2018).

Ion exchange, co-precipitation, and impregnation methods were tested to substitute hydroxyapatite with Cu aiming to adjust the location of active sites. The synthesized copper substituted solids were evaluated in formaldehyde oxidation. Results showed that the substitution improved the synthesized catalyst by achieving complete oxidation of formaldehyde at 180ºC (using all three methods) when compared to normal HAP, which needed 240ºC to achieve the same conversion. Comparing the synthesis methods, the highest conversion achieved at 150ºC was 86% using the ion exchange method (with 1.4 wt.%). Under the same reaction conditions, the co-precipitation (4.7 wt.%) and impregnation methods (4.6 wt.%) achieved conversion of 63% and 33%, respectively. It is suggested that the reason for the higher conversion using the ion exchange method might be due to the presence of small Cu (II) clusters highly dispersed on the surface. However, co-precipitation and impregnation methods resulted in a lower amount of Cu (II) clusters, causing a decrease in the conversion with it (Qu et al., 2014).

Iron and copper substituted hydroxyapatite have been utilized as a heterogeneous catalyst for Fenton-like oxidation of potentially harmful organic pollutants with 30% H_2O_2 used as a green oxidant. Under the optimum reaction conditions (solution pH = 7.5, H_2O_2 dosage = 8 mM, and catalyst loading = 20 mg), almost 92% MB (Methylene Blue) and 90% MG (Malachite Green) were degraded within 45 min. The degradation of the dyes followed the first order kinetics with rate constants for methylene blue and malachite green as follows 0.05113 min^{-1} and 0.0613 min^{-1}, respectively. The Fe-Cu/HAP showed the same conversion after five cycles and exhibited high stability (Kalita et al., 2022).

Strontium

The substitution with strontium was tested in catalytic reactions, with partially and fully substituted strontium. The oxidative dehydrogenation of propane introducing tetrachloromethane into the feed stream, and using strontium and barium hydroxyapatites was studied (Sugiyama et al., 2001). The catalytic experiments were conducted in a fixed-bed continuous flow quartz reactor at atmospheric pressure and with a reaction temperature of 723 K they concluded that the conversion of propane was lower than the case without tetrachloromethane; however, the selectivity towards propylene increased with the increase of the time on stream (Sugiyama et al., 2001). The oxidative dehydrogenation of methane was also studied (Matsumura et al., 1994). The experiments were conducted in a conventional fixed-bed flow reactor under atmospheric pressure and at the reaction temperature of 600 ºC. Substituted strontium hydroxyapatites, with an Sr/P ratio from 1.52 to 1.63, were prepared. As Sr/P ratio decreased from 1.63 to 1.52, the selectivity to CO was from 34 to 42%, and the molar ratio of H_2/CO decreased from 4.4 to 3.3 (Matsumura et al., 1994).

Strontium hydroxyapatite and the solid solution, $Ca_{10-x}Sr_x(PO_4)_6(OH)_2$ were tested in the selective synthesis of 1-butanol from ethanol (Ogo et al. 2011, 2012). All the catalysts were synthesized by hydrothermal method. Compared to calcium hydroxyapatite, it was observed that strontium substituted HAP reported a higher selectivity to 1-butanol with ethanol conversions ranging from 1 to 24%, at 300ºC. It was also concluded that Sr-HAP catalysts with higher Sr/P molar ratios achieved higher selectivity to 1-butanol, the range studied went from 1.58 to 1.70 (Ogo et al., 2011).

Ruthenium

Hydroxyapatite with Ca^{2+} substituted by Ru^{3+} was synthesized and tested on the aerobic oxidation of alcohols (Yamaguchi et al., 2000). The reactions were carried out at 80ºC and oxygen atmosphere. Oxidative dehydrogenation, with benzylic and allylic compounds, achieved yields close to 100%. The oxidation of heterocyclic alcohols, including nitrogen and sulfur atoms, also demonstrated high yields. Thus, ruthenium-modified hydroxyapatite could be reused at least in 3 recycles maintaining the selectivity.

Modified ruthenium hydroxyapatite was also evaluated in the oxidation of organosilanes to silanols in the presence of water under atmospheric oxygen

(Mori et al., 2002a). The results showed that hydroxyapatite modified with ruthenium achieved high conversions and selectivity, to a wide range of organosilanes, resulting in most cases yields close to 100%, even after some recycles. It was also reported that the Ru/HAP catalyst presented higher yields when compared to other typical ruthenium catalysts like RuO_2, Ru/Al_2O_3, Ru/Carbon, $RuCl_2(PPh_3)_3$, $Pr_4N^+.RuO_4^-$ and $RuCl_3.nH_2O$, under the same reaction conditions, for the oxidation of dimethyl phenyl silane.

Cesium

Lactic acid conversion into 2,3-pentanedione over cesium substituted HAP catalysts were studied. A high yield (72.3%) into 2,3 pentanedione was observed. In addition, undesired reactions such as dehydration, decarbonylation, and coking were suppressed, since the surface acid/base balance was achieved by $CsNO_3$ addition to HAP (Li et al., 2017).

Lanthanum

Michael reaction of 1,3-dicarbonyls with enones was carried out using modified hydroxyapatite with lanthanum ions synthesized by the cation-exchange method in an aqueous media (Mori et al., 2005). The modified LaHAP showed promising results for the Michael reaction of 1,3-dicarbonyls with enones, for both the solvent-free and aqueous conditions, achieving high yields for the corresponding 1,5-dioxo compounds, mostly in the range of 90 to 100%.

Lead

Lead hydroxyapatite, $Pb_xCa_{10-x}(PO_4)_6(OH)_2$, with x ranging from 0 to 10, were synthesized by the aqueous precipitation method, resulting in non-stoichiometric lead hydroxyapatites. The catalysts were characterized by FT-IR, XRD, XRF, ICP, and Raman/FT-Raman spectroscopy, and tested in oxidative coupling of methane to produce ethane and ethylene. The highest selectivity to C_2 was 62% with an x value of 2, at a reaction temperature of 1048 K, and a fixed conversion of 35%. It was suggested that the oxidative coupling of methane was affected by the surface basicity, which was supposedly related to the coordination of Pb^{2+} ions in the hydroxyapatite lattice (Park et al., 2012).

Anionic Modified Hydroxyapatites

Considering the general formula of apatites, $Me_{10}(XO_4)_6Y_2$, the anionic substitutions of XO_4, which is usually represented by PO_4^{3-}, it can be replaced by other trivalent groups such as AsO_4^{3-} and VO_4^{3-}, as well as divalent groups (SO_4^{2-}, CO_3^{2-} and HPO_4^{2-}) and tetravalent groups (SiO_4^{4-} and GeO_4^{4-}). Finally, the Y element is usually occupied by monovalent anions, characteristically OH^-, F^- and Cl^-, but in some cases, the presence of divalent anions such as O^{2-}, CO_3^{2-} or even vacancies can also be found (Ibrahim et al., 2020). In this topic, we will review some anionic substitutions of hydroxyapatite.

Carbonate

Carbonate hydroxyapatite (CHAP) was synthesized by the chemical precipitation method and characterized by XRD, XRF, FT-IR, XPS, pyridine-TPD, CO_2 chemisorption, and N_2 adsorption. The results showed the presence of both acid and basic surface sites and also that the carbonate ions (CO_3^{2-}) are preferentially located in the phosphate positions of the apatite structure (Lovón-Quintana et al., 2017). The CHAP was tested in the ethanol conversion, using the optimum reaction conditions of temperature and residence time ($W/F_{ETHANOL}$), (500°C, and 40 g h mol^{-1}) and it was reported a yield of 97% to a mixture of oxygenated hydrocarbon fuels (C_4 to C_{18}+).

Vanadate

Species containing vanadium present some of their main characteristics a high Lewis acidity, a redox potential, and a unique ability to transfer oxygen. Thus, the replacement of phosphate ions by vanadate ions can result in catalysts with different characteristics and applications (Ogo et al., 2007, 2008; Onda et al., 2008; Sugiyama et al., 2005). The partial or complete replacement of PO_4^{3-} ions by VO_4^{3-} ions results in highly stable and isomorphic hydroxyapatites, with their structure identified as $Ca_{10}(PO_4)_{6-x}(VO_4)_x(OH)_2$ (Bauer Boechat et al., 2000; Mayer et al., 1979).

For example, vanadium V═O and V—O—V bonds are suggested to have a role in the activation of propane (Sugiyama et al., 2006). Studies have also revealed that VO_4 and OH groups could result in new highly active catalysts for the oxidative dehydrogenation of propane, and both of these groups are present as active sites on the vanadate hydroxyapatites (Sugiyama et al. 2005). Calcium hydroxyapatites partially substituted with vanadate, achieving a vanadium/phosphorous (V/P) ratio up to 0.15, were synthesized and the highest conversion and selectivity reported was achieved for a V/P of 0.10, with a 16.5% propane conversion and a selectivity for propylene of 54.2%. Under the same reaction conditions, conversion of 5.9% and selectivity of 1.4% with calcium hydroxyapatite were achieved (Sugiyama et al., 2006). Sugiyama et al. (2006) have also reported that under the same reaction conditions, the propane conversion on magnesium pyro-vanadate ($Mg_2V_2O_7$) was 14% and the selectivity for propylene was 50.9%.

The aerobic oxidation of propargylic alcohols under atmospheric pressure of oxygen catalyzed by vanadium-substituted hydroxyapatites also showed promising results, with the catalyst demonstrating a high yield and an easy separation by filtration, being able to be reused up to ten times without conversion decrease (Maeda et al., 2004). Reactions in which carbon-carbon bonds are formed, such as the Michael reaction and aldol reactions in aqueous media, are also examples in which vanadium substituted hydroxyapatites can act with high performance (Hara et al., 2006). The selective synthesis of 1-butanol was also investigated using modified vanadate apatites, like calcium and strontium vanadate apatites, however, both vanadate apatites showed a preference for the production of ethylene (Ogo et al., 2011).

To evaluate the acid-base properties of vanadium substituted hydroxyapatites, a catalytic study, using 2-propanol under an argon atmosphere was carried out, considering that, in the presence of acid catalysts, 2-propanol converts to propylene and in the presence of basics in acetone (Ogo et al., 2008). The results show a selectivity of 69.8% for the production of acetone and 30.2% for propylene using calcium vanadate hydroxyapatite as a catalyst against selectivities of 86.7% and 13.3%, respectively, when using HAP without vanadate, thus suggesting an increase in the presence of acidic sites (Ogo et al., 2008).

Hydroxyapatite as Catalytic Support

The supported metal catalysts have received considerable interest because of their operational advantages compared to their homogeneous counterparts (Saha et al., 2012). Hydroxyapatite can act as a support in many heterogenous reactions (Chakraborty and Roy Chowdhury, 2013). Hydroxyapatite is emerging catalytic support possessing high thermal and mechanical strength, chemically stable with low water solubility along with tunable porosity and acid-basic character (Munirathinam et al., 2020). In the previous section, we have described cation- and anion-substituted hydroxyapatites. In this section, we will give some examples of modified hydroxyapatite as a support and their applications in many reactions.

Manganese

Hydroxyapatite-supported manganese oxide catalysts have been prepared with two different precursors, namely, Mn nitrate (Nit) and Mn acetate (Ac). The catalysts were used in the toluene oxidation reaction. The rate of toluene conversion (in mol h^{-1} mol^{-1}Mn) given at 203°C increases as follow: Mn–Al_2O_3 (0.042) < MnAc-HAP (0.052) < MnNit-HAP (0.245). The differences found in toluene conversion are partly explained by the variations in nature and the dispersion of the active Mn sites. The highest toluene conversion was achieved on MnNit-HAP. The incorporation of NO^{3-} into the HAP by the substitution of Ca^{2+} surface site by Mn^{2+} promotes the Ca^{2+} enrichment of the HAP surface and high dispersion of Mn^{n+} ($n \geq 3$) (Chlala et al., 2016).

When hydroxyapatite-supported Mn catalysts (MnxHAP; x = Mn wt.%: 2.5, 5.0, 10, 20, and 30 based on MnO_2) have been studied on the formaldehyde oxidation, the Mn5HAP showed the highest conversion result. The Mn5HAP achieved a rate of formaldehyde (HCHO) conversion into CO_2 per mole of Mn of 0.042 h^{-1} at 145°C. The well dispersed oxidized manganese species on HAP with a Mn average oxidation state for Mn5HAP catalyst contributed to formaldehyde conversion (Chlala et al., 2020).

Cobalt

Cobalt-based catalysts with a series of supports, HZSM-5, CeO_2, ZrO_2, SiO_2, Al_2O_3, TiO_2, and HAP, were synthesized to study the selective hydrogenation of stearic acid. The Co/HAP exhibited the highest yield to 1-octadecanol (97.1% at 190°C and 4 MPa H_2 in water). The high yield of Co/HAP could be due to cobalt particle dispersion, and stronger metal−support interaction. The absorption of stearic acid on the surface of catalyst promotes the hydrogenation process over the Co species leading to a high yield of 1-octadecanol (Jia et al., 2018).

Cobalt- immobilized on hydroxyapatite with a loading of 0.05 wt.% was studied for alkenes epoxidation. The physical structure of 0.05 wt.% Co/HAP characterized by XRD showed pure hexagonal structure of HAP. No peaks of Co species were observed, suggesting the extremely low loading and highly dispersion of Co species. In addition, no dots or nanoparticles were observed in the dark-field TEM image. These results indicated that Co species in 0.05 wt% Co/HAP might exist as single atoms or clusters. The TOR for epoxidation of cyclohexene was 6261 h^{-1}. The high TOR values obtained was due to: (1) Co species existed as monomeric cations surrounded by oxygen; (2) Co–O bond structure makes Co species electron deficient, favoring bonding with olefin double bonds. The proposed sequence of elementary steps in the cyclohexene epoxidation included: first, the formation of acylperoxy radicals by homo and heterolytic cleavage of oxygen molecules and iso-butyraldehyde; then, Co species react with oxygen atom of acylperoxy radical, which was subsequently coordinated with C=C bond of alkenes; finally, the oxidative addition of acylperoxy radical to C=C bond takes place, yielding epoxide product and butyric acid formation (Mekrattanachai et al., 2018).

Supported cobalt catalysts, namely, Co/Al_2O_3 and Co/HAP, with 10 wt.% loading of cobalt was studied in the Fischer-Tropsch (FT) reaction. The temperature of reduction (350 and 400ºC) and time of reduction (7 h and 16 h) of the catalysts were studied. The conversion of CO on Co/HAP dropped from 50% to 35% as the reduction temperature and time increased. However, CO conversions slightly increased (15% versus 9%) for Co/Al_2O_3 catalysts. The Co/HAP reduced at 400ºC showed larger Co particle size and broader distribution width (13.4 ± 4.2 nm versus 11.6 ± 3.2 nm) compared to Co/HAP reduced at 350ºC. The increase in reduction temperature and reduction time resulted in sintering of Co particles, thus decreasing the number of active sites and CO conversion. At 20 bar with a gas hourly space velocity (GHSV) of 770 mL g^{-1} h^{-1} and at 230°C, the Co/HAP catalyst reduced at 350°C, exhibited C_{5+}

selectivity of 82% with methane selectivity of 11%. However, Co/Al_2O_3 catalyst showed lower C_{5+} selectivity (73%) and relatively higher methane selectivity (14%) even at low CO conversion. Therefore, the increased catalytic efficiency of Co/HAP catalysts, are attributed to the higher reducibility of cobalt precursors over HAP support and absence of small and hard to reduce cobalt oxide precursors (Munirathinam et al., 2020).

Cobalt–molybdenum catalysts supported on calcined bone (Co–Mo/HAP) and pyrolyzed bone (Co–Mo/HAP–C) have been studied for hydrodesulphurization (HDS) of dimethyl disulfide (DMDS). Results indicate that Co–Mo/HAP–C (101.70 $m^2 g^{-1}$) is less efficient to reduce sulfur content in the HDS process of DMDS than Co–Mo/HAP (4.18 $m^2 g^{-1}$), which has a much lower surface area. The Co–Mo/HAP catalyst has a highly dispersed molybdenum oxide species strongly interacting with the support. Despite the higher surface area of the Co–Mo/HAP–C catalyst, the presence of carbon in this catalyst has accumulated cobalt and molybdenum species, and CoMoOx phase was formed on the surface of HAP–C. According to the TPR results, the amount of hydrogen consumed in the Co–Mo/HAP catalyst is higher than consumed hydrogen in the Co–Mo/HAP–C catalyst. The Co–Mo/HAP catalyst after 48 h could decrease the sulfur content of the feed to 9 ppm while the commercial catalyst at the same conditions decreased the sulfur content of the same feed to 107 ppm. The Co–Mo/HAP–C catalyst could reduce sulfur content of feed from 2510 ppm to 110 ppm while the commercial catalyst reduces 930 ppm after 24 h. The lower conversion on the commercial catalyst compared to Co–Mo/HAP and Co–Mo/HAP–C, may be related to the presence of some Mo–O–Al bonds between the MoS_2 and (or) Co–Mo–S and the alumina support. Lower conversion on Co–Mo/HAP–C compared to Co–Mo/HAP might be related to the size of the metal nanoclusters. Hydrodesulphurization occurs by means of parallel hydrogenation (HYD) and direct desulfurization (DS) routes; turnover rates for both routes decreased with decreasing metal species size because the stronger sulfur binding at the surfaces of small species decreases the concentration of the required sulfur vacancies during steady-state HDS catalysis. The sulfur binding strength at the metal surfaces also depends strongly on metal identity (Yazdani et al., 2015).

Bifunctional hydroxyapatite-supported Co catalysts (Co/HAP) were synthesized to produce aromatic alcohols from ethanol. The Co/HAP catalyst converted ethanol directly into methyl benzyl alcohols (MB–OH). The dehydrogenation intermediate, 2-butenal, produced aromatic aldehydes by auto condensation and then the aromatic alcohol by hydrogenation. At 325°C, the MB–OH selectivity over a single-bed Co/HAP catalyst reached 54% at an ethanol conversion of 35%. The selectivity can be further increased to 71% by combining Copper/Carbon and Co/HAP in a dual-bed catalyst system at a lower reaction temperature (225°C) and similar conversion. Promoting ethanol dehydrogenation while preventing hydrogenation of the C = C bonds in unsaturated enal intermediates allows HAP-based materials to catalyze the conversion of ethanol to MB–OH efficiently (Wang et al., 2019).

Nickel

Nickel-based catalysts using two hydroxyapatites with different surface areas, namely, HAP1 ($S_{BET} = 7\ m^2\ g^{-1}$) and HAP2 ($S_{BET} = 60\ m^2\ g^{-1}$), were prepared and used in the dry reforming of methane (DRM). The influence of the thermal treatment (calcination in air at 1000°C for 5 h) of HAP supports were studied. The specific surface area decreased for both supports to $<3\ m^2g^{-1}$. The thermal treatment for Ni/HAP1 catalysts reduced the initial conversion of CO_2 and CH_4 from 80% to 30%, H_2 selectivity dropped from 90% to 45%, and CO selectivity decreased from 100% to 70%, when compared to the untreated catalysts. In the case of the support HAP2, the thermal treatment did not change the performance of the catalysts. Both catalysts before and after thermal treatment showed the similar selectivity in H_2 and CO which reached around 90%, and the initial conversion of both CH_4 and CO_2 was 80%. Between the catalysts studied, Ni/HAP2 presented higher conversion due to the presence of smaller nickel particles (10–20 nm), stronger basicity, higher density of basic sites (0.23 mmol g^{-1}) as well as higher specific surface area of the HAP2 support (Rego de Vasconcelos et al., 2020).

Copper

Copper-supported on Co-substituted HAP were prepared and used for complete oxidation of CO, propene, and toluene as the model component of diesel engine exhaust and toluene as VOC. The results of Co-substituted Cu/HAP were compared with those on Cu/HAP. For the Cu/HAP catalyst, the CO conversion was 50% at 298°C while the propane conversion was 50% at 364°C. On Cu/CoHAP catalyst, however, the CO conversion was 50% at 189°C and the propane conversion was 50% at 302°C. Cobalt substitution in HAP increase the concentration of active oxygen on the surface. From XPS studies, the Cu and O peak observed at lower binding energy in Cu/CoH HAP suggested the increase in concentration of more labile adsorbed oxygen on the catalyst surface. The increase in active oxygen species could be due to the Co substitution and interaction of Cu with Co-HAP. The highly dispersed CuO_x phase over CoHAP support synergistically improve the low temperature oxidation of CO, propene and toluene as the model component of diesel engine exhaust and VOC (More et al., 2019).

Zinc

Zinc bromide supported on hydroxyapatite (Zn-HAP) was prepared to promote the synthesis of 3-substituted indoles from Michael addition of indoles to α,β-unsaturated ketones. The substitution on the indole nucleus occurred exclusively at the 3-position, and no N-alkylation products have been observed. The catalyst showed the potential application of hydroxyapatite as an inorganic macroligand for Lewis acid catalyst. The Zn-HAP catalysts can be easily recovered and reused with only minor decrease in conversion (Tahir et al., 2006).

Hydrogenation of benzene to cyclohexene over Ru and Ru–Zn catalysts supported on hydroxyapatite (HAP) were studied. Compared to the Ru/HAP catalyst, the addition of Zn in the catalyst can increase the selectivity to cyclohexene. When the Ru/Zn molar ratio was 1:1, the HAP supported catalyst exhibited higher selectivity for cyclohexene than those with other Ru/Zn molar ratios. Catalytic tests results showed that 58.2% selectivity and 30.5% yield of cyclohexene were reached at 50°C and H_2 pressure of 5.0 MPa. The conversion of benzene decreased at higher Zn load. It was demonstrated that metallic nanoparticles of less than 2 nm were dispersed uniformly on the surface of the HAP, and the bimetallic Ru–Zn/HAP catalysts showed high conversion and selectivity. The catalysts could be reused at least four times without conversion and selectivity decrease (Zhang et al., 2013).

Molybdenum

Molybdenum-supported on nano-hydroxyapatite were synthesized and used as a catalyst in the conversion of methanol to formaldehyde. Different percentages of molybdenum oxide (1–15% w/w) supported on HAP were studied. The highest yield of formaldehyde (97%) was achieved with 5 wt%

MoO_3/HAP at the reaction temperature of 400°C. In addition, the catalyst stability in the production of formaldehyde was high (90 h). The ions Mo^{6+} as Lewis together with Brønsted acid sites play the main role in the formation of formaldehyde (Said et al., 2016).

Ruthenium

Ruthenium supported on HAP, carbonate apatite (CHAP), and fluor apatites (FAP) has been studied in detail by Tõnsuaadu et al., (2006). The catalytic oxidation ability was higher for Ru-HAP and Ru-CHAP compared to Ru-FAP apatite in the oxidation of benzylic alcohol. Oxidation of benzyl alcohol at 80°C with Ru-HAP and Ru-CHAP catalysts in O_2 atmosphere gave 100% conversion with more than 99% yield of benzaldehyde. In the oxidation reaction with Ru-FAP benzaldehyde was also the only product detected. However, the Ru-FAP resulted in a much lower conversion, achieving 20% after 3 h of reaction. The higher conversion using Ruthenium supported on HAP was directly related to the incorporation of Ru since a nonmodified apatite gave only a residual conversion (below 3%). These results highlight the crucial importance of the Ru-surrounding hydroxy ligands and the bonding of Ru to the apatite in the catalytic reaction (Tõnsuaadu et al., 2006).

In another study, hydroxyapatite was a suitable support for metal Ru in the selective vapor phase hydrogenation of levulinic acid (LA) to γ-valerolactone (gVL). A 99.8% selectivity into gLV and LA conversion of 65.1% over 2 wt%Ru/HAP were obtained at 275°C. The FT-IR analysis of the pyridine adsorbed spectra revealed the presence of Lewis acid sites and very weak Bronsted acid sites on the Ru-HAP surface. The surface metal site along with weak Bronsted sites were active for simultaneous hydrogenation and dehydration of levulinic acid to gVL (Sudhakar et al., 2016).

The Ru-HAP structure has been studied using infrared spectroscopy in the attenuated total reflection mode (ATR-IR) for aerobic oxidation of benzylalcohol. The spectroscopic data supported the existence of a hydrated RuOx -like phase, and is organized as a two-dimensional phase on the apatite. The RuOx -like phase coexists with isolated Ru^{3+} cations likely located in the hardly accessible channels of the apatite structure. The RuOx -like phase is shown to be highly reactive at the solid-liquid interface and exhibits several carbonyl species upon CO adsorption. The likely formation of HCO^{3-} in the absence of the alcohol indicates that benzyl alcohol oxidation and HCO^{3-} formation may occur on similar Ru sites providing evidence that these sites are located in the RuOx phase. The hydrated Ru^{n+} oxide phase is the active phase for benzyl alcohol oxidation, while carbonates and hydrogeno-carbonates are identified as spectator species on the catalyst surface (Mondelli et al., 2008).

Palladium

Palladium supported on hydroxyapatite (Pd/HAP) catalysts were prepared for the combustion of lean methane. The influence of the calcination temperature was studied. The Pd/HAP-773, Pd/HAP-873, Pd/HAP-973 and Pd/HAP-1073 catalysts were pre-treated at 773, 873, 973 and 1073 K, respectively, under a 5%O_2/He flow for 2 h. At 350°C, a systematic increase in TOR can be noted; the Pd/HAP-1073 showed the highest TOR value (0.074 s^{-1}), followed by Pd/HAP-973 (0.071 s^{-1}), Pd/HAP-873 (0.053 s^{-1}) and Pd/HAP-773 (0.049 s^{-1}), respectively. The increase in calcination temperature induces a progressive dehydroxylation of the support and structure evolution of the species containing Pd^{2+}, from tetrahedral (Td) to square planar geometry (D_{4h}), enhancing the metal-support interactions. The increase in the Pd-HAP interaction strength seems to expand the HAP lattice, promote PdO reduction and suppress CO chemisorption. These properties benefit the efficiency and stability of the Pd species in methane oxidation reaction (Boukha et al., 2020).

Hydroxyapatite-supported palladium (0) catalyst and its application for the Suzuki coupling and aerobic oxidation of benzyl alcohols in water were studied. The reaction was carried out using benzyl alcohol (1 mmol), HAP-Pd0 (0.1 g, 0.11 mol% Pd), water (5 mL) and atmospheric air as the source of oxygen under stirring at 100°C for 7 h. It was found that benzaldehyde was obtained in 95% isolated yield with 100% selectivity. The deactivation and recyclability were studied and no change in the yield of HAP-Pd0 up to the 5th use were observed (Jamwal et al., 2008).

Silver

Supported nano-Ag catalyst were studied for CO oxidation. A supported nano-Ag catalyst by employing the strong metal–support interactions (SMSIs) between the Ag nanoparticles (NPs) and hydroxyapatite (HAP) were prepared by hydrothermal or co-precipitation method. The unique confinement effect with different morphology HAP and the protective effect of hydroxyl groups on the HAP surface, highly dispersed Ag NPs were obtained. It was found that the particle size of Ag NPs was strongly related to the morphology and structure of the HAP supports, and it significantly affects the CO oxidation performance. 100% CO conversion can be achieved at 160°C (Lan et al., 2022).

Highly dispersed silver nanoparticles supported on hydroxyapatite were studied for the oxidative dehydrogenation of ethanol to acetaldehyde. The reaction started at 423 K and reached to a conversion of ca. 18% at 550 K, while the selectivity to acetaldehyde was ca. 100%. Moreover, the catalyst was stable, with almost no change in conversion and selectivity after time-on-stream 100 h at 523 K. In comparison with bulk Ag catalyst (1.38×10^{-4} mol s^{-1} g_{Ag}^{-1}) the reaction rate over Ag/HAP (1.78×10^{-3} mol s^{-1} g_{Ag}^{-1}) was improved by almost one order of magnitude under similar reaction conditions. The Ag/HAP exhibited superior conversion, selectivity and stability towards the oxidation of ethanol to acetaldehyde under mild conditions (Xu et al., 2016).

Platinum

Hydroxyapatite-supported Pt catalysts were studied as a catalyst in the removal of harmful pollutants such as formaldehyde (HCHO) from indoor environment at room temperature. The effect of three modifiers (sodium citrate-SC, sodium dodecyl sulfate-SDS, and hexamethylenetetramine-HMT) introduced in the preparation process on the HAP-based catalysts was also studied. The HCHO removal efficiency after 60 min was 75, 73, 54, and 48% for Pt/HAP–SC, Pt/HAP–HMT, Pt/HAP, and Pt/HAP–SDS, respectively, which was higher than that for Pt/TiO_2 (38%). This result suggests that the modifier influences the reaction energy pattern on the catalysts. The conversion of HCHO on Pt/HAP modified by the sodium citrate catalyst was higher than on the other solids. The higher rate of reaction could be attributed to the higher surface Ca/P ratio providing more Lewis acidic sites (Ca^{2+}) and for co-operational capture of HCHO molecules and a more significant amount of active oxygen species. The formation of active oxygen species presumably induced by the strong interaction between Pt nanoparticles and HAP support are responsible for the catalytic performance for HCHO oxidation at room temperature (Xu et al., 2019).

Water gas shift (WGS) reaction over HAP supported Pt were studied, and solids with different Ca/P ratios ranging between 1.58 and 2.00 were prepared.

The WGS reaction rates was 0.040, 0.072. 0.558 and 0.548 $mol_{CO}\ mol_{Pt}^{-1}s^{-1}$ for Pt/HAP with Ca/P ratio 1.58, 1.68, 1,75 and 1.76, respectively. Thus, CO conversion increased with Ca/P ratio. The WGS performance is attributed to the activation of H_2O on ionic oxides due to coordination of H_2O to Lewis acidic Ca^{2+} ions and H bonding to basic O atoms of PO_4^{3-} units. Rendering H_2O molecules highly polarized and thus reactive on apatite surfaces with the ensuing formate-like intermediates being well stabilized through bonding to multiple Ca^{2+} ions, as well (Miao et al., 2016).

Gold

Various cation and anion substituted hydroxyapatites (sHAPs) supported gold catalysts were synthesized and evaluated by alkene π-activation reactions. The solids synthesized were Mg and Ce substituted HAP, fluorapatite, and strontium hydroxyvanadate (Sr/HAP) with Au clusters. The Mg- or Ce-partially substituted HAP and Sr/HAP were the most suitable supports for stabilizing Au clusters, while the use of anion-substituted HAPs led to the aggregation of Au NPs, which suggests that hydroxide and phosphate play an important role in stabilizing Au particles. The partial oxidative strong metal-support interaction (SMSI) manifested on Au/Mg- and Ce-substituted Sr/HAP promoted the formation of positively charged Au species and exhibited higher yield into 14DABE than non-SMSI state catalysts (Nakayama et al., 2022).

Bimetallic Au–Cu/HAP catalyst was prepared through a deposition–precipitation method, and CO oxidation reaction was carried out to evaluate the catalytic performance of the samples. From the results of the O_2-TPD, the addition of copper to Au/HAP gave rise to newer O_2 adsorption sites and bigger O_2 adsorption capacity. At a conversion level of 90%, Au–Cu/HAP calcined at 300°C was found to be the most stable among the catalysts, and it maintained the CO conversion at about 90% over 14 h reaction time. The synergistic interaction between gold and CuOx phase of Au-Cu/HAP created most active sites and enhanced the CO oxidation and stability in comparison with Au/HAP and Cu/HAP (Guo et al., 2017).

The understanding of the essential features of the catalysis of the gold nanoparticles loaded HAP (Au/HAP) in formaldehyde (HCHO) and benzene (C_6H_6) oxidation has been documented by Wang et al., (2015). For Au/HAP calcinated at 200 and 600°C, 100% C_6H_6 conversion was achieved at 250 and 360°C, correspondingly. The relatively small Au particle size of Au/HAP-200°C is a key factor for high C_6H_6 conversion. The nano-gold particles interacted with PO_4^{3-} groups at lower temperatures, while increasing temperature in C_6H_6 oxidation, the nano-gold particles were not stabilized by PO_4^{3-} groups anymore, resulting in the sintering of nano-gold particles HAP significantly improves the stability of Au nanoparticles against sintering and play an important role in catalyzing the combustion of HCHO and C_6H_6 (Wang et al., 2015).

Conclusion

Hydroxyapatites modified by substitution of Ca^{2+}, PO_4^{3-} or OH^- or as support has a vast range of catalytic applications. Stable materials with high performances in catalysis were reported through the synthesis of stoichiometric and non-stoichiometric HAP, with partial or fully substitution of the Ca^{2+} and PO_4^{3-} sites.

Additionally, throughout this review, we have tried highlighting hydroxyapatite versatility as catalytic support. The unique surface acid-base, textural properties, and high thermal and mechanical stability characteristics allowed HAP to be used as support for many catalytic reactions. Different metallic impregnation on the HAP surface resulted in strong metal-support interaction, high dispersion of the metallic species, and the location of metallic species on different sites of the HAP. A high conversion, selectivity, reusability, or stability were observed for metallic catalysts supported on HAP.

To summarize, the properties of apatites can be tuned and behave as powerful catalysts in a wide range of organic reactions. As one of the main strengths of hydroxyapatite is the versatility to facilitate the bond making or bond breaking between reactions, which unlocks a wide range of reaction paths, such as the formation C-C, C-O, C–N, and C-S bonds, hydrogenation, dehydrogenation, oxidation, and dehydration. However, there is still room for improvement in the area aiming to tune even more the conversion, selectivity, and reusability for the catalytic processes.

Disclaimer

None.

References

Boechat CB, Eon JG, Rossi AM, Perez CAC, San Gil RAS. Structure of vanadate in calcium phosphate and vanadate apatite solid solutions. *Phys Chem Chem Phys* (2000) 2(18):4225–4230.

Boukha Z, Choya A, Cortés-Reyes M, de Rivas B, Alemany LJ, et al. Influence of the calcination temperature on the activity of hydroxyapatite-supported palladium catalyst in the methane oxidation reaction. *Appl Catal B* (2020) 277:119280.

Chakraborty R, Roy Chowdhury D. Fish bone derived natural hydroxyapatite-supported copper acid catalyst: Taguchi optimization of semibatch oleic acid esterification. *Chemical Engineering Journal* (2013) 215–216:491–499.

Chlala D, Giraudon JM, Labaki M, Lamonier JF. Formaldehyde Total Oxidation on Manganese-Doped Hydroxyapatite: The Effect of Mn Content. *Catalysts* (2020) 10(12):1422.

Chlala D, Giraudon JM, Nuns N, Lancelot C, Vannier RN, et al. Active Mn species well dispersed on Ca^{2+} enriched apatite for total oxidation of toluene. *Appl Catal B* (2016) 184:87–95.

El-Aal MA, Ali HM, Ibrahim SM. Cu-Doped 1D Hydroxyapatite as a Highly Active Catalyst for the Removal of 4-Nitrophenol and Dyes from Water. *ACS Omega* (2022) 7(30):26777–26787.

Fihri A, Len C, Varma RS, Solhy A. Hydroxyapatite: A review of syntheses, structure and applications in heterogeneous catalysis. *Coord Chem Rev* (2017) 347:48–76.

Goldberg MA, Akopyan AV, Gafurov MR, Makshakova ON, Donskaya NO, et al. Iron-Doped Mesoporous Powders of Hydroxyapatite as Molybdenum-Impregnated Catalysts for Deep Oxidative Desulfurization of Model Fuel: Synthesis and Experimental and Theoretical Studies. *J Phys Chem C* (2021) 125(21):11604–11619.

Gruselle M. Apatites: A new family of catalysts in organic synthesis. *J Organomet Chem* (2015) 793:93–101.

Guo J, Yu H, Dong F, Zhu B, Huang W, Zhang S. High efficiency and stability of Au–Cu/hydroxyapatite catalyst for the oxidation of carbon monoxide. *RSC Adv* (2017) 7(72):45420–45431.

Hara T, Kanai S, Mori K, Mizugaki T, Ebitani K, et al. Highly efficient C-C bond-forming reactions in aqueous media catalyzed by monomeric vanadate species in an apatite framework. *J Org Chem* (2006) 71(19):7455–7462.

Ho CR, Zheng S, Shylesh S, Bell AT. The mechanism and kinetics of methyl isobutyl ketone synthesis from acetone over ion-exchanged hydroxyapatite. *J Catal* (2018) 365:174–183.

Ibrahim M, Labaki M, Giraudon JM, Lamonier JF. Hydroxyapatite, a multifunctional material for air, water and soil pollution control: A review. *J Hazard Mater* (2020) 383:121139.

Jamwal N, Gupta M, Paul S. Hydroxyapatite-supported palladium (0) as a highly efficient catalyst for the Suzuki coupling and aerobic oxidation of benzyl alcohols in water. *Green Chem* (2008) 10(9):999-1003.

Jia W, Xu G, Liu X, Zhou F, Ma H, et al. Direct Selective Hydrogenation of Fatty Acids and Jatropha Oil to Fatty Alcohols over Cobalt-Based Catalysts in Water. *Energy Fuels* (2018) 32(8):8438–8446.

Jun J. Nickel–calcium phosphate/hydroxyapatite catalysts for partial oxidation of methane to syngas: characterization and activation. *J Catal* (2004) 221(1):178–190.

Kalita J, Das B, Dhar SS. Synergistic effect of iron and copper in hydroxyapatite nanorods for Fenton-like oxidation of organic dye. *Colloids and Surfaces A: Physicochemical and Engineering Aspects* (2022) 643:128750.

Kamieniak J, Kelly PJ, Doyle AM, Banks CE. Influence of the metal/metal oxide redox cycle on the catalytic activity of methane oxidation over Pd and Ni doped hydroxyapatite. *Catal Commun* (2018) 107:82–86.

Khachani M, Kacimi M, Ensuque A, Piquemal JY, Connan C, et al. Iron–calcium–hydroxyapatite catalysts: Iron speciation and comparative performances in butan-2-ol conversion and propane oxidative dehydrogenation. *Applied Catalysis A: General* (2010) 388(1–2):113–123.

Lamonier C, Lamonier JF, Aellach B, Ezzamarty A, Leglise J. Specific tuning of acid/base sites in apatite materials to enhance their methanol thiolation catalytic performances. *Catal Today* (2011) 164(1):124–130.

Lan YT, Yang XY, Liu SX, Miao YX, Zhao Z. Highly dispersed silver nanoparticles supported on a hydroxyapatite catalyst with different morphologies for CO oxidation. *New J Chem* (2022) 46(15):6940–6945.

Li X, Sun L, Zou W, Cao P, Chen Z, et al. Efficient Conversion of Bio-Lactic Acid to 2,3-Pentanedione on Cesium-Doped Hydroxyapatite Catalysts with Balanced Acid-Base Sites. *Chem Cat Chem* (2017) 9(24):4621–4627.

Lovón-Quintana JJ, Rodriguez-Guerrero JK, Valença PG. Carbonate hydroxyapatite as a catalyst for ethanol conversion to hydrocarbon fuels. *Applied Catalysis A: General* (2017) 542:136–145.

Maeda Y, Washitake Y, Nishimura T, Iwai K, Yamauchi T, Uemura S. Calcium phosphate-vanadate apatite (CPVAP)-catalyzed aerobic oxidation of propargylic alcohols with molecular oxygen. *Tetrahedron* (2004) 60(41):9031–9036.

Matsumura Y, Sugiyama S, Hayashi H, Shigemota N, Saitoh K, Moffat JB. Strontium hydroxyapatites: Catalytic properties in the oxidative dehydrogenation of methane to carbon oxides and hydrogen. *Journal of Molecular Catalysis* (1994) 92(1):81–94.

Mayer I, Wahnon S, Cohen S. Preparation of hydroxyapatites via the MSO4 sulphates (M = Ca, Sr, Pb and Eu). *Mater Res Bull* (1979) 14(11):1479–1483.

Mekrattanachai P, Cao C, Li Z, Li H, Song W. Cobalt immobilized on hydroxyapatite as a low-cost and highly effective heterogeneous catalyst for alkenes epoxidation under mild conditions. *RSC Adv* (2018) 8(65):37303–37306.

Miao D, Goldbach A, Xu H. Platinum/Apatite Water-Gas Shift Catalysts. *ACS Catal* (2016) 6(2):775–783.

Mondelli C, Ferri D, Baiker A. 2008. Ruthenium at work in Ru-hydroxyapatite during the aerobic oxidation of benzyl alcohol: An in situ ATR-IR spectroscopy study. *J Catal* 258(1):170–176.

More RK, Lavande NR, More PM. Copper supported on Co substituted hydroxyapatite for complete oxidation of diesel engine exhaust and VOC. *Molecular Catalysis* (2019) 474:110414.

Mori K, Oshiba M, Hara T, Mizugaki T, Ebitani K, Kaneda K. Michael reaction of 1,3-dicarbonyls with enones catalyzed by a hydroxyapatite-bound La complex. *Tetrahedron Lett* (2005) 46(25):4283–4286.

Mori K, Tano M, Mizugaki T, Ebitani K, Kaneda K. Efficient heterogeneous oxidation of organosilanes to silanols catalysed by a hydroxyapatite-bound Ru complex in the presence of water and molecular oxygen. *New J Chem* (2002) 26(11):1536–1538.

Moussa S ben, Mehri A, Badraoui B. Magnesium modified calcium hydroxyapatite: An efficient and recyclable catalyst for the one-pot Biginelli condensation. *J Mol Struct* (2020) 1200:127111.

Munirathinam R, Pham Minh D, Nzihou A. Hydroxyapatite as a new support material for cobalt-based catalysts in Fischer-Tropsch synthesis. *Int J Hydrogen Energy* (2020) 45(36):18440–18451.

Nakayama A, Sodenaga R, Gangarajula Y, Taketoshi A, Murayama T, et al. Enhancement effect of strong metal-support interaction (SMSI) on the catalytic activity of substituted-hydroxyapatite supported Au clusters. *J Catal* (2022) 410:194–205.

Ogo S, Onda A, Iwasa Y, Hara K, Fukuoka A, Yanagisawa K. 1-Butanol synthesis from ethanol over strontium phosphate hydroxyapatite catalysts with various Sr/P ratios. *J Catal* (2012) 296:24–30.

Ogo S, Onda A, Kajiyoshi K, Yanagisawa K. Hydrothermal synthesis and particle size control of hydroxyapatite solid solutions with vanadate. *Phosphorus Research Bulletin* (2007) 21:84–87.

Ogo S, Onda A, Yanagisawa K. Hydrothermal synthesis of vanadate-substituted hydroxyapatites, and catalytic properties for conversion of 2-propanol. *Applied Catalysis A: General* (2008) 348(1):129–134.

Ogo S, Onda A, Yanagisawa K. Selective synthesis of 1-butanol from ethanol over strontium phosphate hydroxyapatite catalysts. *Applied Catalysis A: General* (2011) 402(1–2):188–195.

Onda A, Ogo S, Kajiyoshi K, Yanagisawa K. Hydrothermal synthesis of vanadate/phosphate hydroxyapatite solid solutions. *Mater Lett* (2008) 62(8–9):1406–1409.

Othmani M, Bachoua H, Ghandour Y, Aissa A, Debbabi M. Synthesis, characterization and catalytic properties of copper-substituted hydroxyapatite nanocrystals. *Mater Res Bull* (2018) 97:560–566.

Pang Y, Kong L, Chen D, Yuvaraja G, Mehmood S. Facilely synthesized cobalt doped hydroxyapatite as hydroxyl promoted peroxymonosulfate activator for degradation of Rhodamine B. *J Hazard Mater* (2020) 384:121447.

Park JH, Lee DW, Im SW, Lee YH, Suh DJ, et al. Oxidative coupling of methane using non-stoichiometric lead hydroxyapatite catalyst mixtures. *Fuel* (2012) 94:433–439.

Qu Z, Sun Y, Chen D, Wang Y. Possible sites of copper located on hydroxyapatite structure and the identification of active sites for formaldehyde oxidation. *J Mol Catal A: Chem* (2014) 393:182–190.

Rego de Vasconcelos B, Pham Minh D, Martins E, Germeau A, Sharrock P, Nzihou A. Highly-efficient hydroxyapatite-supported nickel catalysts for dry reforming of methane. *Int J Hydrogen Energy* (2020) 45(36):18502–18518.

Saha D, Chatterjee T, Mukherjee M, Ranu BC. Copper(I) hydroxyapatite catalyzed Sonogashira reaction of alkynes with styrenyl bromides. Reaction of cis-styrenyl bromides forming unsymmetric diynes. *J Org Chem* (2012) 77(20):9379–9383.

Said AE-AA, El-Wahab MMMA, Alian AM. Selective oxidation of methanol to formaldehyde over active molybdenum oxide supported on hydroxyapatite catalysts. *Catal Lett* (2016) 146(1):82–90.

Sudhakar M, Kumar VV, Naresh G, Kantam ML, Bhargava SK, Venugopal A. Vapor phase hydrogenation of aqueous levulinic acid over hydroxyapatite supported metal (M = Pd, Pt, Ru, Cu, Ni) catalysts. *Appl Catal B* (2016) 180:113–120.

Sugiyama S, Osaka T, Hashimoto T, Sotowa KI. Oxidative Dehydrogenation of Propane on Calcium Hydroxyapatites Partially Substituted with Vanadate. *Catal Lett* (2005) 103(1–2):121–123.

Sugiyama S, Osaka T, Hirata Y, Sotowa K. Enhancement of the activity for oxidative dehydrogenation of propane on calcium hydroxyapatite substituted with vanadate. *Applied Catalysis A: General* (2006) 312:52–58.

Sugiyama S, Shono T, Nitta E, Hayashi H. Effects of gas- and solid-phase additives on oxidative dehydrogenation of propane on strontium and barium hydroxyapatites. *Applied Catalysis A: General* (2001) 211(1):123–130.

Tahir R, Banert K, Solhy A, Sebti S. Zinc bromide supported on hydroxyapatite as a new and efficient solid catalyst for Michael addition of indoles to electron-deficient olefins. *J Mol Catal A: Chem* (2006) 246(1–2):39–42.

Tõnsuaadu K, Gruselle M, Villain F, Thouvenot R, Peld M, et al. A new glance at ruthenium sorption mechanism on hydroxy, carbonate, and fluor apatites: Analytical and structural studies. *J Colloid Interface Sci* (2006) 304(2):283–291.

Wang QN, Weng XF, Zhou BC, Lv SP, Miao S, et al. Direct, Selective Production of Aromatic Alcohols from Ethanol Using a Tailored Bifunctional Cobalt–Hydroxyapatite Catalyst. *ACS Catal* (2019) 9(8):7204–7216.

Wang Y, Chen B, Crocker M, Zhang Y, Zhu X, Shi C. Understanding on the origins of hydroxyapatite stabilized gold nanoparticles as high-efficiency catalysts for formaldehyde and benzene oxidation. *Catal Commun* (2015) 59:195–200.

Xu J, Xu XC, Yang XJ, Han YF. Silver/hydroxyapatite foam as a highly selective catalyst for acetaldehyde production via ethanol oxidation. *Catal Today* (2016) 276:19–27.

Xu Z, Huang G, Yan Z, Wang N, Yue L, Liu Q. Hydroxyapatite-Supported Low-Content Pt Catalysts for Efficient Removal of Formaldehyde at Room Temperature. *ACS Omega* (2019) 4(26):21998–22007.

Yamaguchi K, Mori K, Mizugaki T, Ebitani K, Kaneda K. Creation of a monomeric ru species on the surface of hydroxyapatite as an efficient heterogeneous catalyst for aerobic alcohol oxidation. *J Am Chem Soc* (2000) 122(29):7144–7145.

Yazdani ME, Monjezi BH, Mokfi M, Bozorgzadeh H, Gil A, Ghiaci M. Synthesis and characterization of new hydrodesulphurization Co–Mo catalysts supported on calcined and pyrolyzed bone. *RSC Adv* (2015) 5(51):40647–40656.

Zhang P, Wu T, Jiang T, Wang W, Liu H, et al. Ru–Zn supported on hydroxyapatite as an effective catalyst for partial hydrogenation of benzene. *Green Chem* (2013) 15(1):152–159.

Chapter 6

A Review of Hydroxyapatite: Sustainable Product Development in Terms of Waste Valorization

Murugiah Krishani[1,*], M. Tech, Hazwani Suhaimi[1], PhD and Nonni Soraya Sambudi[2], PhD

[1]Faculty of Integrated Technologies, Universiti Brunei Darussalam Bandar Seri Begawan, Brunei Darussalam

[2]Department of Chemical Engineering, Universiti Teknologi Petronas, Seri Iskandar, Perak, Malaysia

Abstract

Million tons of shell waste are dumped in oceans and landfills and considered nuisance waste. Rather than disposing of the shells, recycling and reusing are excellent alternatives. In general, shells are rich in calcium carbonate, so sustainable development of calcium-based products may help reduce waste shell accumulation. Hydroxyapatite is a calcium phosphate-based mineral having similar composition and structure to natural human bone and teeth. Excellent biocompatibility, bioactivity, biodegradability, and non-toxicity; are some properties that enable hydroxyapatite to use as a superior biomaterial. The demand to obtain hydroxyapatite in a simple, efficient, scalable, and environmentally friendly way is increasing daily. Not only from shells but hydroxyapatite can also be synthesized from animal bones, fish scales, plants, and algae. This review discussed the highlights of hydroxyapatite and its natural sources, synthesis, and biomedical

* Corresponding Author's Email: 20h8400@ubd.edu.bn.

In: What to Know about Hydroxyapatite
Editor: Isaac Lavoie
ISBN: 979-8-88697-523-9
© 2023 Nova Science Publishers, Inc.

applications. Moreover, it also explained the importance of hydroxyapatite synthesized from biogenic waste materials. From this review, researchers may get an idea of the synthesis of hydroxyapatite from various sources in the development of biomedical applications and the conservation of food and agricultural waste management in terms of waste valorization.

Keywords: hydroxyapatite, waste valorization, biomaterial, waste management

Introduction

Recycling and reusing food and agricultural waste should be encouraged for sustainable development in waste management. It helps to prevent the dumping of nuisance waste which creates several unhygienic issues and reduces pollution by avoiding developing chemical-dependent products. Though food and agricultural waste are already utilized in various applications as a renewable energy source, it adds more beauty and is valuable when used in biomedical applications (Abdulrahman et al., 2014).

In the early days, humans used ceramics only in pottery. For the past few decades, ceramics have been developed and applied to treat damaged body parts to restore their function. The ceramic material used to improve the quality of life for biomedical applications is said to be "Bioceramics." Good osteoconductive properties and tissue responses made the calcium phosphate-based bioceramic materials applied in various orthopedic and maxillofacial applications (Laonapakul 2015).

The first successful biomedical application of calcium phosphate-based material used in a human was reported in the year 1920. After a short time, these bio-ceramic materials get to boom in various applications such as replacing teeth, knees, hips, and ligaments and treatment of periodontal disease, jaw bone stabilization and augmentation, spinal fusion, and maxillofacial reconstruction (Kalita, Bhardwaj, and Bhatt 2007). Hydroxyapatite (HAp) is a calcium phosphate-based bioceramic material with structural and chemical similarities with natural bone minerals used in biomedical applications. Excellent biocompatibility, osteoconductivity, and bioactivity are some features of HAp that make it used in various biomedical applications (Laonapakul 2015). Biogenic sources such as aquatic, mammalian, shell, plant, and corals contain calcium in higher amounts, so it is well suited for HAp extraction. The HAp extracted from a biogenic source is safer and has better cell adhesion and proliferation (Arokiasamy et al., 2022).

More than 2.2 million people undergo bone grafting surgeries to repair bone defects. Healing from bone defects remains a challenge for clinicians in orthopedic applications. Restricted supply of autologous bone and infection threats on using allografts cells and development of long-lasting bone implants made an idea to obtain bone segments from another species technically termed xenogenous. The material obtained from xenogenous bone is similar to the structure of human bones. Moreover, the scalable processing and readily available trace amounts in large quantities are some criteria that are beneficial to using xenogenous bone (Bahrololoom et al., 2009); (Szcześ, Hołysz, and Chibowski 2017). HAp is produced by sintering it at high temperatures, and its mechanical properties are controlled by varying sintering temperature, the composition of the material used, and the duration of sintering. The biogenic sources such as sea shells, egg shells, plants, corals, and mammalian and fish bones act as a reservoir for synthesizing hydroxyapatite. The presence of Al^{3+}, Zn^{2+}, Sr^{2+}, Mg^{2+}, K^{+}, and Na^{+} cations and Co_3^{2-}, SO_4^{2-}, Cl^{-} and F^{-} anions in the hydroxyapatite synthesized from natural materials helps in rapid bone regeneration. Especially the presence of Mg^{2+} and Na^{+} ions helps in bone and teeth development, while their deficiency may lead to bone fragility and loss (Akram et al., 2014). From this study, readers can get an idea about the structure, synthesis, and biomedical applications of HAp and how biogenic wastes help synthesize HAp for converting wastes into a valuable product.

The History behind the Development of Hydroxyapatite

Plaster of Paris ($CaSO_4 . H_2O$) is the first bioceramic material widely used to repair bone defects. This bioceramic material shows an attractive physiochemical performance and poor mechanical properties. In the early 1920s, tricalcium phosphate $Ca_3(PO_4)_2$ was found to repair bone defects in rabbits. Researchers reported that tricalcium phosphate helps to improve bone defects in 31 days instead of 41 days. Later, many studies revealed these calcium salts also stimulate immature bone formation. In 1963, Smith successfully studied cerosium - a ceramic bone substitute containing porous ceramic aluminate infused with epoxy resin. Those days were considered a modern era of bioceramics. It was regarded as a successful bone replacement material in those periods because of its excellent biocompatibility, porosity, and good physicochemical properties similar to the bone (Shackelford 1999);

(Pokhrel 2018). Later in the 1970s, the widespread interest in bioceramics made a way to initiate the dental application of $CaPO_4$ in surgically developed periodontal defects. In addition, dense hydroxyapatite cylinders are used for instant tooth root replacement, and composites based on hydroxyapatite were used to repair significant bone defects. Sooner hydroxyapatite has become an extensive commercial bioceramic material in dental and orthopedic applications (Shackelford 1999); (Dorozhkin 2017).

Hydroxyapatite and Calcium Phosphate-Based Biomaterials

Calcium phosphate class minerals are commonly named apatite with a general formula of $A_4B_6(MO_4)_6X_2$ where A and B represent calcium, MO_4 indicates the phosphate group, and X represents the OH^- group (Brown W. E. and Chow., 1976). These calcium phosphate-based materials contain Ca/P molar ratios of 0.5 to 2. The list of calcium phosphate-based biomaterials widely used in biomedical applications with their formula, Ca/P molar ratio, and symbol are tabulated in Table 1. These materials are successfully used in orthopedics and dentistry to replace damaged hard tissue. Hydroxyapatite is a vital calcium phosphate-based material with a stoichiometric Ca/P ratio of 1.67. Mostly the hydroxyapatite synthesized from natural sources is non-stoichiometric due to the presence of trace elements. Still, it is biocompatible, bioactive, and used in various *in-vitro* applications.

Table 1. List of calcium phosphate-based biomaterials used in biomedical applications (Zaman et al., 2020)

Name	Symbol	Formula	Ca/P Ratio
Amorphous Calcium Phosphate	ACP	$Ca_x(PO_4)_y.nH_2O$	1.2-2.2
β- Tri Calcium Phosphate	β-TCP	$Ca_3(PO_4)_2$	1.5
α- Tri Calcium phosphate	α-TCP	$Ca_3(PO_4)_2$	1.5
Octa Calcium phosphate	OCP	$Ca_8(HPO_4)_2(PO_4)_4.5H_2O$	1.33
Di Calcium Phosphate Anhydrous (monetite)	DCPA	$CaHPO_4$	1
Di Calcium Phosphate Dihydrate (brushite)	DCPD	$CaHPO_4.2H_2O$	1
Mono Calcium Phosphate Anhydrous	MCPA	$Ca(H_2PO_4)_2$	0.5
Name	Symbol	Formula	Ca/P Ratio
Mono Calcium Phosphate Monohydrate	MCPM	$Ca(H_2PO_4)2.H_2O$	0.5
Hydroxyapatite	HAp	$Ca_{10}(PO_4)_6(OH)_2$	1.67

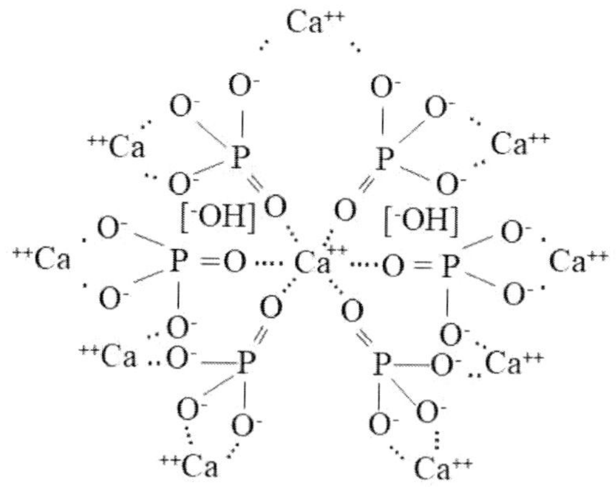

Figure 1. The molecular structure of hydroxyapatite.

Structure and Composition of Hydroxyapatite

$Ca_{10}(PO_4)_6OH_2$ is the basic formula of hydroxyapatite structure, and its stoichiometric ratio (Ca/P) is 1.67 and contains 3.38% of OH, 18.5% P, and 39% Ca by weight. Hydroxyapatite crystalline is hexagonal in structure, having space group $P6_{3/m}$ with lattice parameters a = b = 9.418 Å and c = 6.884 Å with reflection plane and rotational symmetry (Zaman et al., 2020). In hydroxyapatite unit cells, the crystallographic positions of Ca atoms are referred to as M1 and M2. At the M1 place, nine O atoms of phosphate moieties surround four Ca atoms, whereas, at the M2 position, the remaining six O atoms of phosphate moieties surround the other six Ca atoms. The molecular structure of hydroxyapatite is shown in Figure 1. Other than that, hydroxyapatite contains some trace impurities like hydroxyl ions (OH^-), Flouride ions (F^-), Chloride ions (Cl^-), and Phosphite ions (PO_3^{3-}). Studies revealed that the presence of OH^- and F^- ions enhance the strength of hydroxyapatite, whereas a Cl^- and PO_3^{3-} ion weakens the apatite structure (Haider et al., 2017).

Synthesis of Hydroxyapatite

Hydroxyapatite can either synthesized from both organic and inorganic components. The hydroxyapatite synthesized from inorganic components is stoichiometric, while the organic element is commonly found to be greater than 1.67, i.e., non-stoichiometric. But it is also biocompatible, bioactive, and widely used in various biomedical applications. Moreover, hydroxyapatite obtained from organic sources is easy, efficient, cheap, and environmentally friendly. Hence the demand to produce hydroxyapatite from organic sources is increasing daily (Akram et al., 2014); (Ma 2019). In general, hydroxyapatite can be synthesized by various techniques and has been divided into four broad categories such as 1) dry methods, 2) wet methods, 3) hydrothermal methods, and 4) high-temperature processes (Figure 2). The methods like the solid-state reaction method and the mechanochemical method do not require a solvent for hydroxyapatite production and are comes under the category of dry methods. Wet methods are methods in which hydroxyapatite is produced using the solvent medium. The methods like Sol-gel, hydrolysis and chemical precipitation fall under the wet methods. The usage of aqueous solution with high temperature and voltage is said to be a hydrothermal method. The sonochemical, emulsion and microemulsion methods are the best examples of the hydrothermal method. The methods with a high-temperature process like combustion and pyrolysis come under high-temperature methods. These various synthesizing methods result in multiple morphologies, and crystal growth control is the main challenge in the hydroxyapatite synthesis process because the size, shape, and distribution size of crystal can affects the bioactivity, biocompatibility, surface chemistry, mechanical properties, and processing conditions (Gomes et al., 2019).

Dry Methods

As the name suggests, this method does not require solvent, and these methods can produce a large scale of powders with a simple and low-cost procedure. In solid-state synthesis, HAp is produced by mixing calcium and phosphate-containing precursors and is calcinated at a high temperature for a specific time duration, as shown in Figure 3. The resulting powder is heterogeneous and contains secondary phases. HAp synthesized by the solid-state reaction is not produced under precise control processing parameters, so it is not well suited for biomedical applications. HAp production by the solid-state reaction is based upon the following two reactions with water vapor at 1200°C.

$$3CaP_2O_7 + 7CaCO_3 + H_2O \rightarrow Ca_{10}(PO_4)_6(OH)_2 + CO_2 + 7.5\,O_2 \quad (1)$$

$$3\,Ca(PO_4)_2 + 7\,CaCO_3 + H_2O \rightarrow Ca_{10}(PO_4)_6(OH)_2 + CO_2 + 9O_2 \quad (2)$$

Straight forward procedures, high-temperature calcination yields proper crystallized structure, and low cost are some advantages of the solid-state method, while the high temperature is required, inefficient, and the low phase purity of products are some disadvantages of this method (Pu'ad et al., 2019); (Sadat-Shojai et al., 2013).

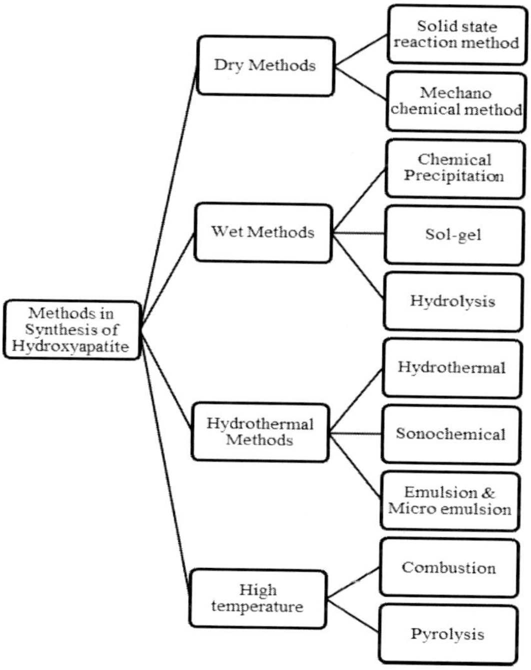

Figure 2. Methods in the synthesis of Hydroxyapatite.

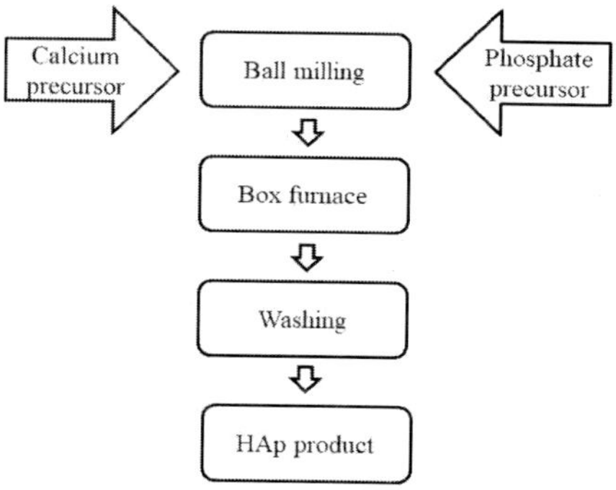

Figure 3. Schematic representation of the solid-state reaction method.

The mechanochemical method is simple and produces well-defined structured powders. In this method, the precursor materials are grounded, and mechanical energy promotes chemical reactions and structural change, as shown in Figure 4. The main reactions involved in mechanochemical synthesis are

$$6CaHPO_4.2H_2O + 4\ CaO \rightarrow Ca_{10}(PO_4)_6\ (OH)_2 + 4\ H_2O \quad (3)$$

$$10CaCO_3 + 6(NH_4)H_2PO_4 \rightarrow Ca_{10}(PO_4)_6\ (OH)_2 + 8H_2O + 10CO_2 + 6NH_3 \quad (4)$$

$$3Ca_3\ (PO_4)_2.xH_2O + Ca(OH)_2 \rightarrow Ca_{10}(PO_4)_6\ (OH)_2 + xH_2O \quad (5)$$

$$10Ca(OH)_2 + 3P_2O_5 \rightarrow Ca_{10}(PO_4)_6\ (OH)_2 + 9\ H_2O \quad (6)$$

(Ferro and Guedes 2019) synthesized HAp from biogenic wastes such as chicken eggshells and cuttlefish bone using mechanochemical reaction by increasing the milling energy up to 123. 0 kWh/g. The resultant reaction sequence obtained for cuttlefish bone was $CaCO_3 \rightarrow$ DCPD $\rightarrow$ HAp and $CaCO_3 \rightarrow$ DCPD $\rightarrow$ DCPA $\rightarrow$ HAp for chicken egg shell.

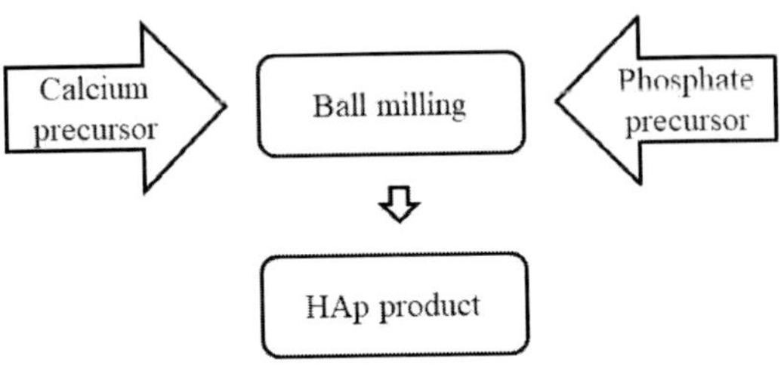

Figure 4. Schematic representation of the mechanochemical method.

Wet Methods

In wet methods, hydroxyapatite powders are produced in the presence of the solvent medium. The powder was obtained using wet methods, having controlled mean size and morphology with low crystallinity and various CaP phase production. Chemical precipitation is a widely used method for synthesizing a large amount of hydroxyapatite powder at a simple and low cost. This technique consists of several processes, as mentioned in Figure 5.

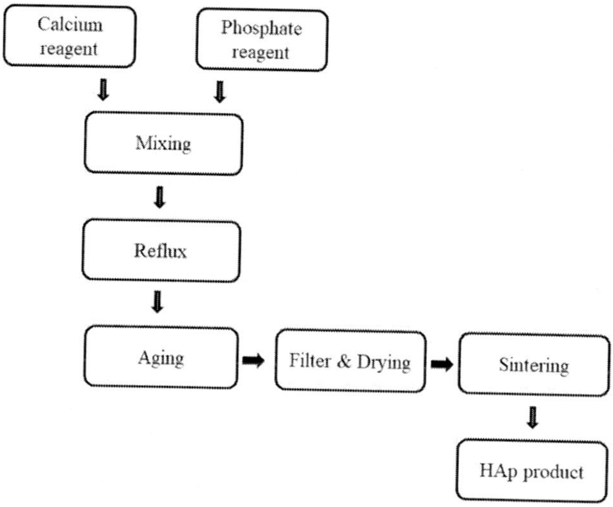

Figure 5. Schematic representation of the wet chemical precipitation method.

At first, the precursor materials containing calcium and phosphate ions are mixed based on the molar ratio of hydroxyapatite. Then the pH is adjusted to an alkaline range, and the solution is allowed for an aging process for around 24 hours and then filtered, dried, and crushed to obtain hydroxyapatite. Usually, calcium hydroxide or calcium nitrate is used as a calcium source, and diammonium hydrogen phosphate and orthophosphoric acid are used as phosphate sources (Pu'ad et al., 2019). Researchers use biogenic sources such as egg shells, clam shells, and seashells as a precursor to synthesize hydroxyapatite. Krishani et al., used blood cockle clam and Asiatic hard clam as a calcium precursor material to synthesize hydroxyapatite using the wet precipitation method (Murugiah et al., 2021). Usually, the powder obtained using the chemical precipitation method is non-stoichiometric and poorly crystallized with an irregular shape (Sadat-Shojai et al., 2013). The precursors in this method should be used according to the hydroxyapatite molar ratio, or another compound will be formed. pH plays a vital role in hydroxyapatite production, and a decrease in pH results in a reduction in the amount of Ca in hydroxyapatite precipitate (Pu'ad et al., 2019).

Hydrolysis is a rarely used wet method for synthesizing hydroxyapatite. A water ionization process causes the diffusion of hydrogen and hydroxide ions. Hydrolysis is carried out to transform calcium and phosphate precursors into high purity hydroxyapatite. The hydrolysis method may take several hours (Pu'ad et al., 2019).

At first, in a conventional sol-gel method, a 3D inorganic network is prepared by mixing suitable precursors or alkoxides in either the organic or aqueous phase, then allowed for aging at room temperature followed by gelation, drying in a hot plate to remove the organic phase. The resultant gel allows calcination to produce hydroxyapatite, as shown in Figure 6. This process takes more time, especially in the solution phase; the reaction between calcium and phosphate precursors happens more gradually. Further, it leads to a longer duration in the aging process to form the apatitic phase.

Moreover, pH and temperature are maintained during the process, and solvent nature and gelation rate are mainly based on the chemical used during the sol-gel process. Poor aging, uncontrolled heat treatment, and gelation may lead to various impurities like CaO, $Ca_3(PO_4)_2$, $CaCO_3$, and $Ca_2P_2O_7$. Usually, calcium nitrate or calcium diethoxide and triethyl phosphite or triethyl phosphate is used as calcium and phosphate precursors in this process, and the general reaction used in this sol-gel process is mentioned below (Sadat-Shojai et al., 2013)

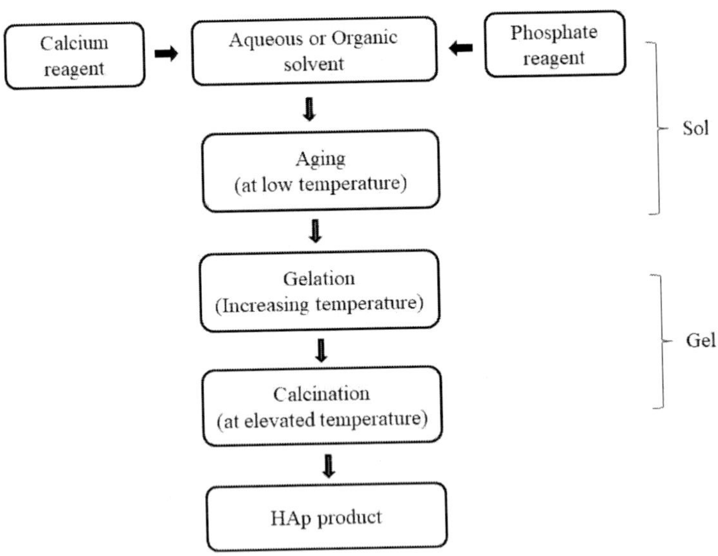

Figure 6. Schematic representation of the sol-gel method.

$$6Ca(NO_3)_2 + 6\,(C_2H_5O)_3\,P(O) \rightarrow Ca_{10}(PO_4)_6\,(OH)_2 + \text{byproducts}$$

Yields HAp of excellent dimensions structure, high transparency, better surface area, low synthesis temperature, high product purity, and homogeneous molecular mixing are some benefits of the sol-gel method. The drying process can result in volume shrinkage and cracking, and low crystallinity and high raw materials costs are some drawbacks of this method (Pai et al., 2020); (Bakan, Laçin, and Sarac 2013); (Cox 2015).

Hydrothermal Methods

The hydrothermal method is one of the commonly used methods to synthesize hydroxyapatite. In the hydrothermal method, chemical reactions occur in aqueous media containing calcium and phosphate precursors at elevated pressure and temperature. Moreover, it is also similar to the chemical precipitation method. Still, the aging process in the hydrothermal method is conducted at above ambient temperature and pressure like inside a pressure vessel or autoclave. Temperature and pressure increase reactivity and the condensation effects create chemical bonds to form a nucleus that produces highly crystalline and stoichiometric hydroxyapatite (Pu'ad et al., 2019); (Sadat-Shojai et al., 2013). The hydrothermal treatment enhances the mechanical and morphological properties of HAp nanoparticles with relatively higher crystallinity, but this process requires high temperature and pressure above ambient and high cost (Sadat-Shojai et al., 2013); (Pai et al., 2020).

Sonochemical methods widely produce nanosized hydroxyapatite particles, in which powerful ultrasound radiation activates the chemical reaction. Formation, elongation, and collision of microbubbles based on the mechanism of acoustic cavitation in an aqueous phase is the central concept in synthesizing hydroxyapatite by the sonochemical method, as shown in Figure 7.

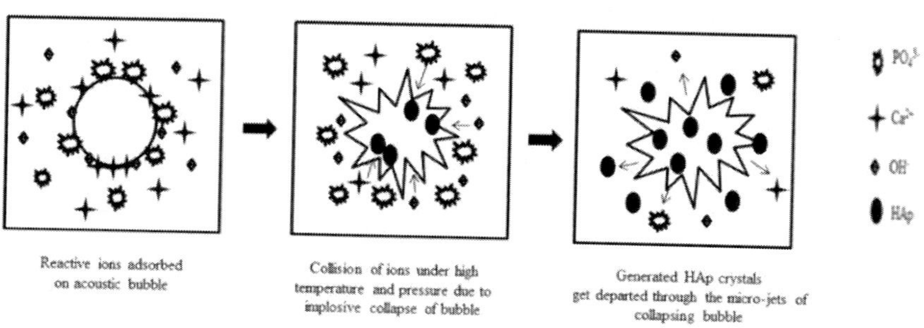

Figure 7. Schematic representation of mechanism behind the sonochemical method.

Hydroxyapatite particles synthesized by sonochemical methods are smaller, uniform, and purer crystals with less agglomeration. Recent studies revealed that an increase in sonication is directly proportional to the rate of hydroxyapatite crystal growth, even up to 5.5 times. Faster reaction and better energy efficiency, low cost, and high purity yield are some advantages, and the sonochemical reduction rate is proportional to the ultrasonic frequency is the only disadvantage of this method (Pu'ad et al., 2019); (Sadat-Shojai et al., 2013); (Pai et al., 2020).

Dispersion of two immiscible liquids by surfactants is the principle behind the microemulsion method. Surfactant is an amphiphilic molecule with a hydrophobic tail and hydrophilic head. Surfactant reduces the surface tension of immiscible liquids resulting in a dispersed phase confined in nanometer regimes. When the reaction is bound to nanosized domains, the resultant emulsion can provide nanosized particles. The emulsion process is enormously dependent on the surfactant found in the medium. The emulsion method can be carried out by three main routes: water in oil emulsion, oil in water emulsion, and water in oil in water emulsion (Sadat-Shojai et al., 2013).

High-Temperature Methods

The rapid exothermic and self-sustained redox reaction between oxidants and organic fuel in an aqueous phase is the primary mechanism behind the synthesis of hydroxyapatite by combustion method. At first, an aqueous solution of $(NH_4)_2$ HPO_4 and Ca $(NO_3)_2$ are mixed to form a white precipitate. Then concentrated HNO_3 was added to dissolve the precipitate. Single or mixtures of fuels are incorporated into the resulting solution mixture. Then the mixture is heated in the furnace at a low temperature of around 300°C to initiate the reaction, followed by a sudden increase in temperature to maximum value for the combustion process. The mixture was finally cooled to induce nucleation and avoid further particle growth, as shown in Figure 8. Yields, high crystalline HAp, low cost, and high phase purity are some advantages, and not being environmentally friendly is the main drawback of this method (Pu'ad et al., 2019); (Sadat-Shojai et al., 2013).

Sometimes pyrolysis is known as "Spray pyrolysis," in which precursor solutions are sprayed into an electric furnace's flame or hot zone area to synthesize the required product. Aizwa et al., synthesized HAp using HNO_3 solutions containing various concentrations of $(NH_4)_2HPO_4$ and $Ca(NO_3)_2$ with a Ca/P molar ratio of 1.67. The setup comprises two electric furnaces: the lower furnace maintains the temperature of 300-500°C for evaporating solvent from the droplet, and the upper furnace maintains the temperature of 750-900°C for pyrolysis of the precipitated salts. Above mentioned solutions were sprayed into the heating zone of the furnace (Sadat-Shojai et al., 2013).

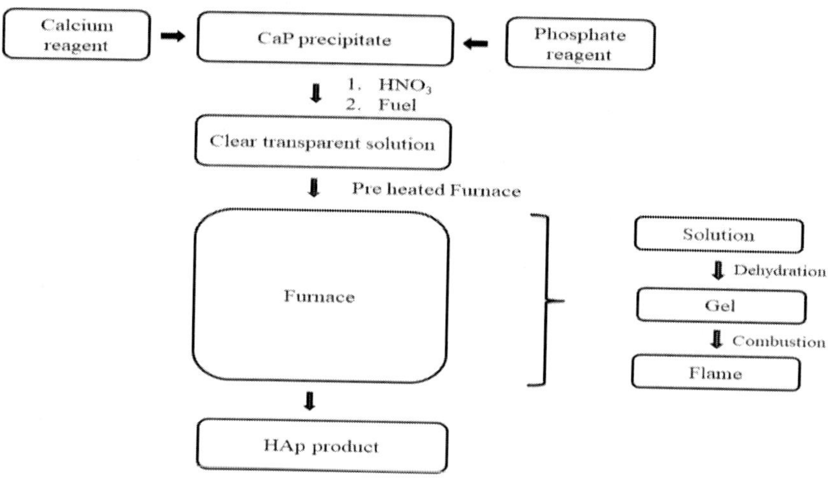

Figure 8. Schematic representation of combustion method.

Natural Sources for Hydroxyapatite Synthesis

Hydroxyapatite Extraction from Mammalian Bones

Mammalian bone contains 20% of collagenous fibers, 69% of apatite minerals, and 9% of water and other organic substances like polysaccharides, lipids, and proteins in minimal quantities. Extraction of hydroxyapatite from mammalian bone is simple and scalable (Kalita, Bhardwaj, and Bhatt 2007). Hydroxyapatite is developed in a biological system's organic matrix of protein molecules. These organic parts of the bone aid crystallite size, size distribution, and lattice orientation during the biomineralization process and simultaneously control the apatite crystal's thickness. Apatite crystals from biological systems endure smaller crystallite sizes with a larger surface area, which helps absorb large amounts of ions (Urist 1965).

Moreover, hydroxyapatite extracted from mammalian bone exhibits different forms, such as a sphere, rod, needle, and plate, and their molecular ratio is also non-stoichiometric, i.e., greater than 1.67. Na^+, Mg^{2+}, K^+, Ti^{2+},

Zn^{2+}, and CO_3^{2-} are some trace elements commonly present in hydroxyapatite synthesized from mammalian bones. At first, mammalian bones are washed and boiled in distilled water and then treated with hypochlorite or NaOH to degrade proteins and remove all other dirt on the bone surface. After washing, the bone was crushed into small pieces using a ball milling process and then calcinated at a temperature ranging from 600-1400°C to remove all organic matter and pathogens in the bone. This process may also help avoid disease transmission from mammals to patients (Oladele et al., 2018).

Hydroxyapatite Extraction from Aquatic Sources

The demand for fish consumption will never fade, and nearly 970-2700 billion tons of fish have been caught, among which humans will consume around 450-1000 billion. Fish bones and scales are wastes from the fish processing industry (Shi et al., 2018). Fish bones and scales are rich in calcium, phosphate, and carbonate, so they are well suited for hydroxyapatite production. To produce hydroxyapatite from fish bones, the bones should be washed, boiled, and treated with alkaline to remove all biomolecules and other impurities, followed by calcination at a high temperature. XRD studies revealed that untreated HAp shows particle size distribution and excellent crystallinity (Oladele et al., 2018). Fish scale is rich in type I collagen and hydroxyapatite, and the extracted collagen and hydroxyapatite from fish scales show higher mechanical properties, biocompatibility, improved cell adhesion, and proliferation (Lim et al., 2020). Kara et al., reported that the fish scales extracted from *Carassius auratus* improved the biomineralization tendency, mechanical properties, alkaline phosphatase activity, cell viability, and type I collagen.

Moreover, the composite scaffolds produced from fish bones are ideal for bone tissue engineering (Kara et al., 2020). Niyasha et al., developed a scaffold based on hydroxyapatite derived from Salmon and Red snapper fish bones. They concluded both scaffolds could play a significant role in periodontal tissue regeneration. Nevertheless, red snapper-based scaffolds display superior features to salmon fish (Wijedasa et al., 2020). Based on the research, Akpan E. S. et al., concluded that hydroxyapatite obtained from catfish bones has various attractive properties in biomedical applications (Akpan et al., 2020).

Hydroxyapatite Extraction from Plant Source

Researchers tried synthesizing hydroxyapatite from plant sources such as wood, algae, calcite-rich plants, and corals. Wood comprises organized parallel hollow tubes with cellular microstructure having good strength, stiffness, and toughness. a) Carbon template production, b) carbon template to calcium carbide conversion, c) calcium oxide formation, d) calcium carbonate formation, and e) HAp production using hydrothermal phosphatization are the five steps involved in the conversion of wood to HAp. Stalks and leaves of plants contain HAp as a significant part and calcium hydroxide as trace elements, e.g., Basil leaves (Akram et al., 2014).

Algae and coral contain $CaCO_3$ with high porosity and interconnectivity that resembles human bone. Usually, the calcite from red algae is converted into HAp using the hydrothermal method or pyrolysis, followed by the calcination method in the presence of diammonium hydrogen phosphate (DAHP) and magnesium nitrate. *Halimeda* spp., like *Halimeda macroloba* and *Halimeda cylindracea,* are some green algae commonly found in the Atlantic, Caribbean, and Indo-Pacific oceans that precipitate $CaCO_3$ into the sediments of their skeleton. Recent research shows that the hydrothermal conversion process produces hydroxyapatite from the skeleton of *Halimeda cylindracea* (Karacan, Ben-Nissan, and Sinutok 2019). Red algae such as *Corallina officinalis* and *Phymatolithon calcareum* are considered the best sources to extract hydroxyapatite (Saleem, Rasheed, and Yougen 2020). Hydroxyapatite is produced from these green and red algae and is applied in bone tissue engineering as a composite bone scaffold (Barua et al., 2018).

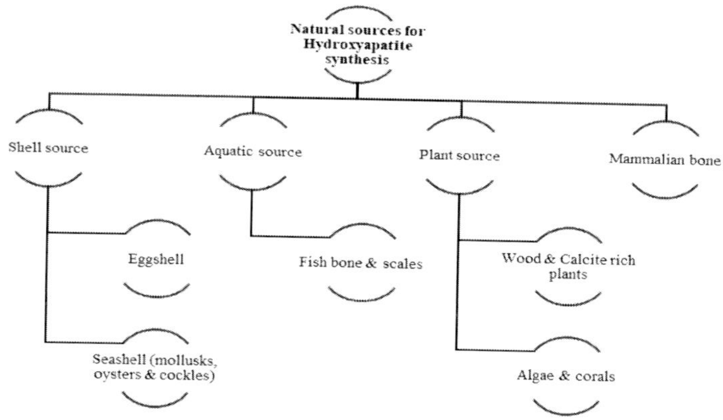

Figure 9. Natural sources for hydroxyapatite.

Corals are the organisms that inhabit the ocean with a rough exoskeleton containing $CaCO_3$. Due to their porous nature and physical properties, corals are used for scaffolding dental and bone growth (Matsuda et al., 2017). Moreover, the pores of the corals are similar to human bone and have superior osteoconductivity and biodegradability. Thus stem cells seeded on the coral-based scaffolds are used to treat various bone defects, including cranial bone defects in canines (Wang et al., 2021). Corals belonging to *Goniastrea* sp., *Porites* sp., and *Tubipora musica* are some corals currently used in complex tissue engineering (Karacan, Ben-Nissan, and Sinutok 2019). An outline of natural sources used for hydroxyapatite synthesis was mentioned in Figure 9.

Hydroxyapatite Extraction from a Shell Source

Shells are the global source of $CaCO_3$ and have produced hydroxyapatite (Ehrlich et al., 2020). Some marine organisms like mollusks, oysters, cockles, and bird's eggs are widely used to develop hydroxyapatite-based scaffolds by combining with some porogen agents. These seashells and eggshells are classified under the category of biogenic sources. Eggs and sea-shells waste are generally dumped into landfills, creating many unhygienic issues like odor and microbial growth. Eggshells are thrown from bakeries, hatcheries, restaurants, and houses annually in a million tons. The eggshell generally contains 94% calcium carbonate, 1% magnesium carbonate, 1% calcium phosphate, and 4% organic substances. Due to enormous availability and cheaper cost, egg shells are an excellent option to use as a calcium precursor for synthesizing hydroxyapatite (Rivera et al., 1999). Bivalves, including clams and mussels, come under the phylum Mollusca; the second largest phylum contains 80,000 invertebrate species. Annually, 5% of bivalves were harvested among the total weight of world harvest for human consumption. The waste shell includes 95-99% of $CaCO_3$ and is usually destined for landfills. Over the past few decades, Many methods have been developed to synthesize HAp from biowaste (Suresh Kumar et al., 2020). Many researchers tried to convert the waste into helpful biomaterial like that (Holloman et al., 2020) developed hydroxyapatite-based scaffolds using chicken eggshell and clam sea shell wastes. (Sari et al., 2021) synthesized the hydroxyapatites from shells of abalone mussels and fabricated them with a honeycomb to make a pore. (Afriani et al., 2019) prepared hydroxyapatite composites to develop scaffolds in bone tissue engineering by using cockle shell waste.

Biomedical Applications of Hydroxyapatite

Drug Delivery System

Over the past decade, hydroxyapatite has been used as a drug carrier to carry antibiotics to hard tissues and also helps provide growth factors, proteins, hormones, vitamins, anti-osteoporotic substances, anti-inflammatory drugs, and anticancer substances to the targeted tissue (Gomes et al., 2019). Drug molecules can be incorporated into HAp either chemically or physically to retain the drug until it reaches the target site. Moreover, it can also degrade and deliver the drug in a controlled manner. Hence HAp is a promising candidate for drug delivery (Jafari 2015). Physiochemical properties such as porous structure, surface composition, size and morphology, excellent biocompatibility, low production cost, low toxicity, pH-dependent dissolution, dormancy to microbial degradation, and excellent storage properties are some properties of HAp to serve as a carrier in drug delivery system. The decrease in the pH value from alkaline to acidic conditions increases the degradation rate of HAp. It results in enhancing the release of drug molecules from the HAp particles. Usually, 7.4 is the pH observed in normal tissues. Still, in the extracellular regions of solid tumors, especially within the endosome-lysosome cell compartment, the pH value is 5, which aids in releasing drugs from HAp. It is an interesting example of pH-dependent dissolution property (Lin and Chang 2015).

Hard Tissue Regeneration

Hard tissue regeneration is one of the significant challenges in the clinical industry due to its complex biocomposites (Kumar et al., 2019). But HAp acts as an aspirant candidate to repair hard tissue due to its similarity with human bone and teeth. Clinically, in repairing hard tissues, HAp are used as 1) granules or powders in bone or teeth defect fillers, 2) Ingredients in bone cement and toothpaste, 3) small implants, 4) porous scaffold material, 5) coatings on metal implants 6) mechanical reinforcement in polymer-bioactive ceramic composites. Studies showed that inserted HAp grafts integrated with host bone and enhanced the formation of new bones, further leading to the recovery of damaged bone (Lin and Chang 2015). Enamel has unique mechanical strength and is self-repaired by life forms after considerable mineral loss. Enamel consists of 97% of apatite nanocrystals. Researchers investigated that 20 nm-sized HAp is an aspirant candidate to repair eroded enamel. These nanosized haps were adsorbed into the surface of enamel and induced the remineralization process. HAp nanocrystals also marked their footprints in cartilage tissue regeneration as a promising material by stimulating the adhesion and proliferation of chondrocytes without using any growth factors. Besides the bioactivity, it also improves the mechanical properties such as anti-friction properties in cartilage tissue regeneration.

Scaffolds in Bone Tissue Engineering

The combination of cells, scaffolds, and signaling molecules to construct a 3D material to repair or restore tissue function is said to be Tissue Engineering (Amini, Laurencin, and Nukavarapu 2012). The scaffold is a base material in which cells and growth factors are embedded to imitate the extracellular matrix to restore and maintain tissue function (Chocholata, Kulda, and Babuska 2019). Scaffold development plays a challenging role in bone tissue engineering because the bone is highly dynamic and diverse tissue both structurally and functionally (Amini, Laurencin, and Nukavarapu 2012). Biocompatibility, biodegradability, mechanical properties, high porosity, and linkages between pores are the essential properties that must be considered while designing the scaffold (Biswal 2020). HAp is a potential candidate with all the above requirements and can be widely used in bone tissue engineering (Lin and Chang 2015).

Moreover, it also has osteoconductive properties, which enhance osteoblast cell attachment, growth, and proliferation (Szcześ, Hołysz, and Chibowski 2017). A solitary polymer cannot achieve every single property of a scaffold, so the desired property can be attained by mixing it with various polymers or materials (Dehghan, Mehrizi, and Nikukar 2021). Usually, natural-based biopolymer has poor mechanical properties so it can be fabricated with hydroxyapatite and used in bone tissue engineering. Hydroxyapatite produced from these green and red algae is applied in bone tissue engineering as a composite bone scaffold. Hydroxyapatite derived from pig and bovine bones is also used in scaffold development (Barua et al., 2018).

Coatings on Metal Implants

Metals like titanium and their alloys are widely used as implants due to their excellent mechanical properties, tolerance and stability in body fluids, and low density. Though implants show better mechanical strength and hardness, they fail to display osteointegration and biocompatibility. To enhance these features, metal implants are coated with HAp. Electrophoretic deposition, plasma spraying, sol-gel deposition, pulsed layer deposition, sputtering techniques, spin coating, and ion beam-assisted deposition are widely used for coating implants with HAp. Good adhesion of hydroxyapatite to the substrate, optimal porosity, high degree of crystallinity, and proper stoichiometry are the characteristics of good hydroxyapatite coatings on metal implants. Later, it was found that bacterial cells started to adhere, produce biofilms, and cause infection after placing the implants on life forms. To prevent bacteria, researchers began to develop HAp coatings with antibacterial properties. HAp - Zn coatings & HAp - Ag coatings are the findings for the above mentioned problem. These coatings show excellent biocompatibility, osteoconductivity, and no cytotoxicity with antibacterial properties (Gomes et al., 2019).

Bioimaging and Diagnosis

Fluorescence imaging is a standard method for imaging a biological matter in a high resolution and safe manner. Various nanostructured materials are used for fluorescence bioimaging, in that Ca-based materials, mainly; hydroxyapatite, gained an increasing interest. Safety, drug loading capacity, easy preparation, pH-dependent solubility, tunable biodegradability, and superior biocompatibility are a few attractive features of hydroxyapatite. Due to low intrinsic optical response, HAp is doped with organic fluorophores, lanthanide ions, composites, or other inorganic materials and is used in cell imaging. Machado et al., successfully developed hydroxyapatite nanorods with good biocompatibility and incorporated in into the human body without any trace of cellular damage (Machado et al., 2019). Deshmukh et al., created a new class of fluorescence HAp particles using the sol-gel method and internalized them into prokaryotic and eukaryotic systems, which exhibit no cytotoxicity. This resultant fluorescence HAp nanoparticles provide exciting opportunities in biomedical applications as treatment in conjunction and bioimaging (Deshmukh et al., 2016).

Downstream Processing

Chromatography is a crucial purification process widely used in all biopharmaceutical industries. HAp is commonly used in chromatography for protein purification, Antigen fractionation, nucleic acids separation and purification, and enzyme purification. No toxicity, resistance to biochemical degradation, high selectivity, high adsorption capacity, and high stability are the advantages of HAp to apply as a purification agent and adsorbent. Ashokan et al., developed HAp chromatographic matrices from waste eggshells regarding waste valorization. The developed matrix showed high surface area, better flow characteristics, and good column efficiency and resolution. Moreover, a large yield can be obtained with lesser time and cheaper cost (Ashokan et al., 2021). An outline of biomedical applications of hydroxyapatite is shown in Figure 10.

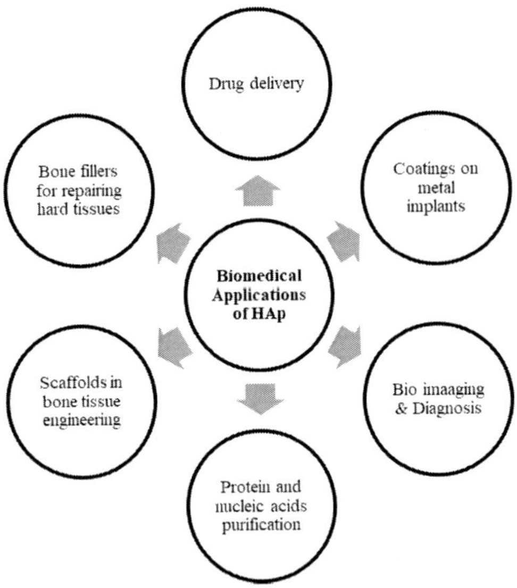

Figure 10. Biomedical applications of hydroxyapatite.

Conclusion

This article primarily focused on the methods involved in synthesizing hydroxyapatite, followed by natural sources and biomedical applications of hydroxyapatite. As mentioned in the article, hydroxyapatite can be synthesized from various biogenic sources such as animal bones, sea shells, eggshells, fish bones, plants, algae, and corals. These waste materials are commonly dumped into landfills and create pollution. I hope this article gives readers some idea about developing hydroxyapatite and converting waste into a valuable product.

References

Abdulrahman, I., Hamzat Ibiyeye Tijani, Bashir Abubakar Mohammed, Haruna Saidu, Hindatu Yusuf, Mohammed Ndejiko Jibrin, and Sulaiman Mohammed. 2014. "From Garbage to Biomaterials: An Overview on Egg Shell Based Hydroxyapatite." *Journal of Materials* 2014: 1-6. https://doi.org/10.1155/2014/802467.

Afriani, F., Y. Tiandho, J. Evi, A. Indriawati, and R. A. Rafsanjani. 2019. "Synthesis and Characterization of Hydroxyapatite/Silica Composites Based on Cockle Shells Waste and Tin Tailings." *IOP Conference Series: Earth and Environmental Science* 353 (1). https://doi.org/10.1088/1755-1315/353/1/012032.

Akpan, E. S., M. Dauda, L. S. Kuburi, D. O. Obada, and D. Dodoo-Arhin. 2020. "A Comparative Study of the Mechanical Integrity of Natural Hydroxyapatite Scaffolds Prepared from Two Biogenic Sources Using a Low Compaction Pressure Method." *Results in Physics* 17 (December 2019): 103051. https://doi.org/10.1016/j.rinp.2020.103051.

Akram, M., Rashid Ahmed, Imran Shakir, Wan Aini Wan Ibrahim, and Rafaqat Hussain. 2014. "Extracting Hydroxyapatite and Its Precursors from Natural Resources." *Journal of Materials Science* 49 (4): 1461-75. https://doi.org/10.1007/s10853-013-7864-x.

Amini, A. R., Cato T. Laurencin, and Syam P. Nukavarapu. 2012. "Bone Tissue Engineering: Recent Advances and Challenges." *Critical Reviews in Biomedical Engineering* 40 (5): 363-408. https://doi.org/10.1615/CritRevBiomedEng.v40.i5.10.

Arokiasamy, P., Mohd Mustafa Al Bakri Abdullah, Shayfull Zamree Abd Rahim, Salmabanu Luhar, Andrei Victor Sandu, Noorina Hidayu Jamil, and Marcin Nabiałek. 2022. "Synthesis Methods of Hydroxyapatite from Natural Sources: A Review." *Ceramics International* 48 (11): 14959-79. https://doi.org/10.1016/j.ceramint.2022.03.064.

Ashokan, A., Vivek Rajendran, T. S. Sampath Kumar, and Guhan Jayaraman. 2021. "Eggshell Derived Hydroxyapatite Microspheres for Chromatographic Applications by a Novel Dissolution - Precipitation Method." *Ceramics International* 47 (13): 18575-83. https://doi.org/10.1016/j.ceramint.2021.03.183.

Bahrololoom, M. E., M. Javidi, S. Javadpour, and J. Ma. 2009. "Characterisation of Natural Hydroxyapatite Extracted from Bovine Cortical Bone Ash." *J. Ceram. Process. Res* 10 (2): 129-38.

Bakan, F., Oral Laçin, and Hanifi Sarac. 2013. "A Novel Low Temperature Sol-Gel Synthesis Process for Thermally Stable Nano Crystalline Hydroxyapatite." *Powder Technology* 233: 295-302. https://doi.org/10.1016/j.powtec.2012.08.030.

Barua, E., Ashish B. Deoghare, Payel Deb, and Sumit Das Lala. 2018. "Naturally Derived Biomaterials for Development of Composite Bone Scaffold: A Review." *In IOP Conference Series: Materials Science and Engineering*. Vol. 377. https://doi.org/10.1088/1757-899X/377/1/012013.

Biswal, T. 2020. "Biopolymers for Tissue Engineering Applications: A Review." *Materials Today: Proceedings*. https://doi.org/10.1016/j.matpr.2020.09.628.

Brown, W. E., and L. C. Chow. 1976. "Chemical Properties of Bone Mineral." *Annu Rev Mater Sci* 6: 213-236.

Chocholata, P., Vlastimil Kulda, and Vaclav Babuska. 2019. "Fabrication of Scaffolds for Bone-Tissue Regeneration." *Materials* 12 (4). https://doi.org/10.3390/ma12040568.

Cox, S. C. 2015. "Synthesis Method of Hydroxyapatite." *Ceram,* no. January 2014: 1-7. www.ceram.com.

Dehghan, M., Mohammad Khajeh Mehrizi, and Habib Nikukar. 2021. "Modeling and Optimizing a Polycaprolactone/Gelatin/Polydimethylsiloxane Nanofiber Scaffold for Tissue Engineering: Using Response Surface Methodology." *Journal of the Textile Institute* 112 (3): 482-93. https://doi.org/10.1080/00405000.2020.1766317.

Deshmukh, K., M. Monsoor Shaik, Sutapa Roy Ramanan, and Meenal Kowshik. 2016. "Self-Activated Fluorescent Hydroxyapatite Nanoparticles: A Promising Agent for Bio-Imaging and Bio-Labeling Self-Activated Fluorescent Hydroxyapatite Nanoparticles: A Promising Agent for Bio-Imaging and Bio-Labeling." *ACS Biomaterials Science & Engineerin* 2 (8): 1257-64. https://doi.org/10.1021/acsbiomaterials.6b00169.

Dorozhkin, S. V. 2017. "A History of Calcium Orthophosphates (CaPO4) and Their Biomedical Applications." *Morphologie* 101 (334): 143-53. https://doi.org/10.1016/j.morpho.2017.05.001.

Ehrlich, H., Rajko Martinović, Danijela Joksimović, Iaroslav Petrenko, Stefano Schiaparelli, Marcin Wysokowski, Dmitry Tsurkan, Allison L. Stelling, Armin Springer, Michael Gelinsky & Aleksandar Joksimović. 2020. "Conchixes: Organic Scaffolds Which Resemble the Size and Shapes of Mollusks Shells, Their Isolation and Potential Multifunctional Applications." *Applied Physics A: Materials Science and Processing* 126 (7): 1-13. https://doi.org/10.1007/s00339-020-03728-7.

Ferro, A. C., and Mafalda Guedes. 2019. "Mechanochemical Synthesis of Hydroxyapatite Using Cuttlefish Bone and Chicken Eggshell as Calcium Precursors." *Materials Science and Engineering C* 97 (October 2018): 124-40. https://doi.org/10.1016/j.msec.2018.11.083.

Gomes, D. S., A. M. C. Santos, G. A. Neves, R. R. Menezes, Campina Grande, and Campina Grande. 2019. "A Brief Review on Hydroxyapatite Production and Use in Biomedicine." *Ceramica* 65: 282-302.

Haider, A., Sajjad Haider, Sung Soo Han, and Inn Kyu Kang. 2017. "Recent Advances in the Synthesis, Functionalization and Biomedical Applications of Hydroxyapatite: A Review." *RSC Advances* 7 (13): 7442-58. https://doi.org/10.1039/c6ra26124h.

Hembrick-Holloman, V., Temesgen Samuel, Zaheeruddin Mohammed, Shaik Jeelani, and Vijaya K. Rangari. 2020. "Ecofriendly Production of Bioactive Tissue Engineering Scaffolds Derived from Egg- And Sea-Shells." *Journal of Materials Research and Technology* 9 (6): 13729-39. https://doi.org/10.1016/j.jmrt.2020.09.093.

Jafari, S. 2015. "Application of Hydroxyapatite Nanoparticle in the Drug Delivery Systems." *Journal of Molecular Pharmaceutics & Organic Process Research* 03 (01): 1-3. https://doi.org/10.4172/2329-9053.1000e118.

Kalita, S. J., Abhilasha Bhardwaj, and Himesh A. Bhatt. 2007. "Nanocrystalline Calcium Phosphate Ceramics in Biomedical Engineering." *Materials Science and Engineering* C 27 (3): 441-49. https://doi.org/10.1016/j.msec.2006.05.018.

Kara, A., Oylum C. Gunes, Aylin Z. Albayrak, Gokcen Bilici, Guven Erbil, and Hasan Havitcioglu. 2020. "Fish Scale/Poly(3-Hydroxybutyrate-Co-3-Hydroxyvalerate) Nanofibrous Composite Scaffolds for Bone Regeneration." *Journal of Biomaterials Applications* 34 (9): 1201-15. https://doi.org/10.1177/0885328220901987.

Karacan, I., Besim Ben-Nissan, and Sutinee Sinutok. 2019. *Marine-Based Calcium Phosphates from Hard Coral and Calcified Algae for Biomedical Applications*. Springer Singapore. https://doi.org/10.1007/978-981-13-8855-2_7.

Kumar, A., Saeid Kargozar, Francesco Baino, and Sung S Han. 2019. "Additive Manufacturing Methods for Producing Hydroxyapatite and Hydroxyapatite-Based Composite Scaffolds: A Review." *Frontiers in Materials* 6. https://doi.org/10.3389/fmats.2019.00313.

Laonapakul, T. 2015. "Synthesis of Hydroxyapatite from Biogenic Wastes." *Kku Engineering Journal* 42 (3): 269-75.

Lim, S. S., Chye Jian Oon, Kit Wayne Chew, Jing Ying Yap, May Teng Lim, Derwin Su, Lai Yee Lee, and Hwei San Loh. 2020. "Improved Physical Properties and *in Vitro* Biocompatibility of Chitosan Composite Scaffolds Incorporated with a Green Filler on Bone Cells." *Clean Technologies and Environmental Policy* 22 (3): 701-12. https://doi.org/10.1007/s10098-020-01815-0.

Lin, K., and J. Chang. 2015. Structure and Properties of Hydroxyapatite for Biomedical Applications. *Hydroxyapatite (Hap) for Biomedical Applications*. Vol. 4214. Elsevier Ltd. https://doi.org/10.1016/b978-1-78242-033-0.00001-8.

Ma, Guoqing. 2019. "Three Common Preparation Methods of Hydroxyapatite." *IOP Conference Series: Materials Science and Engineering* 688 (3): 0-12. https://doi.org/10.1088/1757-899X/688/3/033057.

Machado, T. R., I. S. Leite, N. M. Inada, M. S. Li, E. Cordoncillo, and E. Longo. 2019. "Designing Biocompatible and Multicolor Fl Uorescent Hydroxyapatite Nanoparticles for Cell-Imaging Applications." *Materials Today Chemistry* 14. https://doi.org/10.1016/j.mtchem.2019.100211.

Matsuda, Y., Tetsunari Nishikawa, Tomoharu Okamura, Kazuya Tominaga, Masahiro Wato, Hajime Tabata, Makoto Umeda, Nobutaka Okusa, Koichi Imai, Akio Tanaka, Isao Tamura 2017. "Comparative Study of Tissue Affinity, Chemical Characteristics of Cultured and Natural Coral as a Bioabsorbable Scaffold." *Journal of Oral Tissue Engineering* 14 (3): 164-70. https://doi.org/10.11223/jarde.14.164.

Murugiah, K., M. I. Zakaria, H. Suhaimi, W. Caesarendra, and N. S. Sambudi. 2021. "Synthesis and Characterisation of Hydroxyapatite (HAp) from Asiatic Hard Clam

(Meretrix Meretrix) and Blood Cockle Clam (Anadara Granosa) Using Wet Precipitation Process." In *2021 IEEE National Biomedical Engineering Conference (NBEC)*, 1-6. https://doi.org/10.1109/NBEC53282.2021.9618744.

Oladele, I. O., O. G. Agbabiaka, O. G. Olasunkanmi, A. O. Balogun, and M. O. Popoola. 2018. "Non-Synthetic Sources for the Development of Hydroxyapatite." *Journal of Applied Biotechnology & Bioengineering* 5 (2): 88-95. https://doi.org/10.15406/jabb.2018.05.00122.

Pai, S., S. M Kini, R. Selvaraj, and A. Pugazhendhi. 2020. "A Review on the Synthesis of Hydroxyapatite, Its Composites and Adsorptive Removal of Pollutants from Wastewater." *J. Water Process Eng* 38. https://doi.org/10.1016/j.jwpe.2020.101574.

Pokhrel, S. 2018. "Hydroxyapatite: Preparation, Properties and Its Biomedical Applications." *Advances in Chemical Engineering and Science* 8: 225-40. https://doi.org/10.4236/aces.2018.84016.

Pu'ad, N. A. S. Mohd, R. H. Abdul Haq, H. Mohd Noh, H. Z. Abdullah, M. I. Idris, and T. C. Lee. 2019. "Synthesis Method of Hydroxyapatite: A Review." *Materials Today: Proceedings* 29: 233-239. https://doi.org/10.1016/j.matpr.2020.05.536.

Rivera, E. M., Miguel Araiza, Witold Brostow, Victor M. Castaño, J. R. Díaz-Estrada, R. Hernández, and J. Rogelio Rodríguez. 1999. "Synthesis of Hydroxyapatite from Eggshells." *Materials Letters* 41 (3): 128-34. https://doi.org/10.1016/S0167-577X(99)00118-4.

Sadat-Shojai, M., Mohammad Taghi Khorasani, Ehsan Dinpanah-Khoshdargi, and Ahmad Jamshidi. 2013. "Synthesis Methods for Nanosized Hydroxyapatite with Diverse Structures." *Acta Biomaterialia* 9 (8): 7591-7621. https://doi.org/10.1016/j.actbio.2013.04.012.

Saleem, M., Sidra Rasheed, and Chen Yougen. 2020. "Silk Fibroin/Hydroxyapatite Scaffold: A Highly Compatible Material for Bone Regeneration." *Science and Technology of Advanced Materials* 21 (1): 242-66. https://doi.org/10.1080/14686996.2020.1748520.

Sari, M., Puspa Hening, Chotimah, Ika Dewi Ana, and Yusril Yusuf. 2021. "Porous Structure of Bioceramics Carbonated Hydroxyapatite-Based Honeycomb Scaffold for Bone Tissue Engineering." *Materials Today Communications* 26 (October 2020): 102135. https://doi.org/10.1016/j.mtcomm.2021.102135.

Shackelford, J. F. 1999. "Bioceramics - an Historical Perspective." *Materials Science Forum* 293: 1-4. https://doi.org/10.4028/www.scientific.net/msf.293.1.

Shi, P., Meng Liu, Fengjiao Fan, Cuiping Yu, Weihong Lu, and Ming Du. 2018. "Characterization of Natural Hydroxyapatite Originated from Fish Bone and Its Biocompatibility with Osteoblasts." *Materials Science and Engineering C* 90 (March): 706-12. https://doi.org/10.1016/j.msec.2018.04.026.

Suresh Kumar, C., K. Dhanaraj, R. M. Vimalathithan, P. Ilaiyaraja, and G. Suresh. 2020. "Hydroxyapatite for Bone Related Applications Derived from Sea Shell Waste by Simpleprecipitation Method." *Journal of Asian Ceramic Societies* 8 (2): 416-29. https://doi.org/10.1080/21870764.2020.1749373.

Szcześ, A., Lucyna Hołysz, and Emil Chibowski. 2017. "Synthesis of Hydroxyapatite for Biomedical Applications." *Advances in Colloid and Interface Science* 249 (April): 321-30. https://doi.org/10.1016/j.cis.2017.04.007.

Urist, M. R. 1965. "Bone: Formation by Autoinduction." *Science* (New York, N.Y.) 150 (3698): 893-99. https://doi.org/10.1126/science.150.3698.893.

Wang, Z., L. Han, T. Sun, W. Wang, X. Li, and B. Wu. 2021. "Construction of Tissue-Engineered Bone with Differentiated Osteoblasts from Adipose-Derived Stem Cell and Coral Scaffolds at an Ectopic Site." *British Journal of Oral and Maxillofacial Surgery* 59 (1): 46-51. https://doi.org/10.1016/j.bjoms.2020.07.006.

Wijedasa, N. P., Sarah M. Broas, Rachel E. Daso, and Ipsita A. Banerjee. 2020. "Varying Fish Scale Derived Hydroxyapatite Bound Hybrid Peptide Nanofiber Scaffolds for Potential Applications in Periodontal Tissue Regeneration." *Materials Science and Engineering C* 109 (August 2019): 110540. https://doi.org/10.1016/j.msec.2019.110540.

Zaman, S. U., Muhammad Khaliq, U Zaman, and Nawshad Muhammad. 2020. "Overview of Hydroxyapatite; Composition, Structure, Synthesis Methods and Its Biomedical Uses." *Biomedical Letters* 6 (1): 84-99. https://www.researchgate.net/publication/346657427.

Chapter 7

Deposition of HA-Based Coatings on Bio-Inert Substrates by Various Deposition Methods for Improved Bio-Mechanical Properties

Shahid Hussain and Kazi Sabiruddin*

Department of Mechanical Engineering, Indian Institute of Technology Indore, Simrol, Indore, India

Abstract

Hydroxyapatite (HA) is an attractive bioactive material in the biomedical field due to its chemical composition and structure which are similar to the mineral components of natural bone. It is a highly bio-compatible and osteo-conductive material. However, the low bending strength and fracture toughness of pure HA restrict it from implant applications. Although bio-inert materials such as titanium alloy and bio-grade stainless steel have excellent mechanical properties, they show poor osteo-conductivity. Coating of HA applied to these metallic materials can be a solution to the problem. The HA-coated metallic implant can have superior mechanical and biological characteristics. HA powder can be synthesized from chemical reagents or biogenic materials by using various routes. Chemical composition and other properties of biogenic material are preserved in the synthesized biogenic-derived hydroxy-apatite. The HA powder is deposited as a coating on bio-inert metals for biomedical implant applications by using several deposition methods. In this chapter, several techniques such as plasma spray, electrophoretic deposition, sputtering deposition, pulsed laser deposition, spin coating, ion beam assisted deposition, sol-gel deposition, electrodeposition, cold

* Corresponding Author's Email: skazi@iiti.ac.in (Kazi Sabiruddin).

In: What to Know about Hydroxyapatite
Editor: Isaac Lavoie
ISBN: 979-8-88697-523-9
© 2023 Nova Science Publishers, Inc.

spray, and suspension spray techniques which were used to deposit HA coatings are discussed thoroughly. The characteristics of the HA coatings in terms of their microstructure, phases, mechanical properties, and biological behavior are described in detail.

Keywords: hydroxyapatite; biogenic; plasma spray process; characterization

Introduction

Hydroxyapatite (HA) is considered one of the excellent bio-ceramic material among the various calcium phosphate members. It is applied in biomedical areas, especially orthopedic, dental, and maxillofacial fields, due to its bioactivity, biocompatibility, and osteoconductive characteristics. The bioactivity (ability to generate bone apatite-like material), biocompatibility (capability to interact with the living system without creating detrimental effect), and osteoconductivity (formation of suitable scaffold for bone development) are required properties of HA during biomedical applications. The properties of HA are fundamentally determined by its composition, phase, crystal size, particle size, morphology, and crystallinity, which in turn depend on synthesis precursors and processing methods. In some cases, heating temperature and the duration of its application mainly control the properties of synthesized HA, particularly microstructure and mechanical characteristics [1–3]. Titanium-based alloys, cobalt-based alloys, and stainless steel (316L) are some of the most widely used metallic biomaterials to replace or repair damaged bone due to their high durability, toughness, and strength [4]. However, these bio-inert metallic implants show poor osteoconductivity property that causes a weak interface when surrounded by bone tissues and results in aseptic loosening [5]. As a solution, bioactive materials such as HA can be deposited on these metallic implants to obtain both high mechanical and excellent biological properties [6]. Synthetic calcium and different biogenic materials can be employed as a starting material for the preparation of HA. The properties of biogenic-derived HA are different than those of synthetic calcium-based HA in terms of bioactivity. Various biogenic resources, namely seashell [7, 8], eggshell [9, 10], bovine bone [11], fish bone [12], chicken bone [13], oyster shell [14], fish scale [15], plant [16], coral [17], pig bone [18], snail shell [19], mussel shell [20], abalone shell [21] etc., have been employed in the past by many researchers for the preparation of HA. It is necessary to utilize these biogenic materials for easy waste management and to produce low-cost HA. In this way, the harmful effects of raw biogenic materials on the environment can be reduced.

HA Deposition Methods

Like any other coating material, the properties of HA coating depend on the coating deposition technique. Also, for a specific deposition technique, the character of HA coating may vary with the varying process parameters. Based on the application area, the technique to deposit HA coating is decided. Over the years, different coating methods have been developed and adopted to deposit HA powders on different bio-inert substrate materials. The commonly used coating methods are plasma spray, electrophoretic, sputter deposition, pulsed laser deposition, spin coating, ion beam-assisted deposition, sol-gel deposition, electrodeposition, cold spray, and suspension spray technique. The application of these methods for the deposition of HA coating is described in the following sections.

Plasma Spray Deposition

Plasma spraying is considered one of the most versatile and accepted techniques for manufacturing HA coatings. This method has also been approved by US Food and Drug Administration (FDA). The main component of the plasma spraying setup is a gun which consists of a tungsten cathode and a copper anode. Non-reactive gas such as helium (He), argon (Ar), hydrogen (H_2), nitrogen (N_2), or a mixture of these, is fed across the high-energy arc to ionize and produce a plasma plume. After gas ionization, electrons move to their lower energy state, and recombination of ions takes place in an open atmosphere. During this process, a huge amount of energy is released in the form of a flame. The maximum temperature of a plasma flame may reach up to 25000 °C based on the electric power and gas mixture employed. Coating material in the form of powder/wire is fed to the flame for melting, and then deposition on the substrate is kept in front of the flame at a sufficient distance. Plasma jet provides particle velocity in the range of 80 to 300 m/s [22, 23]. Singh et al., [24] studied the HA-graphene nanoplatelets composite coating deposited on titanium alloy substrate by using the atmospheric plasma spraying (APS) method. The amount of graphene nanoplatelets was kept as 0,

1, and 2 wt % in composite coating. The plasma power of the coating setup varied from 15 kW to 35 kW. Characterization of the composite coatings was carried out in terms of porosity, crystallinity, elastic modulus, fracture toughness, microstructure, phase, micro-hardness, and biocompatibility. The improvement in crystallinity of HA-based coating was observed with the incorporation of graphene nanoplatelets into it. When plasma power was increased from 15 kW to 35 kW, an increment in fracture toughness, elastic modulus, and hardness of the coating was noticed. The addition of graphene nanoplatelets to the HA feedstock helped to improve the toughness and reduce the porosity of the coating. Chambard et al., [25] investigated the effect of characteristics of the precursor powder on the HA coatings fabricated by APS and RF-suspension plasma spray techniques. The feedstock HA powders are varied in terms of density, Ca/P ratio, and powder size. The X-ray diffractometer (XRD), field-emission scanning electron microscope (FESEM), Fourier transform infrared (FTIR) spectroscope, and Raman spectroscope were employed to characterize the precursor powders and their respective coatings. Unlike the APS coated HA, the particle size of the feedstock was not found to have any significant effect on the microstructure of HA coating deposited by the RF-suspension plasma spray method. Singh et al., [26] developed HA-calcium silicate composite coating on Ti-6Al-4V alloy with varying calcium silicate content through the atmospheric plasma spray technique. The amount of calcium silicate in the composite coating was chosen as 0, 10, and 20% by weight. The corrosion and mechanical behaviour of coated samples were analyzed. Surface roughness, microhardness, porosity, crystallinity, and corrosion resistance were found to increase with the increasing content of calcium silicate in HA coating. In another study, the plasma spray process was employed to deposit pure HA and HA reinforced with Al_2O_3 as well as TiO_2 on titanium alloy substrates [27]. The reinforced HA coating showed improved mechanical properties with enhanced corrosion resistance behaviour. Chen et al., [28] compared plasma sprayed pure HA and plasma sprayed HA-graphene nanosheet composite coatings in terms of the mechanical and biological properties. The mechanical properties such as yield strength, elastic modulus, and fracture toughness, along with the biological properties such as cell adhesion strength, differentiation, and proliferation, are found to be improved in the case of graphene nanosheet reinforced HA. The graphene nanosheet pull-out, crack branching, crack deflection, load transfer, inter-layer sliding of graphene nanosheet, and graphene nanosheet bridging were the mechanisms responsible for the enhanced mechanical performance of the composite coating.

Electrophoretic Deposition

In electrophoretic deposition (EPD) method, the coating material is used in the form of a suspension. The substrate is kept as the cathode among the two electrodes used in the process. Stainless steel or carbon rod is used as the anode. The deposition area, cathode to anode distance, applied voltage, and the deposition time are the variable process parameters of EPD. The suspended particles are dispersed homogeneously by using a magnetic stirrer. Drevet et al., [29] deposited the HA through EPD by varying two important process parameters such as deposition time and applied voltage. The best results are achieved by applying 10 V of voltage for a time duration of 10 min. Further, the mechanical properties such as Young's modulus and hardness of coated samples are improved via thermal treatment carried out in a controlled atmosphere. Khanmohammadi et al., [30] studied the EPD deposited HA-based bioglass composite coatings on titanium with changing composition of the composite. The HA content was kept as 0, 25, 50, 75, and 100 wt % in the composite. Different characterization techniques were employed to analyze the coated samples. The coating obtained from 50 wt % HA in bioglass showed the maximum bonding strength and the highest bioactivity property among all. Zhang et al., [31] fabricated the antibacterial-functionalized Ag-loaded HA coating on titanium substrate with the help of the EPD technique by changing the Ag-ion concentration. The Ag-ion addition with HA was noticed to improve the adhesion strength of the coating. Mahmoodi et al., [32] used the EPD method to deposit HA-graphene oxide composite coating on tantalum substrates. The bio-tribological and in-vitro corrosion behavior of the coatings were studied. The deposition was done at a fixed voltage of 30 V applied for two different time durations: 5 and 10 min. The improvement in fracture toughness, resistance to plastic deformation, and bio-tribological behavior was noticed for the composite coated sample as compared to the uncoated sample. Sun et al., [33] reported a study on Zn-substituted HA coating deposited on stainless steel substrate by the EPD process. Before coating, they synthesized the Zn-substituted HA through the co-precipitation method. The effect of Zn concentration on the microstructure and morphology of the fabricated coating was investigated. When the Zn concentration was increased from 1 to 10%, the deposition current density was found to be enhanced significantly. The coated samples were analyzed by using different techniques such as Fourier transform infrared spectroscopy, scanning electron microscopy, and X-ray diffraction. The results indicated that uniform and dense Zn-substituted HA coating could be obtained with a Zn/Ca ratio of 5%.

Sputter Deposition

In sputter deposion, the atoms/ions of coating material are plucked from the solid target by applying high energy to it and deposited on a solid substrate kept in a vacuum environment. Highly pure coating with excellent surface characteristics can be achieved by this technique. The target material, deposition time, reactive gas flow rate, applied power, target to substrate distance, and deposition temperature are some of the essential process parameters that control the coating properties obtained by this technique. Parau et al., [34] studied the influence of the deposition temperature on the characteristics of HA coating deposited on AZ31B alloy substrate by radio frequency (RF) magnetron sputtering technique. The deposition temperature varied from room temperature to 400 °C, and it was noticed that with an increase in temperature, the grain size of HA increased greatly. However, the reduction in hardness was observed by the increasing temperature. Improvement in anti-corrosion and mechanical properties were observed up to 200 °C. Lenis et al., [35] deposited HA coating on Ti-alloy through the RF magnetron sputtering technique. Coated samples were heat-treated in the Ar atmosphere at 400 and 700 °C temperatures. Further, the morphology, mechanical properties, phase composition, and microstructure of the coatings were analyzed by scanning electron microscopy, micro-Raman spectroscopy, X-ray diffraction, energy dispersive x-ray spectroscopy, atomic force microscopy, Fourier transform infrared spectroscopy and nanoindentation methods. The coated sample, which was thermally treated at 400 °C, showed an enhanced crystallinity, elastic modulus, hardness, and adhesion strength compared to the as-deposited coating. Vladescu et al., [36] fabricated the composite coatings of HA on Ti-6Al-4V alloy substrate with variable Ti content by using RF magnetron sputtering technique. With an increase in Ti content, the biological and mechanical properties such as elastic modulus and hardness of the composite coatings were found to be increased. However, the average surface roughness of the coating was decreased with the quantity of Ti addition. Cho et al., [37] employed the RF magnetron sputtering technique to deposit HA with Sr on Ti alloy nanotube with variable process parameters. Atomic force microscope, water contact angle goniometer, field emission scanning electron microscope, and X-ray diffractometer were used to characterize the coated samples in terms of roughness, surface topography, wettability, morphology, and phase identification, respectively. The average surface roughness of the as-deposited coating was found to be decreased with an increase in applied voltage. Bramowicz et al., [38] correlated the surface morphology with the mechanical properties of HA coatings prepared by the RF magnetron sputtering technique. The deposition temperature ranging from 400 to 800 °C was varied during the process. With an increase in deposition temperature, both the elastic modulus and hardness of the coating were found to be decreased.

Pulsed Laser Deposition

In pulsed laser deposition (PLD) technique, a high energy laser beam in pulsed mode is used to focus over the source/coating material to evaporate in a controlled manner. The vaporized species in the form of molecule, atom, ion, or cluster of all are deposited on the substrate in a vacuum or gas environment. Pereiro et al., [39] deposited Sr-substituted HA successfully by pulsed laser deposition (PLD) technique. Commercially available HA and $SrCO_3$ powder were mixed and subsequently compacted to obtain the initial targets for the PLD. The effect of varying Sr content on the properties of resulting coatings was analyzed. Khandelwal et al., [40] used the PLD technique to deposit HA coating on stainless steel. They also analyzed the influence of varying laser energy on the characteristics of coated samples. At higher laser energy, HA particles were found to agglomerate, which led to reduce the average surface roughness of the coating. Also, with an increase in the laser energy, the Ca/P ratio is found to decrease. Duta et al., [41] deposited bovine and bovine-derived HA over titanium substrate with the help of PLD. Further, bovine-derived HA coating was compared with commercial HA coating in terms of various properties. Hidalgo-Robatto et al., [42] fabricated the Fluor-carbonated HA coating through the PLD method. They also studied how the crystalline structure and composition of the coating were influenced by the water vapor pressure employed in the method. Bao et al., [43] reported the effects of deposition time and target properties on the HA coating produced by using the PLD technique. With an increase in deposition time and target sintering temperature, the Ca/P ratio of the deposited HA was found to be increased.

Spin Coating

Spin coating is a thin film deposition method where a few drops of coating material preferably in liquid form kept on a flat substrate rotating with a very high speed. Under the centrifugal force, the droplet/s the coating material settles down in an uniform manner on the substrate by evaporating the solvent part. Factors such as speed and duration of rotation play important roles to achieve specific properties of the coated sample obtained by this method. Yuan et al., [44] developed the Zn-doped HA coating on stainless steel substrate by using the spin-coating method. The quantity of Zn varied as 0, 5, 10, 15, and 20 mol% in the Zn-doped HA coating. The coating was heat-treated at 600 °C in a normal atmospheric condition, and further, it was analyzed in terms of morphological and compositional properties. The pure HA coating showed nano rod-like morphology in its microstructure. On the other hand, the crystallite size of the Zn-doped HA coating was found to be reduced. Shokri et al., [45] employed the spin coating technique to deposit HA coating on NiTi substrate. Calcination of prepared HA coatings was performed at 400 to 600 °C temperature to achieve a highly crystalline HA phase in the resulting coating. The HA coating calcined at 600 °C showed the maximum crystallinity.

Ion Beam Assisted Deposition

Choi et al., [46] developed HA coating on the Ti-alloy substrate by using the ion-beam assisted deposition (IBAD) method. Kim et al., [47] used the IBAD method to deposit HA on alumina substrate. The prepared coating was annealed at 500 °C for a 2 h of time duration, and subsequently, it was immersed in deionized water at room temperature for 72 h. Luo et al., [48] studied the degradation behaviour of HA coating deposited by IBAD in both in-vitro and in-vivo modes. Rabiei et al., [49] deposited the functionally graded HA coating through IBAD with in situ heat treatment conditions. The IBAD-coated HA samples showed higher hardness and modulus of elasticity values compared to the same of sputter-deposited film. Blalock et al., [50] reported a study on HA coatings deposited on hot titanium and silicon-based substrates with the help of the IBAD technique. The process parameters and substrate temperature were varied to find their effects on the composition and microstructural properties of the fabricated coatings. With an increase in substrate temperature, the Ca/P ratio of the HA was found to be enhanced for both the coatings deposited on silicon and titanium substrates.

Sol-Gel Method

Domínguez-Trujillo et al., [51] employed a sol-gel deposition technique to coat HA on a porous titanium substrate. Cai et al., [52] deposited an HA coating containing both Mg and F ions in it on Ti-alloy through the sol-gel method. The resulting coatings showed improved bioactivity by promoting bone formation in a simulated body fluid (SBF) medium. Priyadarshini et al., [53] studied the biological, morphological, and structural properties of the Ce-doped HA coating prepared on Ti-alloy substrate via sol-gel technique. The corrosion resistance and bioactivity properties of the HA coating were found to improve by the addition of Ce. Ahmadi et al., [54] deposited a composite coating of HA, TiO_2, and Al_2O_3 on stainless steel substrate by using the sol-gel method. The weight quantity of TiO_2 and Al_2O_3 varied with 20, 30, and 40% in the composite coatings. Further, the biocompatibility, electrochemical reactivity, and bond strength of the coatings were analyzed in vitro conditions.

Electrochemical Deposition/Electroplating

Lin et al., [55] prepared HA coating on a Ti sheet by using the electrochemical deposition method. Gopi et al., [56] reported a study on HA-carbon nanotube composite coating developed by electrodeposition technique. Nizami et al., [57] used an electrochemical deposition technique to deposit HA-graphene oxide composite coating on the medical-grade stainless steel substrate. Mechanical peeling resistance, bioactivity, and corrosion resistance properties of the coating were found to be improved with the incorporation of graphene oxide in HA. Zeng et al., [58] compared electroplated HA-graphene oxide composite coating with the same electroplated pure HA coating deposited on Ti substrates based on their properties. The HA-graphene oxide composite coating showed higher crystallinity, bonding strength, and bioactivity. Qiu et al., [59] studied the HA-TiO_2 composite coating deposited on NiTi alloy by the electrochemical deposition method. Further, the mechanical and corrosion behaviors of the fabricated composite coating were analyzed.

Cold Spray Method

Chen et al., [60] studied the mechanical, structural, and morphological properties of HA coating deposited on the stainless-steel substrate by using the cold spraying technique. Tang et al., [61] used the cold spray technique to deposit a composite of HA and Ta on a Ti-alloy substrate. Ta was added to HA to avoid the dissociation of HA and to improve its bioactivity property. In a different approach, Chatelain et al., [62] heat-treated the HA powder at 700 and 900 °C, and then it was sprayed with a cold spray method to obtain the HA coatings. Vilardell et al., [63] fabricated micro-crystalline HA, nano-crystalline HA, and porous titanium coatings by using the cold spray technique. All three coatings were studied and compared in terms of crystallinity and in vitro bioactivity properties.

Suspension Plasma Spray Technique

Ročňáková et al., [64] produced HA coatings by using the suspension plasma spray (SPS) technique. The influence of torch speed on the crystalline composition, microstructure, and thickness of the coatings was studied. Xu et al., [65] used the SPS technique to deposit the nanostructured HA coating on a stainless-steel substrate. The suspension was prepared through the wet chemical method, and hydrogen gas was used to create the plasma during the spray. The hydrogen gas flow rate was varied to observe its effect on the properties of the deposited coating. The results indicated that with an increase in hydrogen flow rate, the secondary HA phases were formed with an increased crystallite size. Chen et al., [66] developed composite coatings of HA and nanodiamond with varying composition by the SPS method. Mechanical and microstructural properties of the coatings were studied. Improved mechanical properties and uniform microstructure observed for nanodiamond reinforced HA coatings were due to the reduced porosity of the coating. Zheng et al., [67] studied the crystallinity of HA coating deposited by the SPS technique. The arc current was varied to analyze its effect on the crystallinity of the coating. With an increase in arc current from 400 to 600 A, the crystallinity of the coating was found to be increased from 68.68 to 76.84%. Abir et al., [68] employed the SPS technique to prepare composite coatings of HA and gray titania with varying composition. The effect of the gray titania addition to HA on the antibacterial and mechanical properties of the composite coating was studied.

Conclusion

Different coating methods to deposit pure HA and HA-based coatings on various substrate materials are discussed in this chapter. Ten commonly used coating methods include plasma spraying, sputtering deposition, spin coating, sol-gel deposition, cold spray, electrophoretic deposition, pulsed laser deposition, ion beam-assisted deposition, electrochemical deposition, and suspension plasma spray techniques have been reviewed. In these works, various materials were added to the pure HA precursor/feedstock to increase the bio-mechanical properties of the deposited coatings with the aim of making them suitable for biomedical applications. These deposition methods are capable of producing HA/HA-based coatings with different microstructures, phases, and properties. Also, every coating method has certain benefits and limitations. Hence, based on the specific application and requirement, one should select the process (with proper parameters and conditions) and the materials for depositing the HA-based coatings.

References

[1] Nasker P., A. Samanta, S. Rudra, A. Sinha, A. K. Mukhopadhyay, M. Das, Effect of fluorine substitution on sintering behaviour, mechanical and bioactivity of hydroxyapatite, *J. Mech. Behav. Biomed. Mater.* 95 (2019) 136–142. https://doi.org/10.1016/j.jmbbm.2019.03.032.

[2] Nikpour M. R., S. M. Rabiee, M. Jahanshahi, Synthesis and characterization of hydroxyapatite/chitosan nanocomposite materials for medical engineering applications, *Compos. B. Eng.* 43 (2012) 1881–1886. https://doi.org/10.1016/j.compositesb.2012.01.056.

[3] Turnbull G., J. Clarke, F. Picard, P. Riches, L. Jia, F. Han, B. Li, W. Shu, 3D bioactive composite scaffolds for bone tissue engineering, *Bioact. Mater.* 3 (2018) 278–314. https://doi.org/10.1016/j.bioactmat.2017.10.001.

[4] Harun W S W, R I M Asri, J. Alias, F H Zulkifli, K. Kadirgama, S. A. C. Ghani, J. H. M. Shariffuddin, A comprehensive review of hydroxyapatite-based coatings adhesion on metallic biomaterials, *Ceram. Int.* 44 (2018) 1250–1268. https://doi.org/10.1016/j.ceramint.2017.10.162.

[5] Lahiri D., S. Ghosh, A. Agarwal, Carbon nanotube reinforced hydroxyapatite composite for orthopedic application: *A review, Materials Science and Engineering: C.* 32 (2012) 1727–1758. https://doi.org/10.1016/j.msec. 2012.05. 010.

[6] Lima R. S., B. R. Marple, Thermal Spray Coatings Engineered from Nanostructured Ceramic Agglomerated Powders for Structural, Thermal Barrier and Biomedical

Applications: *A Review, Journal of Thermal Spray Technology*. 16 (2007) 40–63. https://doi.org/10.1007/s11666-006-9010-7.

[7] Vecchio K. S., X. Zhang, J. B. Massie, M. Wang, C. W. Kim, Conversion of bulk seashells to biocompatible hydroxyapatite for bone implants, Acta Biomater. 3 (2007) 910–918. https://doi.org/10.1016/j.actbio.2007.06.003.

[8] Hussain S., K. Sabiruddin, Effect of heat treatment on the synthesis of hydroxyapatite from Indian clam seashell by hydrothermal method, *Ceram. Int*. 47 (2021) 29660–29669. https://doi.org/10.1016/j.ceramint.2021.07.137.

[9] Wu S. C., H. C. Hsu, S. K. Hsu, Y. C. Chang, W. F. Ho, Effects of heat treatment on the synthesis of hydroxyapatite from eggshell powders, *Ceram. Int.* 41 (2015) 10718–10724. https://doi.org/10.1016/j.ceramint.2015.05.006.

[10] Hussain S., K. Sabiruddin, Synthesis of eggshell based hydroxyapatite using hydrothermal method, *IOP Conf. Ser. Mater. Sci. Eng.* 1189 (2021) 012024. https://doi.org/10.1088/1757-899x/1189/1/012024.

[11] Niakan A., S. Ramesh, P. Ganesan, C. Y. Tan, J. Purbolaksono, H. Chandran, S. Ramesh, W. D. Teng, Sintering behaviour of natural porous hydroxyapatite derived from bovine bone, *Ceram. Int.* 41 (2015) 3024–3029. https://doi.org/ 10.1016/ j.ceramint.2014.10.138.

[12] Ozawa M., S. Suzuki, Microstructural development of natural hydroxyapatite originated from fish-bone waste through heat treatment, *Journal of the American Ceramic Society*. 85 (2002) 1315–1317. https://doi.org/10.1111/j.1151-2916. 2002.tb00268.x.

[13] Bee S.-L., Z. A. A. Hamid, Characterization of chicken bone waste-derived hydroxyapatite and its functionality on chitosan membrane for guided bone regeneration, *Compos. B. Eng.* 163 (2019) 562–573. https://doi.org/10.1016/ j.compositesb.2019.01.036.

[14] Wu S. C., H. C. Hsu, Y. N. Wu, W. F. Ho, Hydroxyapatite synthesized from oyster shell powders by ball milling and heat treatment, *Mater. Charact.* 62 (2011) 1180–1187. https://doi.org/10.1016/j.matchar.2011.09.009.

[15] Mondal B., S. Mondal, A. Mondal, N. Mandal, Fish scale derived hydroxyapatite scaffold for bone tissue engineering, *Mater. Charact.* 121 (2016) 112–124. https://doi.org/10.1016/j.matchar.2016.09.034.

[16] Wan Y Z., L. Hong, S.R. Jia, Y. Huang, Y. Zhu, Y.L. Wang, H.J. Jiang, Synthesis and characterization of hydroxyapatite-bacterial cellulose nanocomposites, *Compos. Sci. Technol.* 66 (2006) 1825–1832. https://doi.org/ 10.1016/j.compsci tech.2005.11.027.

[17] Sivakumar M., T. S. S. Kumar, K. L. Shantha, K. P. Rao, Development of hydroxyapatite derived from Indian coral, *Biomaterials*. 17 (1996) 1709–1714. https://doi.org/10.1016/0142-9612(96)87651-4.

[18] Ofudje E. A., A. Rajendran, A.I. Adeogun, MA Idowu, S.O. Kareem, D.K. Pattanayak, Synthesis of organic derived hydroxyapatite scaffold from pig bone waste for tissue engineering applications, *Advanced Powder Technology*. 29 (2018) 1–8. https://doi.org/10.1016/j.apt.2017.09.008.

[19] Zhou H., M. Yang, M. Zhang, S. Hou, S. Kong, L. Yang, L. Deng, Preparation of Chinese mystery snail shells derived hydroxyapatite with different morphology

using condensed phosphate sources, *Ceram. Int.* 42 (2016) 16671–16676. https://doi.org/10.1016/j.ceramint.2016.07.101.

[20] Edralin E J M, J. L. Garcia, F. M. dela Rosa, E.R. Punzalan, Sonochemical synthesis, characterization and photocatalytic properties of hydroxyapatite nano-rods derived from mussel shells, *Mater. Lett.* 196 (2017) 33–36. https://doi.org/10.1016/j.matlet.2017.03.016.

[21] Chen J., Z. Wen, S. Zhong, Z. Wang, J. Wu, Q. Zhang, Synthesis of hydroxyapatite nanorods from abalone shells via hydrothermal solid-state conversion, *Mater Des.* 87 (2015) 445–449. https://doi.org/10.1016/j.matdes. 2015.08.056.

[22] Singh S., K. K. Pandey, A. Islam, A. K. Keshri, Corrosion behaviour of plasma sprayed graphene nanoplatelets reinforced hydroxyapatite composite coatings in simulated body fluid, *Ceram. Int.* 46 (2020) 13539–13548. https://doi.org/10.1016/j.ceramint.2020.02.139.

[23] Henao J., C. Poblano-Salas, M. Monsalve, J. Corona-Castuera, O. Barceinas-Sanchez, Bio-active glass coatings manufactured by thermal spray: A status report, *Journal of Materials Research and Technology.* 8 (2019) 4965–4984. https://doi.org/10.1016/j.jmrt.2019.07.011.

[24] Singh S., K. K. Pandey, O. S. Asiq Rahman, S. Haldar, D. Lahiri, A. K. Keshri, Investigation of crystallinity, mechanical properties, fracture toughness and cell proliferation in plasma sprayed graphene nano platelets reinforced hydroxyapatite coating, *Mater. Res. Express.* 7 (2020) 015415. https://doi.org/ 10.1088/2053-1591/ab6c23.

[25] Chambard M., O. Marsan, C. Charvillat, D. Grossin, P. Fort, C. Rey, F. Gitzhofer, G. Bertrand, Effect of the deposition route on the microstructure of plasma-sprayed hydroxyapatite coatings, *Surf. Coat. Technol.* 371 (2019) 68–77. https://doi.org/10.1016/j.surfcoat.2019.01.027.

[26] Singh J., S. S. Chatha, H. Singh, Characterization and corrosion behavior of plasma sprayed calcium silicate reinforced hydroxyapatite composite coatings for medical implant applications, *Ceram. Int.* 47 (2021) 782–792. https://doi.org/10.1016/j.ceramint.2020.08.189.

[27] Singh G., S. Singh, S. Prakash, Surface characterization of plasma sprayed pure and reinforced hydroxyapatite coating on Ti6Al4V alloy, *Surf. Coat. Technol.* 205 (2011) 4814–4820. https://doi.org/10.1016/j.surfcoat. 2011.04.064.

[28] Chen Y., J. Ren, Y. Sun, W. Liu, X. Lu, S. Guan, Efficacy of graphene nanosheets on the plasma sprayed hydroxyapatite coating: Improved strength, toughness and in-vitro bioperformance with osteoblast, *Mater. Des.* 203 (2021) 109585. https://doi.org/10.1016/j.matdes.2021.109585.

[29] Drevet R., N. ben Jaber, J. Fauré, A. Tara, A. ben Cheikh Larbi, H. Benhayoune, Electrophoretic deposition (EPD) of nano-hydroxyapatite coatings with improved mechanical properties on prosthetic Ti6Al4V substrates, *Surf. Coat. Technol.* 301 (2016) 94–99. https://doi.org/10.1016/ j.surfcoat.2015.12.058.

[30] S. Khanmohammadi, M. Ojaghi-Ilkhchi, M. Farrokhi-Rad, Evaluation of bioglass and hydroxyapatite based nanocomposite coatings obtained by electrophoretic deposition, *Ceram. Int.* 46 (2020) 26069–26077. https:// doi.org/10.1016/j.ceramint.2020.07.100.

[31] Zhang C., T. Uchikoshi, L. Liu, K. Iwanami-Kadowaki, M. Uezono, K. Moriyama, M. Kikuchi, Antibacterial-functionalized Ag loaded-hydroxyapatite (HAp) coatings fabricated by electrophoretic deposition (EPD) process, *Mater. Lett.* 297 (2021) 129955. https://doi.org/10.1016/j.matlet. 2021.129955.

[32] Mahmoodi M., M. H. Hydari, L. Mahmoodi, L. Gazanfari, M. Mirhaj, Electrophoretic deposition of graphene oxide reinforced hydroxyapatite on the tantalum substrate for bone implant applications: In vitro corrosion and bio-tribological behavior, *Surf. Coat. Technol.* 424 (2021) 127642. https://doi.org/10.1016/j.surfcoat.2021.127642.

[33] Sun G., J. Ma, S. Zhang, Electrophoretic deposition of zinc-substituted hydroxyapatite coatings, *Materials Science and Engineering*: C. 39 (2014) 67–72. https://doi.org/10.1016/j.msec.2014.02.023.

[34] Parau A. C., C. M. Cotrut, P. Guglielmi, A. Cusanno, G. Palumbo, M. Dinu, G. Serratore, G. Ambrogio, D.M. Vranceanu, A. Vladescu, Deposition temperature effect on sputtered hydroxyapatite coatings prepared on AZ31B alloy substrate, *Ceram. Int.* 48 (2022) 10486–10497. https://doi.org/10.1016/ j.ceramint.2021. 12.258.

[35] Lenis J A, F. M. Hurtado, M. A. Gómez, F. J. Bolívar, Effect of thermal treatment on structure, phase and mechanical properties of hydroxyapatite thin films grown by RF magnetron sputtering, *Thin. Solid Films.* 669 (2019) 571–578. https://doi.org/10.1016/j.tsf.2018.11.045.

[36] Vladescu A., S. C. Padmanabhan, F. Ak Azem, M. Braic, I. Titorencu, I. Birlik, M.A. Morris, V. Braic, Mechanical properties and biocompatibility of the sputtered Ti doped hydroxyapatite, *J. Mech. Behav. Biomed. Mater.* 63 (2016) 314–325. https://doi.org/10.1016/j.jmbbm.2016.06.025.

[37] Cho H.-R., H.-C. Choe, Morphology of hydroxyapatite and Sr coatings deposited using radio frequency-magnetron sputtering method on nanotube formed Ti-6Al-4V alloy, *Thin Solid. Films.* 735 (2021) 138893. https://doi.org/10.1016/j.tsf.2021. 138893.

[38] Bramowicz M., L. Braic, F. A. Azem, S. Kulesza, I. Birlik, A. Vladescu, Mechanical properties and fractal analysis of the surface texture of sputtered hydroxyapatite coatings, *Appl. Surf. Sci.* 379 (2016) 338–346. https://doi.org/10.1016/j.apsusc.2016.04.077.

[39] Pereiro I., C. Rodríguez-Valencia, C. Serra, E. L. Solla, J. Serra, P. González, Pulsed laser deposition of strontium-substituted hydroxyapatite coatings, *Appl. Surf. Sci.* 258 (2012) 9192–9197. https://doi.org/10.1016/j.apsusc. 2012.04. 063.

[40] Khandelwal H., G. Singh, K. Agrawal, S. Prakash, R. D. Agarwal, Characterization of hydroxyapatite coating by pulse laser deposition technique on stainless steel 316 L by varying laser energy, *Appl. Surf. Sci.* 265 (2013) 30–35. https://doi.org/10.1016/j.apsusc.2012.10.072.

[41] Duta L., FN Oktar, G. E. Stan, G. Popescu-Pelin, N. Serban, C. Luculescu, I. N. Mihailescu, Novel doped hydroxyapatite thin films obtained by pulsed laser deposition, *Appl. Surf. Sci.* 265 (2013) 41–49. https://doi.org/10.1016/ j.apsusc. 2012.10.077.

[42] Hidalgo-Robatto B. M., J. J. Aguilera-Correa, M. López-Álvarez, D. Romera, J. Esteban, P. González, J. Serra, Fluor-carbonated hydroxyapatite coatings by pulsed laser deposition to promote cell viability and antibacterial properties, *Surf. Coat. Technol.* 349 (2018) 736–744. https://doi.org/10.1016/J.SURFCOAT.2018.06.047.
[43] Bao Q., C. Chen, D. Wang, J. Liu, The influences of target properties and deposition times on pulsed laser deposited hydroxyapatite films, *Appl. Surf. Sci.* 255 (2008) 619–621. https://doi.org/10.1016/j.apsusc.2008.06.080.
[44] Yuan Q., J. Wu, C. Qin, A. Xu, Z. Zhang, S. Lin, X. Ren, P. Zhang, Spin-coating synthesis and characterization of Zn-doped hydroxyapatite/polylactic acid composite coatings, *Surf. Coat. Technol.* 307 (2016) 461–469. https://doi.org/10.1016/j.surfcoat.2016.09.021.
[45] Shokri N., M S Safavi, M. Etminanfar, F. C. Walsh, J. Khalil-Allafi, Enhanced corrosion protection of NiTi orthopedic implants by highly crystalline hydroxyapatite deposited by spin coating: The importance of pre-treatment, *Mater. Chem. Phys.* 259 (2021) 124041. https://doi.org/10.1016/j.match emphys.2020.124041.
[46] Choi J.-M., H.-E. Kim, I.-S. Lee, Ion-beam-assisted deposition (IBAD) of hydroxyapatite coating layer on Ti-based metal substrate, *Biomaterials.* 21 (2000) 469–473. https://doi.org/10.1016/S0142-9612(99)00186-6.
[47] Kim T. N., Q. L. Feng, Z. S. Luo, F. Z. Cui, J. O. Kim, Highly adhesive hydroxyapatite coatings on alumina substrates prepared by ion-beam assisted deposition, *Surf. Coat. Technol.* 99 (1998) 20–23. https://doi.org/10.1016/ S0257-8972(97)00121-7.
[48] Luo Z. S., F. Z. Cui, Q. L. Feng, H. D. Li, X D. Zhu, M. Spector, In vitro and in vivo evaluation of degradability of hydroxyapatite coatings synthesized by ion beam-assisted deposition, *Surf. Coat Technol.* 131 (2000) 192–195. https://doi.org/10.1016/S0257-8972(00)00824-0.
[49] Rabiei A., B. Thomas, C. Jin, R. Narayan, J. Cuomo, Y. Yang, J.L. Ong, A study on functionally graded HA coatings processed using ion beam assisted deposition with in situ heat treatment, *Surf. Coat Technol.* 200 (2006) 6111–6116. https://doi.org/10.1016/j.surfcoat.2005.09.027.
[50] Blalock T., X. Bai, A. Rabiei, A study on microstructure and properties of calcium phosphate coatings processed using ion beam assisted deposition on heated substrates, *Surf Coat Technol.* 201 (2007) 5850–5858. https://doi.org/ 10.1016/ j.surfcoat.2006.10.039.
[51] Domínguez-Trujillo C., E. Peón, E. Chicardi, H. Pérez, J. A. Rodríguez-Ortiz, J. J. Pavón, J. García-Couce, J. C. Galván, F. García-Moreno, Y. Torres, Sol-gel deposition of hydroxyapatite coatings on porous titanium for biomedical applications, *Surf. Coat. Technol.* 333 (2018) 158–162. https://doi.org/ 10.1016/ j.surfcoat.2017.10.079.
[52] Cai Y., S. Zhang, X. Zeng, Y. Wang, M. Qian, W. Weng, Improvement of bioactivity with magnesium and fluorine ions incorporated hydroxyapatite coatings via sol–gel deposition on Ti6Al4V alloys, *Thin Solid Films.* 517 (2009) 5347–5351. https://doi.org/10.1016/j.tsf.2009.03.071.

[53] Priyadarshini B., S. Ramya, E. Shinyjoy, L. Kavitha, D. Gopi, U. Vijayalakshmi, Structural, morphological and biological evaluations of cerium incorporated hydroxyapatite sol–gel coatings on Ti–6Al–4V for orthopaedic applications, *Journal of Materials Research and Technology.* 12 (2021) 1319–1338. https://doi.org/10.1016/j.jmrt.2021.03.009.

[54] Ahmadi R., A. Afshar, In vitro study: Bond strength, electrochemical and biocompatibility evaluations of TiO2/Al2O3 reinforced hydroxyapatite sol–gel coatings on 316L SS, *Surf. Coat Technol.* 405 (2021) 126594. https://doi.org/10.1016/j.surfcoat.2020.126594.

[55] Lin D.-Y., X.-X. Wang, Preparation of hydroxyapatite coating on smooth implant surface by electrodeposition, *Ceram. Int.* 37 (2011) 403–406. https://doi.org/10.1016/j.ceramint.2010.08.018.

[56] Gopi D., E. Shinyjoy, M. Sekar, M. Surendiran, L. Kavitha, T.S. Sampath Kumar, Development of carbon nanotubes reinforced hydroxyapatite composite coatings on titanium by electrodeposition method, *Corros. Sci.* 73 (2013) 321–330. https://doi.org/10.1016/j.corsci.2013.04.021.

[57] Nizami MZI., B. D. L. Campéon, Y. Nishina, Electrodeposition of hydroxyapatite and graphene oxide improves the bioactivity of medical grade stainless steel, *Materials Today Sustainability.* 19 (2022) 100193. https://doi.org/10.1016/j.mtsust.2022.100193.

[58] Zeng Y., X. Pei, S. Yang, H. Qin, H. Cai, S. Hu, L. Sui, Q. Wan, J. Wang, Graphene oxide/hydroxyapatite composite coatings fabricated by electrochemical deposition, *Surf Coat Technol.* 286 (2016) 72–79. https://doi.org/10.1016/j.surfcoat.2015.12.013.

[59] Qiu D., L. Yang, Y. Yin, A. Wang, Preparation and characterization of hydroxyapatite/titania composite coating on NiTi alloy by electrochemical deposition, *Surf Coat Technol.* 205 (2011) 3280–3284. https://doi.org/ 10.1016/j.surfcoat.2010.11.049.

[60] Chen Q.-Y., Y.-L. Zou, X. Chen, X.-B. Bai, G.-C. Ji, H.-L. Yao, H.-T. Wang, F. Wang, Morphological, structural and mechanical characterization of cold sprayed hydroxyapatite coating, *Surf Coat Technol.* 357 (2019) 910–923. https://doi.org/10.1016/j.surfcoat.2018.10.056.

[61] Tang J., Z. Zhao, H. Liu, X. Cui, J. Wang, T. Xiong, A novel bioactive Ta/hydroxyapatite composite coating fabricated by cold spraying, *Mater Lett.* 250 (2019) 197–201. https://doi.org/10.1016/j.matlet.2019.04.123.

[62] Chatelain D., A. Denoirjean, V. Guipont, F. Rossignol, N. Tessier-Doyen, Influence of the thermal treatment of a hydroxyapatite powder on the characteristics of coatings deposited by cold gas spraying, *Surf Coat Technol.* (2022) 128697. https://doi.org/10.1016/j.surfcoat.2022.128697.

[63] Vilardell A. M., N. Cinca, N. Garcia-Giralt, S. Dosta, I. G. Cano, X. Nogués, J. M. Guilemany, Functionalized coatings by cold spray: An in vitro study of micro- and nanocrystalline hydroxyapatite compared to porous titanium, *Materials Science and Engineering:* C. 87 (2018) 41–49. https://doi.org/ 10.1016/j.msec.2018.02.009.

[64] Ročňáková I., K. Slámečka, E. B. Montufar, M. Remešová, L. Dyčková, A. Břínek, D. Jech, K. Dvořák, L. Čelko, J. Kaiser, Deposition of hydroxyapatite and

tricalcium phosphate coatings by suspension plasma spraying: Effects of torch speed, *J. Eur. Ceram. Soc.* 38 (2018) 5489–5496. https://doi.org/ 10.1016/ j.jeurceramsoc.2018.08.007.

[65] Xu H., X. Geng, G. Liu, J. Xiao, D. Li, Y. Zhang, P. Zhu, C. Zhang, Deposition, nanostructure and phase composition of suspension plasma-sprayed hydroxyapatite coatings, *Ceram. Int.* 42 (2016) 8684–8690. https://doi.org/10.1016/j.ceramint. 2016.02.102.

[66] Chen X., B. Zhang, Y. Gong, P. Zhou, H. Li, Mechanical properties of nanodiamond-reinforced hydroxyapatite composite coatings deposited by suspension plasma spraying, *Appl. Surf. Sci.* 439 (2018) 60–65. https://doi.org 10.1016/j.apsusc.2018.01.014.

[67] Zheng B., Y. Luo, H. Liao, C. Zhang, Investigation of the crystallinity of suspension plasma sprayed hydroxyapatite coatings, *J. Eur. Ceram. Soc.* 37 (2017) 5017–5021. https://doi.org/10.1016/j.jeurceramsoc.2017.07.007.

[68] Abir M M M, Y. Otsuka, K. Ohnuma, Y. Miyashita, Effects of composition of hydroxyapatite/gray titania coating fabricated by suspension plasma spraying on mechanical and antibacterial properties, *J. Mech. Beha.v Biomed. Mater.* 125 (2022) 104888. https://doi.org/10.1016/j.jmbbm.2021.104888.

Chapter 8

3D Printed HA Ceramic and Its Composite Scaffolds for Tissue Engineering

Qing Zhou and Rujie He
Institute of Advanced Structure Technology, Beijing Institute of Technology, Beijing, China

Abstract

Hydroxyapatite (HA) ceramic has gained tremendous attention in tissue engineering owing to its excellent mechanical properties, biocompatibility, and osteoconductivity. In most tissue engineering cases, complex-shaped HA scaffolds are usually needed, which brings great challenges to their manufacturing. Recently, additive manufacturing, also known as 3D printing, has been widely studied for its fabrication ability of complex-shaped HA ceramic and its composite scaffolds. Herein, this chapter introduces the latest advances of 3D printed HA ceramic and its composite scaffolds, including HA, HA/TCP, and various HA/polymer composite scaffolds. The performances of these scaffolds, including their mechanical properties, and *in vitro* and *in vivo* biological behaviors, are further summarized. Finally, the challenges and opportunities involved in this field are discussed and forecasted. It is believed this chapter can give some guidance for researchers or graduate students in the field of 3D printing of bioceramic scaffolds for tissue engineering.

Keywords: hydroxyapatite, scaffold, additive manufacturing, 3D printing, tissue engineering

In: What to Know about Hydroxyapatite
Editor: Isaac Lavoie
ISBN: 979-8-88697-523-9
© 2023 Nova Science Publishers, Inc.

Introduction

External trauma to bone can be susceptibly inflicted by traffic accidents, natural disasters, sports injuries, and so on. Even pathological diseases like cancer and infections can also result in bone fragments. The raw bone tissue healing process is usually not satisfactory in such situations because it is difficult for articular cartilage to heal naturally when encountering lesions caused by external trauma and disorders. Furthermore, it is highly susceptible to bone separation, delayed healing, healing failure, and even localized dysfunction in the patient (Frisbie et al., 2003). With the increasing maturity of medical technologies and bone repair products, bone repair materials for filling defects and inducing tissue regeneration have gradually extended to the clinic, showing significant outcomes, and steadily emerging as the prevailing tendency of clinical bone scaffolds (Ramesh, Moratti, and Dias, 2018). Common treatments for repairing bone defects include autologous bone grafting and scaffold reconstruction (Batstone et al., 2018).

Autologous bone grafting is usually accompanied by problems of limited sources and causing secondary injuries, whereas scaffold restoration has dramatically reduced operative time and procedures and is becoming the primary method of restoration bone engineering. Present scaffolds for bone regeneration include metallic scaffolds (Nemcakova et al., 2022, Zhao et al., 2017), polymeric scaffolds (Yuan et al., 2018, Bedell et al., 2020), and bioceramic scaffolds (Wen et al., 2017, Zhang et al., 2018). However, metallic scaffolds tend to conduct heat which interferes with treatment. Polymeric scaffolds have superior biological properties but insufficient mechanical properties which induce inflammation. Fortunately, bioceramic scaffolds, represented by hydroxyapatite (HA) ceramic, are extensively accepted for bone regeneration because of their reasonably secure ingredients and suitable mechanical properties, and have demonstrated favorable bone repair capabilities, presenting restorative potential in bone scaffold applications.

Normally, HA ceramic causes an osseous bond between the material and the broken end scaffolded by inducing specific physiological reactions at the interface. HA ceramic usually contains large amounts of calcium (Ca) and phosphorus (P), which are the main inorganic components of bone tissue. When the bioactive ceramic is dissolved in micro amounts, the calcium and phosphorus ions will dissolve, which can promote cell adhesion, proliferation, and secretion of the matrix, and regulate the fluid environment, all of which will favor the regeneration of bone tissues. HA ceramic is incredibly biocompatible, but the high brittle, low flexural strength and easily broken-ability make them difficult to manufacture, limiting the wide applications for restoration bone engineering. Combined with 3D printing, it can regulate pore size, pore shape, pore arrangement, and morphology to prepare multi-scale porous HA ceramic structures suitable for cell growth.

3D printing allows the maximum flexibility in design, enabling the preparation of various components from micro-scaled to macro-scaled, which tends to the trial of HA bioceramic. Typically, 3D printing technologies can be classified into selective laser sintering (SLS), selective laser melting (SLM), laminated object manufacturing (LOM), binder jetting (BJ), vat photopolymerization (VP), fused deposition modeling (FDM), direct ink writing (DIW), and so on (Zhang, et al., 2022). Among these 3D printing technologies, VP, DIW, and FDM are the most commonly used methods for fabricating bioceramic green bodies. In addition, after obtaining the bioceramic green bodies, 3D printing technologies are usually hybridized with degreasing and sintering technologies to fabricate the final bioceramic scaffolds (except for polymers and composite implants, which generally do not require sintering and degreasing).

Consequently, 3D printed HA ceramic and its composites have been broadly adopted as bone regeneration materials, including HA, biphasic calcium phosphate (HA/TCP), and HA/polymer, for autologous bone graft substitutes due to their remarkably bioactive and osteoinductive qualities. Table 1 lists the advantages and disadvantages, technologies, mechanical properties, and biological experiments of several major types of 3D printed HA and its composite scaffolds.

3D Printed HA Scaffolds for Tissue Engineering

HA ($Ca_{10}(PO_4)_6(OH)_2$) is well known as a ceramic material with excellent biocompatibility and bone guidance. HA is widely distributed in the bone tissues of animals. In human bones, the content of HA is as high as 70% (Sommerfeldt and Rubin, 2001). After HA implantation, the new bone can grow along with the bone tissues around the defects and the internal pores of the scaffolds (Sommerfeldt and Rubin, 2001). 3D printed HA scaffolds have been developed to customize porous structures and complex shapes suitable for bone regeneration. The non-cytotoxic HA was obtained through SLA by Chen (Chen et al., 2019). The cross-section of the hole was hexagonal, the inner tangent circle was 500μm, and the whole scaffold had a honeycomb structure. Then the HA structure was scaffolded into the rabbit parietal. Based on the micro-CT images after the implantation of HA, rabbits showed no adverse physiological reactions such as infection. The wafer formed a strong bone connection with the defect indicating HA had a good biosafety *in vivo*.

Similarly, Gibbons (Gibbons et al., 2015) prepared HA scaffolds by SLS and characterized them for mechanical properties, microstructure, and porosity. Liu (Liu et al., 2021) prepared HA scaffolds by DLP. The DLP-prepared HA scaffold was cylindrical with a pore size of 500 μm and had a high surface quality. Mechanical and biological behavioral analysis of the HA scaffold were performed, and the scaffold showed a good stability and biosafety. Du. (Du et al., 2021) printed HA scaffolds with high resolution by DLP, for spherical star units with circular channels and octahedral unit structures consisting of cylinders, respectively. These samples had a channel width of 380 μm, and the pores were intacted and undistorted. Kang (Kang et al., 2022) prepared HA scaffolds by SLA for mandibular repair. *In vitro* cell cultivation experiments suggested that the cells were attached to the HA scaffold. During DIW, the adhesive is a critical factor for the 3D printing process, as it determines the nature and structure of the printed bone scaffold and the reliability of the overall printing system. Borvornwu (Borvornwu et al., 2022) verified the restoration effects of 3D printed nano-HA in alveolar bone, where scaffolds were filled with the new bone around both the root tip and crown after implantation and showed a significant degradation and resorption after some time.

Overall, the biosafety of 3D printed HA scaffolds has been widely demonstrated and the 3D printed HA scaffolds has shown great potential in bone repair engineering. Although 3D printing technologies of HA are diverse and comparatively sophisticated, high resolution and safer non-toxic photosensitive resins need to be continue to be found.

3D Printed HA-TCP Scaffolds for Tissue Engineering

Beta-tricalcium phosphate (β-TCP) ceramic is also biocompatible and osteoinductive due to its main component of calcium phosphate and has a faster degradation rate compared to HA. It has a relative density of 99.5%, higher compressive strength than typical scaffolds, and provides cytocompatibility and osteogenicity (Schmidleithner et al., 2019). It is necessary to point out that a higher degradation rate *in vivo* leads to more osteogenic induction potentials (Rh Owen, Dard, and Larjava, 2018).

Table 1. Advantages and disadvantages, technologies, mechanical properties, and biological experiments of 3D printed HA and its composite scaffolds

Scaffolds	Advantages/ Disadvantages	Technology	Mechanical properties	In vivo/in vitro results	Ref.
HA	•*Ad* Good biocompatibility, bioactivity, and osteoconductivity •*Dis* Poor osteoinduction and high brittleness, low degradation rate	DIW	Compressive strength was 0.88±0.02 MPa	--	(Cox et al. 2015)
		SLA	Bending strength was 7.2 MPa	After 21 days, MC3T3-E1 cells were proliferated and differentiated on the scaffold.	(Kang et al. 2022)
		DIW	Compressive strength was ~5 MPa	After 12 weeks of in vivo culture, the perimeter of the scaffold was filled with new bone.	(Luo et al. 2015)
HA/TCP	•*Ad* Excellent biocompatibility, osteoconductivity, resorbability, an internal structure similar to natural bone •*Dis* High brittleness	DLP	Compressive strength was 6.20MPa	Granulation of the scaffold at 4 weeks of implantation.	(Lim et al. 2020)
		DLP	Compressive strength was 1.060 MPa	Greater elevation of osteogenic gene expression 7 days after stent implantation.	(Liu et al. 2022)
HA/ Polymer	GelMA-gelatin/ HA: •*Ad* Good bioactivity, does not cause inflammatory reactions, degradable •*Dis* Low strength	VP	--	After 28 days, MC3T3-E1 cells proliferated significantly, ALP assay activity and cell viability were significantly enhanced.	(Allen et al. 2022)

Table 1. (Continued)

Scaffolds	Advantages/ Disadvantages		Technology	Mechanical properties	In vivo/in vitro results	Ref.
	PTMC/ HA	•*Ad* Easy cell adhesion on the surface, high biocompatibility •*Dis* Structural instability	SLA	--	MSCs, macrophages, preosteoblasts, and osteoblasts adhered well to the surface of the material and expanded morphologically at 6 h after seeding.	(Teotia et al. 2020)
	PCL/ HA	•*Ad* Easy to process, non-toxic, good biocompatibility;	FDM	--	The stent had high expression of Runx-2, Osterix, and osteogenic proteins at day 60, significantly	(Helaehil et al. 2021)
		•*Dis* Hydrophobic surface, cells do not adhere easily			promoting the regeneration of bone tissue.	
	PEGDA/ HA	•*Ad* Better flexibility, biocompatibility, swellability •*Dis* Low strength	DLP	Compressive strength was 0.045-1.063 MPa	New bone production near the scaffold 6 weeks after implantation in rats.	(Tikhonov et al. 2020)

Ad means advantages.
Dis means disadvantages.

The faster degradation rate of β-TCP results in better osteoinductive performance compared to HA when in contact with body, and can quickly dissolve and release calcium and phosphate ions (Fiume et al., 2021). However, the rate of its degradation for β-TCP still does not match the rate of new bone production in the human body, and the mechanical properties are also very poor. On the contrary, HA is known for its better biocompatibility and stability in contact with body fluids, but its degradation performance is poor. Based on the advantages and disadvantages of HA and β-TCP, HA/TCP composite has demonstrated its superior osteogenic properties, appropriate degradation rate, and suitable mechanical strength (Rh Owen, Dard, and Larjava, 2018, Le Gars Santoni et al., 2022, Ballouze et al., 2021).

The interaction between HA/TCP composite and cells leads to cell-initiated enzymatic degradation and acidic degradation of absorbed HA/TCP composite, resulting in the formation of a HA layer on the HA/TCP surface which is important for bone regeneration. This promotes the adhesion and proliferation of osteoblasts and differentiation, which eventually leads to the gradual maturation of osteoblasts. The HA/TCP composite scaffold was prepared by DLP. The HA/TCP composite scaffold had the same composition as natural bone, while providing structural stability, and had also shown excellent biocompatibility in New Zealand white rabbits *in vivo*, and remarkably observed bone enhancement adjacent to the scaffold (Lim et al., 2020). Ahn (Ahn et al., 2021) prepared HA/TCP scaffolds by binder jetting (BJ), and the scaffolds had a regular geometry. The as-prepared HA/TCP scaffolds significantly promoted the initial response of cells and exhibited sufficient proliferation and differentiation. Liu (Liu et al., 2022) prepared HA/TCP scaffolds by DLP, and the unmodified HA/TCP scaffold showed a regular porous structure. The results of *in vitro* experiments showed that the human umbilical vein endothelial cells adhered well to the scaffolds, significantly enhanced proliferation in the cells after a period of incubation with the scaffolds, and promoted vascularization. Prabh (Prabh et al., 2021) prepared HA/TCP scaffolds with different filling densities to mimic natural tissues, and *in vitro* experiments with rabbit adipose-derived mesenchymal stem cells indicated that the scaffolds were biocompatible, promoting differentiation to bone/chondrocytes, and induced expression of relevant genes.

3D Printed HA/Polymer Scaffolds for Tissue Engineering

Hydrogels used for bone repair include methacrylate gelatin (GelMA), methacrylate hyaluronic acid (MeHA), sodium alginate (SA), collagen, alginate, chitosan, etc. Most of these hydrogels are naturally occurring, have excellent biocompatibility, are safe and non-toxic, and provide a more natural growing environment for cells. It has been proved that GelMA, MeHA, and GelMA-MeHA blend hydrogels exhibited higher potential for osteoblasts regeneration, which has been widely proven in the field of craniomaxillofacial prosthetics (Miao et al., 2022, Janarthanan et al., 2019, Temple et al., 2014).

Specifically, SA is a natural polymer material that has great bioactivity, biocompatibility, biodegradability, and non-toxicity of degradation products. To improve the mechanical properties, Liu (Liu et al., 2019) combined SA with HA through additive manufacturing, and a final HA/SA composite porous layered scaffold was prepared by DIW. The scaffold showed better mechanical properties than pure SA scaffold, with a large number of cell adhesion and good proliferation, indicating excellent direct biocompatibility. To evaluate the biocompatibility of the scaffold, *in vitro* cell culture experiments were performed by seeding mouse bone mesenchymal stem cells on the scaffold and observing the results related to cell proliferation after 1, 3, and 5 days, which showed a significant increase in the number of bone marrow mesenchymal stem cells. In the polymer matrix which promoted cell attachment and proliferation, HA helped to promote the scaffold to adsorb proteins in culture thus further promoting cell proliferation, in addition, the increase of HA enhanced the roughness of the cell surface, which facilitated cell attachment.

Lee (Lee et al., 2022) prepared HA/gelatin scaffolds by DIW followed by photocuring and characterized the morphology of 3D printed HA/gelatin/GO scaffolds. The surface morphology of the HA/gel/graphene oxide scaffold became smooth with the increase of graphene oxide concentration. Then the 3D printed parameters, mechanical properties, and biological behavior for scaffolds were further characterized. The scaffold accelerated cell growth and even osteogenic fragmentation. Allen (Allen et al., 2022) printed composite scaffolds by mixing GelMA, gelatin, and HA by DIW. The survival and proliferation of cells in the composite scaffolds were evaluated by 28-day in vitro experiments for histological analysis, ALP assay, and osteogenic protein expression levels. The scaffolds displayed well-organic biodegradability, strength, and tolerance osteoconduction, while the addition of inorganic substances reduced the swelling of hydrogels, the scaffolds changed from transparent to opaque. They illustrated the expression levels of Bla7 (bone morphogenetic protein 7) and Bglap (osteocalcin) genes in scaffolds with different HA contents, and these results suggested that the presence of hydrogels enhances osteogenic.

PLGA is also attracting attention for the control of biodegradability and convenient biocompatibility. Zhang (Zhang, et al., 2022) demonstrated good adhesion and proliferation of HepG2 cells by mixing silk protein, PLGA, and HA ceramics. 3D printing technology enabled scaffolds with homogeneous microstructure, appropriate porosity, proper mechanical properties, controlled degradation rate, stable pH environment, and good cell culture compatibility. Li (Li et al., 2021) used DIW to prepare PLGA/HA scaffolds with different HA contents and evaluated their structure and mechanical properties. All the scaffolds had new bone production. The composite scaffold containing zirconia had a larger area of new bone compared to the pure PLGA scaffold. According to the bone and cell staining results, the composite scaffolds had a better osteogenic ability, and the degradability of the scaffolds with different contents was similar, but there were significant differences in osteogenic integrity and osteogenic behavior, and the specific pattern needs to be verified by further *in vivo* experiments.

Future Perspectives for 3D Printed HA and Composite Scaffolds

3D printed bioceramics have received much attention as the promising materials for bone repair. However, enhancement of biological properties requires multi-materials complementing each other and catalyzed by active factors.

To enhance the cell expansion and adhesion on the scaffold and to provide a more suitable environment for cell proliferation, scaffolds are usually loaded with biologically active agents, including proteins, extracellular matrix, rare minerals, cells, etc. Another way is to modify the scaffold surface to enhance the roughness of the scaffold surface or to attract cell adhesion for growth. Ranga (Ranga et al., 2020) demonstrated that the incorporation of Zn^{2+}, Mg^{2+}, and Sr^{2+} supported cell regeneration and enhanced the mechanical properties of bioceramic scaffolds while improving antimicrobial properties. The addition of Mg^{2+} to HA/TCP produced desirable physicochemical properties that enhance hydroxyapatite formation compared to pristine HA/TCP, and in vitro results showed that the Mg-HA/TCP was biologically active and supported cell growth. Ballouze (Ballouze et al., 2021) added Mg to HA/TCP and performed in vitro culture experiments. The observation of morphological alterations in MSCs revealed that the cells trans-colonized to an osteoblast-like shape. The HA/TCP scaffold containing Mg was scaffolded into the skull defect of mice and gradual new bone production was observed at the defect site.

Cell-laid extracellular matrix (ECM) plays a critical role in cellular responses which has been demonstrated (Owida, Kuiper, and Yang, 2021). Combining AM and bioceramics can drastically stimulate the expansion of cells. Kumar (Kumar et al., 2016) prepared HA scaffolds encapsulated in decellularized extracellular matrix (dECM) by 3D printing. The addition of dECM established a microenvironment favorable for cell growth. The HA scaffold loaded with dECM exhibited stronger cell adhesion compared to the pure HA scaffold. The expression of osteogenic proteins was determined by staining, cells grown on dECM-covered HA scaffolds demonstrated a 3D morphology with high levels of both actin and vinculin. Chen (Chen et al., 2021) coated nano-DBM (demineralized bone matrix) on top of 3D printed PCL/β-TCP scaffolds. Through comparing the cell growth after 1, 3, and 7 days of in vitro culture, cells on nano-DBM coated PCL/β-TCP scaffold showed better activity than those on PCL/β-TCP scaffold.

Summary

3D printed HA scaffolds have gained significant attention as the most effective material for bone prosthetics. Several types of 3D printing techniques used to prepare bioceramics, including VP, DIW, and FDM, of which VP has the advantage of high precision and speed and is the most widely used technique today. HA/TCP combines the advantages of HA and β-TCP and has an internal structure similar to that of natural bone, showing tremendous potential for bone repair. HA/polymer composite scaffolds combines the bioactivity of polymers with the osteoconductivity of HA and will be an important direction in the future. Composite scaffolds have shown excellent biological properties and promising potential. However, the specific performance of the various composite scaffolds needs to be further validated by in vivo experiments.

Biological property is the key factor limiting the development of bioceramic scaffolds. Doping with biofactors, enhancing the adhesion of implants to cells, and loading biological factors are the main methods to enhance the biological properties of scaffolds. In particular, implant-loaded bioactive factors have expressed huge potential for bone repair. With the increasing development of additively manufactured bioceramic scaffolds which are being explored for other functional applications such as shape memory effects and drug release.

This offers the possibility of minimally invasive and targeted therapies in bone restoration.

At present, 3D printed HA ceramic and its composite scaffolds have received remarkable achievements in developing ideal treatment options for bone regeneration engineering. With the rapid evolution of technology and updated materials, we believe that 3D printed bioceramics will contribute to the development of restoration medicine and the quality of human life. Scaffolds will become more functional and diverse.

Disclaimer

None.

References

Ahn Ji-Ho, Kim J, Han G, Kim DE, Cheon K-H, Lee H, Kim H-E, Kim Y-J, Jang TS and Jung H-D. 2021. "3D-printed biodegradable composite scaffolds with significantly enhanced mechanical properties via the combination of binder jetting and capillary rise infiltration process." *Additive Manufacturing,* 41. doi: 10.1016/j.addma.2021.101988.

Allen Nicholas B, Abar B, Johnson L, Burbano J, Danilkowicz RM and Adams SB. 2022. "3D-bioprinted GelMA-gelatin-hydroxyapatite osteoblast-laden composite hydrogels for bone tissue engineering." *Bioprinting,* 26:1690-1698. doi: 10.1016/j.bprint.2022.e00196.

Ballouze R, Marahat MH, Mohamad S, Saidin NA, Kasim SR and Ooi JP. 2021. "Biocompatible magnesium-doped biphasic calcium phosphate for bone regeneration." *J Biomed Mater Res B Appl Biomater,* 109 (10):1426-1435. doi: 10.1002/jbm.b.34802.

Batstone MD. 2018. "Reconstruction of major defects of the jaws." *Aust Dent J,* 63 Suppl 1:S108-S113. doi: 10.1111/adj.12596.

Bedell ML, Navara AM, Du Y, Zhang S and Mikos AG. 2020. "Polymeric Systems for Bioprinting." *Chem Rev,* 120 (19):10744-10792. doi: 10.1021/acs.chemrev.9b00834.

Chen Q, Zou B, Lai Q, Wang Y, Xue R, Xing H, Fu X, Huang C and Yao P. 2019. "A study on biosafety of HAP ceramic prepared by SLA-3D printing technology directly." *J Mech Behav Biomed Mater,* 98:327-335. doi: 10.1016/j.jmbbm.2019.06.031.

Cox SC, Thornby JA, Gibbons GJ, Williams MA and Mallick KK. 2015. "3D printing of porous hydroxyapatite scaffolds intended for use in bone tissue engineering applications." *Mater Sci Eng C Mater Biol Appl,* 47:237-47. doi: 10.1016/j.msec.2014.11.024.

Du Yao, Hu T, You J, Ye Y, Zhang B, Bao B, Li M, Liu Y, Wang Y and Wang T. 2021. "Study of falling-down-type DLP 3D printing technology for high-resolution hydroxyapatite scaffolds." *International Journal of Applied Ceramic Technology,* 19 (1):268-280. doi: 10.1111/ijac.13915.

Fiume Elisa, Magnaterra G, Rahdar A, Verné E and Baino F. 2021. "Hydroxyapatite for Biomedical Applications: A Short Overview." *Ceramics,* 4 (4):542-563. doi: 10.3390/ceramics4040039.

Frisbie David D, Oxford JT, Southwood L, Trotter GW, Rodkey WG, Steadman JR, Goodnight JL and Wayne McIlwraith C. 2003. "Early Events in Cartilage Repair After Subchondral Bone Microfracture." *Clinical Orthopaedics and Related Research®,* 407:215-227.

Helaehil JV, Lourenco CB, Huang B, Helaehil LV, de Camargo IX, Chiarotto GB, Santamaria-Jr M, Bartolo M and Caetano GF. 2021. "In Vivo Investigation of Polymer-Ceramic PCL/HA and PCL/beta-TCP 3D Composite Scaffolds and Electrical Stimulation for Bone Regeneration." *Polymers,* (Basel) 14 (1). doi: 10.3390/polym14010065.

Janarthanan Gopinathan, Pillai MM, Kulasekaran SS, Rajendran S and Bhattacharyya A. 2019. "Engineered knee meniscus construct: understanding the structure and impact of functionalization in 3D environment." *Polymer Bulletin,* 77 (5):2611-2629. doi: 10.1007/s00289-019-02874-0.

Kang Jin-Ho, Sakthiabirami K, Jang K-J, Jang J-G, Oh G-J, Park C, Fisher JG and Park S-W. 2022. "Mechanical and biological evaluation of lattice structured hydroxyapatite scaffolds produced via stereolithography additive manufacturing." *Materials & Design,* 214. doi: 10.1016/j.matdes.2021.110372.

Kijartorn P, Wongpairojpanich J, Thammarakcharoen F, Suwanprateeb J and Buranawat B. 2022. "Clinical evaluation of 3D printed nano-porous hydroxyapatite bone graft for alveolar ridge preservation: A randomized controlled trial." *J Dent Sci,* 17 (1):194-203. doi: 10.1016/j.jds.2021.05.003.

Kumar A, Nune KC and Misra RD. 2016. "Biological functionality of extracellular matrix-ornamented three-dimensional printed hydroxyapatite scaffolds." *J Biomed Mater Res A,* 104 (6):1343-51. doi: 10.1002/jbm.a.35664.

Le Gars Santoni B, Niggli L, Dolder S, Loeffel O, Sblendorio GA, Heuberger R, Maazouz Y, Stahli C, Dobelin N, Bowen P, Hofstetter W and Bohner M. 2022. "Effect of minor amounts of beta-calcium pyrophosphate and hydroxyapatite on the physico-chemical properties and osteoclastic resorption of beta-tricalcium phosphate cylinders." *Bioact Mater,* 10:222-235. doi: 10.1016/j.bioactmat.2021.09.003.

Lee Hoyeol, Yoo JM, Ponnusamy NK and Nam SY. 2022. "3D-printed hydroxyapatite/gelatin bone scaffolds reinforced with graphene oxide: Optimized fabrication and mechanical characterization." *Ceramics International,* 48 (7):10155-10163. doi: 10.1016/j.ceramint.2021.12.227.

Lim HK, Hong SJ, Byeon SJ, Chung SM, On SW, Yang BE, Lee JH and Byun SH. 2020. "3D-Printed Ceramic Bone Scaffolds with Variable Pore Architectures." *Int J Mol Sci,* 21 (18). doi: 10.3390/ijms21186942.

Liu G, Zhang B, Wan T, Zhou C, Fan Y, Tian W and Jing W. 2022. "A 3D-printed biphasic calcium phosphate scaffold loaded with platelet lysate/gelatin methacrylate to promote vascularization." *J Mater Chem B,* 10 (16):3138-3151. doi: 10.1039/d2tb00006g.

Liu Ruoxi, Ma L, Liu H, Xu B, Feng C and He R. 2021. "Effects of pore size on the mechanical and biological properties of stereolithographic 3D printed HAp bioceramic scaffold." *Ceramics International,* 47 (20):28924-28931. doi: 10.1016/j.ceramint.2021.07.053.

Liu Shuifeng, Hu Y, Zhang J, Bao S, Xian L, Dong X, Zheng W, Li Y, Gao H and Zhou W. 2019. "Bioactive and Biocompatible Macroporous Scaffolds with Tunable Performances Prepared Based on 3D Printing of the Pre-Crosslinked Sodium Alginate/Hydroxyapatite Hydrogel Ink." *Macromolecular Materials and Engineering,* 304 (4). doi: 10.1002/mame.201800698.

Luo Y, Zhai D, Huan Z, Zhu H, Xia L, Chang J and Wu C. 2015. "Three-Dimensional Printing of Hollow-Struts-Packed Bioceramic Scaffolds for Bone Regeneration." *ACS Appl Mater Interfaces,* 7 (43):24377-83. doi: 10.1021/acsami.5b08911.

Miao F, Liu T, Zhang X, Wang X, Wei Y, Hu Y, Lian X, Zhao L, Chen W and Huang D. 2022. "Engineered bone tissues using biomineralized gelatin methacryloyl/sodium alginate hydrogels." *J Biomater Sci Polym Ed,* 33 (2):137-154. doi: 10.1080/09205063.2021.1980360.

Natarajan ABM, Sivadas VPD and Nair PD. 2021. "3D-printed biphasic scaffolds for the simultaneous regeneration of osteochondral tissues." *Biomed Mater,* 16 (5). doi: 10.1088/1748-605X/ac14cb.

Nemcakova I, Litvinec A, Mandys V, Potocky S, Plencner M, Doubkova M, Nanka O, Olejnickova V, Sankova B, Bartos M, Ukraintsev E, Babcenko O, Bacakova L, Kromka A, Rezek B and Sedmera D. 2022. "Coating Ti6Al4V implants with nanocrystalline diamond functionalized with BMP-7 promotes extracellular matrix mineralization in vitro and faster osseointegration in vivo." *Sci Rep,* 12 (1):5264. doi: 10.1038/s41598-022-09183-z.

Owida HA, Kuiper NL and Yang Y. 2021. "Maintenance and Acceleration of Pericellular Matrix Formation within 3D Cartilage Cell Culture Models." *Cartilage,* 13 (2_suppl):847S-861S. doi: 10.1177/1947603519870839.

Ramesh N, Moratti SC and Dias GJ. 2018. "Hydroxyapatite-polymer biocomposites for bone regeneration: A review of current trends." *J Biomed Mater Res B Appl Biomater,* 106 (5):2046-2057. doi: 10.1002/jbm.b.33950.

Ranga N, Gahlyan S and Duhan S. 2020. "Antibacterial Efficiency of Zn, Mg and Sr Doped Bioactive Glass for Bone Tissue Engineering." *J Nanosci Nanotechnol,* 20 (4):2465-2472. doi: 10.1166/jnn.2020.17336.

Rh Owen G, Dard M and Larjava H. 2018. "Hydoxyapatite/beta-tricalcium phosphate biphasic ceramics as regenerative material for the repair of complex bone defects." *J Biomed Mater Res B Appl Biomater,* 106 (6):2493-2512. doi: 10.1002/jbm.b.34049.

Schmidleithner C, Malferarri S, Palgrave R, Bomze D, Schwentenwein M and Kalaskar DM. 2019. "Application of high resolution DLP stereolithography for fabrication of tricalcium phosphate scaffolds for bone regeneration." *Biomed Mater,* 14 (4):045018. doi: 10.1088/1748-605X/ab279d.

Sommerfeldt DW and Rubin CT. 2001. "Biology of bone and how it orchestrates the form and function of the skeleton." *Eur Spine J,* 10 Suppl 2:S86-95. doi: 10.1007/s005860100283.

Temple JP, Hutton DL, Hung BP, Huri PY, Cook CA, Kondragunta R, Jia X and Grayson WL. 2014. "Engineering anatomically shaped vascularized bone grafts with hASCs and 3D-printed PCL scaffolds." *J Biomed Mater Res A,* 102 (12):4317-25. doi: 10.1002/jbm.a.35107.

Teotia AK, Dienel K, Qayoom I, van Bochove B, Gupta S, Partanen J, Seppala J and Kumar A. 2020. "Improved Bone Regeneration in Rabbit Bone Defects Using 3D Printed Composite Scaffolds Functionalized with Osteoinductive Factors." *ACS Appl Mater Interfaces,* 12 (43):48340-48356. doi: 10.1021/acsami.0c13851.

Tikhonov A, Evdokimov P, Klimashina E, Tikhonova S, Karpushkin E, Scherbackov I, Dubrov V and Putlayev V. 2020. "Stereolithographic fabrication of three-dimensional permeable scaffolds from CaP/PEGDA hydrogel biocomposites for use as bone grafts." *J Mech Behav Biomed Mater,* 110:103922. doi: 10.1016/ j.jmbbm.2020.103922.

Wei J, Yan Y, Gao J, Li Y, Wang R, Wang J, Zou Q, Zuo Y, Zhu M and Li J. 2021. "3D-printed hydroxyapatite microspheres reinforced PLGA scaffolds for bone regeneration." *Mater Sci Eng C Mater Biol Appl,* 112618. doi: 10.1016/j.msec.2021.112618.

Wen Y, Xun S, Haoye M, Baichuan S, Peng C, Xuejian L, Kaihong Z, Xuan Y, Jiang P and Shibi L. 2017. "3D printed porous ceramic scaffolds for bone tissue engineering: A review." *Biomater Sci,* 5 (9):1690-1698. doi: 10.1039/c7bm00315c.

Yuan B, Wang Z, Zhao Y, Tang Y, Zhou S, Sun Y and Chen X. 2021. "In Vitro and In Vivo Study of a Novel Nanoscale Demineralized Bone Matrix Coated PCL/beta-TCP Scaffold for Bone Regeneration." *Macromol Biosci,* 21 (3):e2000336. doi: 10.1002/mabi.202000336.

Yuan M, Bi B, Huang J, Zhuo R and Jiang X. 2018. "Thermosensitive and photocrosslinkable hydroxypropyl chitin-based hydrogels for biomedical applications." *Carbohydr Polym,* 192:10-18. doi: 10.1016/j.carbpol.2018.03.031.

Zhang Ben and Song J. 2018. "3D-Printed Biomaterials for Guided Tissue Regeneration." *Small Methods,* 2 (9). doi: 10.1002/smtd.201700306.

Zhang Xueqin, Zhang K, Zhang L, Wang W, Li Y and He R. 2022. "Additive manufacturing of cellular ceramic structures: From structure to structure–function integration." *Materials and Design,* 215. doi: 10.1016/j.matdes.2022.110470.

Zhang Y, Liu L, Li N, Wang Y and Yue X. 2022. "3D scaffold fabricated with composite material for cell culture and its derived platform for safety evaluation of drugs." *Toxicology,* 466:153066. doi: 10.1016/j.tox.2021.153066.

Zhao D, Witte F, Lu F, Wang J, Li J and Qin L. 2017. "Current status on clinical applications of magnesium-based orthopaedic implants: A review from clinical translational perspective." *Biomaterials,* 112:287-302. doi: 10.1016/j.biomaterials.2016.10.017.

Index

E

F

H

I

L

V

W